Computing Supplement 10

H. Hagen, G. Farin, H. Noltemeier (eds.)
in cooperation with R. Albrecht

Geometric Modelling

Dagstuhl 1993

Springer-Verlag Wien New York

Prof. Dr. H. Hagen
Fachbereich Informatik
Universität Kaiserslautern
Federal Republic of Germany

Prof. Dr. G. Farin
Department of Computer Science
Arizona State University
Tempe, U.S.A.

Prof. Dr.H. Noltemeier
Lehrstuhl für Informatik I
Universität Würzburg
Federal Republic of Germany

Prof. Dr. R. Albrecht
Institut für Informatik
Universität Innsbruck
Austria

ISBN-13: 978-3-211-82666-9   e-ISBN-13: 978-3-7091-7584-2
DOI: 10.1007/978-3-7091-7584-2

© 1995 Springer-Verlag/Wien

Typesetting: Thomson Press (India) Ltd., New Delhi

Printed on acid-free and chlorine-free bleached paper

With 188 Figures

Library of Congress Cataloging-in-Publication Data

Geometric modelling : Dagstuhl, 1993 / H. Hagen, G. Farin, H.
   Noltemeier, eds., in cooperation with R. Albrecht.
         p.   cm. – (Computing supplement ; 10)
     Based on lectures given at the second Dagstuhl Seminar on
   Geometric Modelling.
     ISBN-13: 978-3-211-82666-9
     1. Curves on surfaces – Mathematical models – Congresses.
   2. Surfaces – Mathematical models – Congresses. I. Hagen, H. (Hans),
   1953-   II. Farin, Gerald E. III. Noltemeier, Hartmut.
   IV. Dagstuhl Seminar on Geometric Modelling (2nd : 1993) V. Series:
   Computing (Springer-Verlag).   Supplementum ; 10.
   QA565.G45   1995
   006.6–dc20   95-7042
                                                               CIP

ISSN 0344-8029

Preface

This book is based on lectures presented at the second Dagstuhl Seminar on Geometric Modelling organized by Gerald Farin (Arizona State University), Hans Hagen (Universität Kaiserslautern), and Hartmut Noltemeier (Universität Würzburg). International experts from academia and industry were selected to speak on the most interesting topics in geometric modelling. The resulting papers, published in this volume, give a state-of-the-art survey of the relevant problems and issues. The following topics are discussed:

- NURBS
- product engineering
- object oriented modelling
- solid modelling
- surface interrogation
- feature modelling
- variational design
- scattered data algorithms
- geometry processing
- blending methods
- smoothing and fairing algorithms
- spline conversion

The discussion between industry and university has proven to be very fruitful. The scientists from industry were able to give many important and practicable impulses for new research; in the opposite direction the university researchers have developed many new technologies, solving industrial problems, which may be transferred back to industry. Everybody was impressed by the quality of the presentations. They acknowledged the importance of such research exchange between the various partners.

We would like to thank all participating speakers, and the audience, for what appears to have been a very successful workshop. Many thanks go to Frank Weller for his help in editing this book.

Tempe, Kaiserslautern, Würzburg,
October 1994

Hans Hagen
Gerald Farin
Hartmut Noltemeier

Contents

Computing Suppl. 10, 1–34 (1995)

Parametric Offset Surface Approximation

R. E. Barnhill and **T. M. Frost**, Tempe

Abstract. Offset surfaces are of interest in a variety of engineering applications. The formulation of a parametric offset surface involves division by the square root of a parametric equation, therefore, the offset surface is typically non-polynomial. Because of this complexity, offset surfaces cannot, in general, be written as members of the same class of functions or their generating or progenitor surface. Approximation of the offset surface is therefore desirable. Three contemporary methods of offset surface approximation are described which serve as models for the development of a new approximation algorithm. An adaptive offset surface approximation method based on a visually smooth triangular interpolant to position and tangent plane data defined in a triangular mesh is then developed. The criteria used to develop the approximation method are discussed, the components of the algorithm are described and the results of an implementation are illustrated. A conclusion about the success and possible refinement of the triangular offset surface approximation method is drawn and ideas for further research are outlined.

Key words: Offset surfaces, parametric surfaces, surface approximation.

1. Introduction

The parametric offset surface of a given parametric surface $s(u, v)$ is defined as $s(u, v) + d\mathbf{n}(u, v)$ where $\mathbf{n}$ is the unit surface normal at each point $s(u, v)$ and d is the offset magnitude. The surface $s(u, v)$ is referred to as the progenitor surface. Offset surfaces arise in engineering applications such as geometric modeling, definition of tolerance zones, description of growth processes and generation of tool paths for numerical control machining.

One of the key properties of the offset surface is the complex mathematical form of the unit surface normal $\mathbf{n}$. Since the unit surface normal is computed by normalizing the vector $s_u(u, v) \times s_v(u, v)$, the unit surface normal is typically non-polynomial. Approximation of offset surfaces is therefore desirable because of this complexity and the restrictions that some modeling systems place on the types of surfaces that can be manipulated efficiently [9].

It is difficult to generate smooth approximations to offset surfaces with reliable accuracy. This is a result of the difficulty of surface approximation in general and the fact that accurate approximation becomes increasingly more difficult to achieve in regions where the offset distance is comparable to the minimum *concave* radius of curvature of the progenitor surface. The minimum *concave* radius of curvature is defined in terms of the *principal curvatures* of a surface and is explained in detail in Section 2.

When the offset distance is less than, but comparable to, the minimum *concave* radius of curvature, the offset surface can develop regions with high curvature values. Regions of high curvature make approximation difficult. Tangent discontinuities such as vertices and ridges result in the offset surface when the offset distance is equal to the minimum *concave* radius of curvature. Discontinuities in the offset surface are extremely difficult to reproduce in a surface approximation scheme and are usually considered inappropriate characteristics in practical environments. Self-intersection in the offset surface results when the offset distance is greater than the minimum *concave* radius of curvature. Degenerate behavior such as vertices, ridges and self-intersection is not acceptable in most applications and must be detected by surface analysis algorithms so that any such inappropriate behavior is illustrated in an informative manner.

Methods have been presented by [13, 15, 17, 20, 26] for the approximation of offset curves which take into account the geometry of the progenitor curve. The more complex problem of developing an adaptive algorithm for approximating offset surfaces has been the goal of this research. Three contemporary offset surface approximation methods were examined to gain insight into the state of the art. The approximation algorithms proposed by [9] and [16] utilize refinement of piecewise continuous rectangular patch meshes and the approximation algorithm proposed by [19] utilizes subdivision of the knot vectors and control polyhedrons of NURBS surfaces. The goal of this research was to develop a new algorithm for the generation of approximate offset surfaces which is adaptive, efficient and robust and does not suffer from some of the problems associated with existing methods.

Contemporary approaches to generation of approximate offset surfaces by [9] (uniform bicubic Hermite mesh), [19] (NURBS surfaces) and [16] (uniform bicubic/biquintic Bézier mesh) were examined with particular attention to surface analysis techniques developed by Farouki, in order to develop a thorough understanding of the state of the art in offset surface approximation.

An adaptive version of Farouki's algorithm was implemented and served as the basis for developing an interactive approximate offset surface generation program. A second algorithm based on triangular Bernstein-Bézier patches was then developed based on a visually smooth triangular interpolant to position and tangent data defined in a triangular mesh [21]. Various modifications to the original work by Piper were incorporated which produced results with decreased numbers of total patches required for approximation. The presentation of the triangular approximation algorithm follows that found in [11].

2. Mathematical Description

The offset surface is defined as the locus of points swept out by the center of a sphere of constant radius as it moves over the progenitor surface [9]. Using this definition, the fundamental differential properties of progenitor surfaces and their offsets can be described.

For an arbitrary, smooth parametric surface $s(u, v)$, a parametric offset surface may be mathematically defined as:

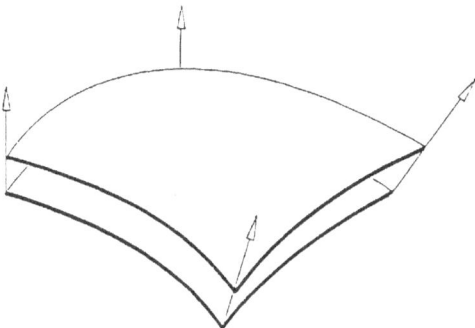

Figure 1. Progenitor surface (below) and its offset (above)

$$o(u, v) = s(u, v) + d\mathbf{n}(u, v),$$

where d is the offset distance,

$$\mathbf{n}(u, v) = \frac{\mathbf{s}_u(u, v) \times \mathbf{s}_v(u, v)}{\| \mathbf{s}_u(u, v) \times \mathbf{s}_v(u, v) \|}$$

is the unit surface normal and $\mathbf{s}_u$ and $\mathbf{s}_v$ are the partial derivatives of $\mathbf{s}$ with respect to u and v. By the definition of an offset surface, we must restrict ourselves to surfaces $s(u, v)$ such that $\| \mathbf{s}_u(u, v) \times \mathbf{s}_v(u, v) \| \neq \mathbf{0}$. This means that s, referred to as the progenitor surface, must have a non-singular surface metric [9]. The definition of the offset surface is illustrated in Fig.1.

The remainder of this section is devoted to developing the differential properties of parametric surfaces and their offsets through examination of the first and second fundamental tensors. [9] presented a detailed analysis of these differential properties and the work presented here is taken from this paper. [19] presented the essential results from [9] and eliminated much of the detailed development. A similar approach is taken in this section, and only the most relevant results are presented.

2.1. Differential Properties of Parametric Surfaces and their Offsets

The shape of a surface is completely characterized by two fundamental tensors associated with each point on the surface, the first and second fundamental tensors. It can be shown that surfaces with identical first and second fundamental tensors exhibit identical shape although their location in space may differ.

The first fundamental tensor defines surface characteristics associated with the tangent plane at a given point and is defined as:

$$\mathbf{G} = \begin{bmatrix} \mathbf{s}_u \cdot \mathbf{s}_u & \mathbf{s}_u \cdot \mathbf{s}_v \\ \mathbf{s}_v \cdot \mathbf{s}_u & \mathbf{s}_v \cdot \mathbf{s}_v \end{bmatrix}$$

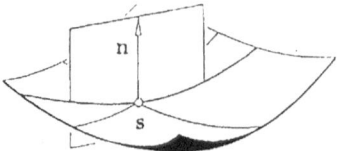

Figure 2. *Normal section* of a surface

and the determinant of the first fundamental form may be written as:

$$|\mathbf{G}| = \det(\mathbf{G}) = |\mathbf{s}_u \times \mathbf{s}_v|^2.$$

The second fundamental tensor defines surface characteristics associated with the curvature at a given point and is defined as:

$$\mathbf{D} = \begin{bmatrix} \mathbf{n} \cdot \mathbf{s}_{uu} & \mathbf{n} \cdot \mathbf{s}_{uv} \\ \mathbf{n} \cdot \mathbf{s}_{vu} & \mathbf{n} \cdot \mathbf{s}_{vv} \end{bmatrix}.$$

To proceed further we need to consider *normal sections* of a surface as illustrated in Fig. 2.

Normal sections are curves $\mathbf{s}(u(t), v(t))$ which are intersections of a surface $\mathbf{s}(u, v)$ with planes containing the surface normal $\mathbf{n}$ at every point $(u(t), v(t))$ where t is a parameter used to define the *normal section*. The curvature $\kappa(u(t), v(t))$ of a *normal section* at a given point $(u(t), v(t))$ is called the *normal curvature* of a surface and is defined as:

$$\kappa = \frac{-q\mathbf{D}q^T}{q\mathbf{G}q^T}, \tag{2.1}$$

where $q = [\dot{u}(t), \dot{v}(t)]$ is the *normal section's* tangent vector written in the coordinate system defined by $\mathbf{s}_u$ and $\mathbf{s}_v$, and $(\cdot)$ denotes derivative with respect to t. The sign convention in Eq. (2.1) ensures that $\kappa > 0$ when the center of curvature is on opposite side of the surface indicated by the surface normal. A similar development is found in [9] and [19].

The *principal curvatures* at a point on the surface are defined as the maximum and minimum *normal curvature* values for all orientations of the normal plane at that point. The *principal curvatures* are solution to the simultaneous equations $\partial\kappa/\partial\dot{u} = 0$ and $\partial\kappa/\partial\dot{v} = 0$ which lead to the quadratic equation:

$$\kappa^2 - 2H\kappa + K = 0, \tag{2.2}$$

where K and H are the Gaussian and mean curvatures given by:

$$K = \frac{|\mathbf{D}|}{|\mathbf{G}|},$$

$$H = \frac{(g_{12}d_{21} + g_{21}d_{12}) - (g_{11}d_{22} + g_{22}d_{11})}{2|\mathbf{G}|}.$$

Let κ_+ and κ_- be the solution of Eq. (2.2) where $\kappa_+ \geq \kappa_-$, then we have:

$$\kappa_+ = H + \sqrt{H^2 - K},$$

$$\kappa_- = H - \sqrt{H^2 - K}, \tag{2.3}$$

$$K = \kappa_+ \kappa_-,$$

$$H = (\kappa_+ + \kappa_-)/2.$$

[9] derived corresponding quantities for the offset surface in terms of quantities derived for the progenitor surface and the offset distance d. The first fundamental tensor of the offset surface is defined as:

$$\mathbf{G}_o = (1 - Kd^2)\mathbf{G} - 2d(1 + Hd)\mathbf{D}$$

and its determinant can be written as:

$$|\mathbf{G}_o| = \det(\mathbf{G}_o) = (Kd^2 + 2Hd + 1)^2|\mathbf{G}|. \tag{2.4}$$

Recall that we have restricted the progenitor surface such that $\mathbf{G} \neq 0$, (i.e. the progenitor surface metric is non-singular for all (u, v) in the domain of interest).

The quadratic term $Kd^2 + 2Hd + 1$ in Eq. (2.4) is of particular interest in the differential geometry of the offset surface. From this quadratic equation we are able to determine for what values of the curvature of the progenitor surface the offset surface metric will be singular. From the set of Eq. (2.3) we get:

$$Kd^2 + 2Hd + 1 = \kappa_+ \kappa_- d^2 + (\kappa_+ + \kappa_-)d + 1$$
$$= (\kappa_+ d + 1)(\kappa_- d + 1)$$

and therefore the offset surface metric $\mathbf{G}_o$ is singular for corresponding points on the progenitor surface where either principal curvature is equal to $-1/d$.

2.2. Classification of Points in the Offset Surface

The qualitative behavior of a point of the offset surface at a point falls into one of several categories which are determined by the principal curvatures of the progenitor surface, κ_+ and κ_-, at that point and the critical curvature $\kappa_c = -1/d$. [9] provides the distinction between *regular* and *singular* points in the offset surface.

Regular points in the offset surface occur when neither κ_+ or κ_- is equal to κ_c. Recalling that $\kappa_+ \geq \kappa_-$ by definition, three cases arise:

(1) $\kappa_+ > \kappa_c, \kappa_- > \kappa_c$,
(2) $\kappa_+ > \kappa_c, \kappa_- < \kappa_c$,
(3) $\kappa_+ < \kappa_c, \kappa_- < \kappa_c$.

In cases (1) and (3), the offset surface normal has the same sense as that of the progenitor surface normal. In case (2), the offset surface normal has the opposite sense as that of the progenitor surface normal.

Singular points in the offset surface occur when either κ_+ or κ_- is equal to κ^c. Again, three cases can arise:

(1) $\kappa_+ > \kappa_c, \kappa_- = \kappa_c,$
(2) $\kappa_+ = \kappa_c, \kappa_- = \kappa_c,$
(3) $\kappa_+ = \kappa_c, \kappa_- < \kappa_c.$

In cases (1) and (3), the point on the offset surface lies on a ridge. In case (2), the point on the offset surface is a cusp and results when the progenitor surface locally approximates a sphere of radius d.

The classification of points in the offset surface based on known properties of the progenitor surface is useful from the standpoint that approximation methods will usually not be able to reproduce such complex characteristics as vertices and ridges. Such features should be identified as early as possible as they are usually undesirable features in any practical application.

3. Offset Surface Approximation: Rectangular Methods

Methods for the approximation of parametric offset surfaces have been developed by [9, 16, 19] which use bicubic Hermite, NURBS, and bicubic/biquintic Bézier surfaces as the basis for approximation respectively. This section briefly describes these rectangular based methods of offset surface approximation.

3.1. Bicubic Hermite Approximation

[9] presented a method for the bicubic Hermite approximation of offset surfaces to parametric progenitor surfaces defined over rectangular domains. The method has three main components. First, a differential analysis of the progenitor surface is performed to determine the maximum offset distance which can be used without degeneracy occurring in the offset surface. Second for each patch of the progenitor surface (assumed piecewise C^1 defined over the unit square $[0, 1] \times [0, 1]$), uniform domain subdivision to some specified level is performed and a bicubic Hermite approximating patch fit to position, first partial derivative and twist information of the offset surface at the corners of each subdomain. Third, a tolerance analysis is performed to check the accuracy of the approximation. As stated by Farouki, the algorithm can be made adaptive by providing feedback between the second and third stages.

The bicubic Hermite patch defined on each subdomain $[u_{min}, u_{max}] \times [v_{min}, v_{max}]$ can be written as:

$$\sum_{i=0}^{3} \sum_{j=0}^{3} h_i(\bar{u}) h_j(\bar{v}) \mathbf{m}_{ij}$$

where

$$\bar{u} = (u - u_{min})/(u_{max} - u_{min}),$$
$$\bar{v} = (v - v_{min})/(v_{max} - v_{min}),$$

and

$$h_0(t) = 2t^3 - 3t^2 + 1,$$
$$h_1(t) = -2t^3 + 3t^2,$$
$$h_2(t) = t^3 - 2t^2 + t,$$
$$h_3(t) = t^3 - t^2,$$

which are characterized by the following well known properties:

$$\begin{pmatrix} h_0(0) & h_0(1) & h_0'(0) & h_0'(1) \\ h_1(0) & h_1(1) & h_1'(0) & h_1'(1) \\ h_2(0) & h_2(1) & h_2'(0) & h_2'(1) \\ h_3(0) & h_3(1) & h_3'(0) & h_3'(1) \end{pmatrix} = \begin{pmatrix} 1 & 0 & 0 & 0 \\ 0 & 1 & 0 & 0 \\ 0 & 0 & 1 & 0 \\ 0 & 0 & 0 & 1 \end{pmatrix}$$

and where the matrix $\mathbf{M}$ is composed of the $\mathbf{m}_{ij}$ and defined as:

$$\begin{vmatrix} \mathbf{o}(u_{min}, v_{min}) & \mathbf{o}(u_{min}, v_{max}) & \dfrac{\mathbf{o}_v(u_{min}, v_{min})}{(\Delta v)} & \dfrac{\mathbf{o}_v(u_{min}, v_{max})}{(\Delta v)} \\[2ex] \mathbf{o}(u_{max}, v_{min}) & \mathbf{o}(u_{max}, v_{max}) & \dfrac{\mathbf{o}_v(u_{max}, v_{min})}{(\Delta v)} & \dfrac{\mathbf{o}_v(u_{max}, v_{max})}{(\Delta v)} \\[2ex] \dfrac{\mathbf{o}_u(u_{min}, v_{min})}{(\Delta u)} & \dfrac{\mathbf{o}_u(u_{min}, v_{max})}{(\Delta u)} & \dfrac{\mathbf{o}_{uv}(u_{min}, v_{min})}{(\Delta u \Delta v)} & \dfrac{\mathbf{o}_{uv}(u_{min}, v_{max})}{(\Delta u \Delta v)} \\[2ex] \dfrac{\mathbf{o}_u(u_{max}, v_{min})}{(\Delta u)} & \dfrac{\mathbf{o}_u(u_{max}, v_{max})}{(\Delta u)} & \dfrac{\mathbf{o}_{uv}(u_{max}, v_{min})}{(\Delta u \Delta v)} & \dfrac{\mathbf{o}_{uv}(u_{max}, v_{max})}{(\Delta u \Delta v)} \end{vmatrix}$$

where $\Delta u = u_{max} - u_{min}$ and $\Delta v = v_{max} - v_{min}$ provide the parameter domain transformation of the given subdomain to the unit square, and

$$\mathbf{o}(u, v) = \mathbf{s}(u, v) + d\mathbf{n}(u, v),$$
$$\mathbf{o}_u(u, v) = \mathbf{s}_u(u, v) + d\mathbf{n}_u(u, v),$$
$$\mathbf{o}_v(u, v) = \mathbf{s}_v(u, v) + d\mathbf{n}_v(u, v),$$
$$\mathbf{o}_{uv}(u, v) = \mathbf{s}_{uv}(u, v) + d\mathbf{n}_{uv}(u, v).$$

While uniform subdivision to some specified level is performed on the domain of each patch of the progenitor surface, a non-uniform subdivision scheme will increase efficiency as noted by [19]. The patch adjacency conditions required by the meshes of bicubic Hermite approximations in order to maintain C^1 continuity across patch boundaries, however, dictate that the subdivision may occur where subdivision is not warranted by the complexity of the exact offset surface. This trend would become increasingly noticeable as the levels of subdivision increase and large numbers of new approximating patches are added in regions which do not warrant further refinement.

This unnecessary refinement can be seen by examining the domain subdivision generated by an adaptive version of this method. These illustrations appear at the end of Section 5. The implementation used to generate the results does not maintain one-to-one patch adjacency. Despite the lack of continuity in the resulting approximation, the unnecessary refinement that would be caused by strictly adhering to uniform subdivision can be observed.

Numerical analysis techniques are applied to determine the extreme values of curvature of the progenitor surface and to determine the maximum error occurring in the offset surface approximation. The Newton-Raphson methods required to solve these problems have a formidable drawback in that reasonable start points must be obtained for all extrema of interest. As [9] points out, convergence of these methods is an open question in all but simple cases.

3.2. NURBS Approximation

[19] presented a method for constructing a NURBS approximation to offset surfaces of NURBS surfaces. The NURBS surface can be written as:

$$\frac{\displaystyle\sum_{i=0}^{k_u}\sum_{j=0}^{k_v} N_i^m(u)N_j^n(v)\mathbf{p}_{ij}w_{ij}}{\displaystyle\sum_{i=0}^{k_u}\sum_{j=0}^{k_v} N_i^m(u)N_j^n(v)w_{ij}},$$

where n and m are the order of the surface in the u and v directions, $N_i^m(u)$ and $N_j^n(v)$ are the B-spline basis functions defined over $u \in [u_0, u_{k_u+m}]$, $v \in [v_0, v_{k_v+n}]$, $\mathbf{p}_{ij} \in \mathbb{R}^3$ are points of the control polyhedron and w_{ij} are their associated weights ($w_{ij} > 0$).

The method of approximation proposed is an iterative scheme which offsets the control points of the control polyhedron of the progenitor surface to obtain an approximation of the offset surface. Each point in the progenitor's control polyhedron is offset along a pseudo-normal defined by the geometry of the points in the control polyhedron. This offset control polyhedron defines the offset approximation. For each value of $\{u_i, v_j\} \in \{u_0, \ldots, u_{k_u+m}\} \times \{v_0, \ldots, v_{k_v+n}\}$ the approximate offset surface is compared with the exact offset position at that point. For each point $\{u_i, v_j\}$ where the approximate is determined to be inaccurate, new knots are added to the left and right midspans in both the u and v knot vectors. If the approximation is determined accurate enough at all $\{u_i, v_j\}$, the approximate is examined at the midspans, one-third spans and so on until some user specified level of interior checking has been accomplished. In both insertion algorithms, multiple knot insertion is not allowed so that the maximum level of continuity is maintained in the approximation.

After this tolerance analysis, two finer knot vectors have been created for the progenitor. The progenitor's control polyhedron is then refined using the Oslo Algorithm [18]. This results in a refined control polyhedron which identically defines the progenitor surface but contains more degrees of freedom with which the offsetting process can create a more accurate approximate offset surface control polyhedron. This process can be shown to converge to the exact offset surface as the pseudo-normals converge to the progenitor normals as the knot refinement proceeds.

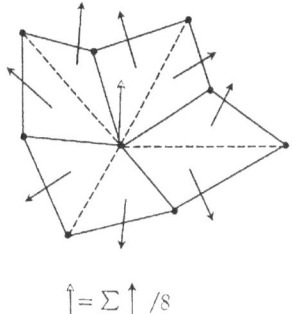

$$\hat{\uparrow} = \Sigma \uparrow / 8$$

Figure 3. Formulation of the pseudo-normal

The method of offsetting the control points of the progenitor surface is accomplished using a pseudo-normal defined by the control polyhedron. For control points located in the interior of the control net, the pseudo-normal is defined as the average of the normals formed by the planes defined by the control point's neighbors taken sequentially two at a time. In this manner, the pseudo-normal is taken to be the average of eight such plane normals. The formulation of the pseudo-normal for a control point in the interior of the control net is illustrated in Fig. 3. For control points on one of the borders (and not in the corner) of the control polyhedron, the pseudo-normal is defined by half of the plane normals used in computing the pseudo-normal for control points located in the interior of the control polyhedron. For corner control points, the exact offset surface normal, which is defined by the corner point and its neighbors in the u and v directions, is used.

The tolerance analysis is accomplished by examining the length of the vector computed by intersecting a line in the direction of the progenitor's surface normal at a given point with the approximate offset surface. Each point $s(u, v)$ of the progenitor surface has a corresponding point $o(\bar{u}, \bar{v})$ on the approximate offset surface determined in this manner. The problem of computing $o(\bar{u}, \bar{v})$ is solved using a Newton-Raphson method to compute the solution to two non-linear equations in the two variables $\bar{u}$ and $\bar{v}$. Once this has been accomplished the measure of accuracy used in the tolerance analysis part of the algorithm becomes:

$$\frac{\| o(\bar{u}, \bar{v}) - s(u, v) \|}{|d|} \leq 1 - \varepsilon,$$

which indicates that the length of the vector $o(\bar{u}, \bar{v}) - s(u, v)$ is approximately equal to the magnitude of the offset distance to within some tolerance ε.

While the subdivision processes involved in this method are non-uniform and knot values are inserted where local inaccuracy is discovered, unwarranted refinement in the approximate offset surface's control polyhedron can occur as insertion of new knots causes a ripple effect in the refinement of the progenitor's control polyhedron (i.e. a new knot causes the generation of a new row or column of control points). This effect will become more noticeable as the number of knots in the u and v directions increases. Similarly, unwarranted refinement can occur in the approximation method proposed by [9]. This problem is described in the previous sub-section.

3.3. Bicubic/Biquintic Bézier Approximation

[16] presented a method for the bicubic and biquintic Bézier approximation of offset surfaces to Bézier surfaces. Although designed for progenitor surfaces in the Bézier form, this method could be extended to handle general parametric progenitor surfaces defined over rectangular domains. Rectangular Bézier patches are of the following form:

$$\mathbf{b}(u, v) = \sum_{i=0}^{m} \sum_{j=0}^{n} B_i^m(u) B_j^n(v) \mathbf{b}_{ij},$$

where m and n are the degree of the surface in the u and v directions, $B_i^m(u)$ and $B_j^n(v)$ are univariate Bernstein polynomials of degree m and n respectively where

$$B_i^n(t) = \binom{n}{i} t^i (1 - t)^{n-i}$$

and $\mathbf{b}_{ij} \in \mathbb{R}^3$ are points of the control polyhedron.

The method of approximation can be briefly described as selecting a suitable rectangular subdivision of the progenitor's domain and then fitting bicubic or biquintic Bézier patches over each subdomain such that the overall surface produced is smooth. Tangent plane continuity (G^1) is required for bicubic approximating patches and *normal section* curvature continuity (G^2) is required for the biquintic approximating patches. Each approximating patch is fit according to the equations that determine G^1 and G^2 continuity. For further details on geometric continuity see [12].

Two surfaces $\mathbf{r}(u, v)$ and $\mathbf{s}(u, v)$ exhibit G^1 continuity if the following conditions hold at a common point of $\mathbf{r}$ and $\mathbf{s}$ (a_{ij} and b_{ij} scalars):

$$\begin{pmatrix} \mathbf{s}_u \\ \mathbf{s}_v \end{pmatrix} = \begin{pmatrix} a_{10} & b_{10} \\ a_{01} & b_{01} \end{pmatrix} \begin{pmatrix} \mathbf{r}_u \\ \mathbf{s}_v \end{pmatrix}.$$

The same two surfaces, $\mathbf{r}$ and $\mathbf{s}$, exhibit G^2 continuity if the conditions for G^1 continuity hold and the following conditions also hold at a common point of $\mathbf{r}$ and $\mathbf{s}$:

$$\begin{pmatrix} \mathbf{s}_{uu} \\ \mathbf{s}_{uv} \\ \mathbf{s}_{vv} \end{pmatrix} = \begin{pmatrix} a_{20} & b_{20} \\ a_{11} & b_{11} \\ a_{02} & b_{02} \end{pmatrix} \begin{pmatrix} \mathbf{r}_u \\ \mathbf{r}_v \end{pmatrix} + \begin{pmatrix} a_{10}^2 & 2a_{10}b_{10} & b_{10}^2 \\ a_{10}a_{01} & a_{10}b_{01} + a_{01}b_{10} & b_{10}b_{01} \\ a_{01}^2 & 2a_{01}b_{01} & b_{01}^2 \end{pmatrix} \begin{pmatrix} \mathbf{r}_{uu} \\ \mathbf{r}_{uv} \\ \mathbf{r}_{vv} \end{pmatrix}.$$

For each approximating bicubic or biquintic patch defined over each subdomain, the algorithm for fitting patches proceeds in two steps. First, the boundaries of the approximating patches are fit such that the appropriate geometric continuity conditions are met at the corner points. Second, the interior of the approximating patch is fit such that the appropriate geometric continuity conditions are met along the boundary curves. In both fitting processes, the scalars a_{ij} and b_{ij} associated with the continuity conditions are used to minimize the length of error vectors sampled at discrete points. The error vectors are computed by subtracting points in the exact offset surface from their corresponding points on the approximate offset surface. In general, the error vectors will not be perpendicular to the approximate offset surface and will not give the best estimate of how close (geometrically) the approximation is to the exact offset surface. Iterative improvement is performed and the error vectors are adjusted until they are approximately perpendicular to the approximate offset and hence the error sum is reduced to better reflect the accuracy achieved by the approximation. An iterative correction formula for performing this adjustment is given by [16]. This correction formula uses second partial derivative information. A similar correction formula involving only first partial derivative information is given in [14] and is further described in Section 4.

The approximation algorithm can be made adaptive by further subdividing the domain of the progenitor surface and processing the successively more refined regions. A method of curvature based subdivision is proposed to facilitate such subdivision. A problem similar to that which occurs in the method of the [9] occurs as the patch adjacency conditions required to maintain continuity across approximating patch boundaries may lead to subdivision in regions which are not complex enough to warrant refined definition.

4. Offset Surface Approximation: Triangular Methods

The main objective of this research was to develop an offset surface approximation algorithm that satisfied criteria for success determined by examination of contemporary research on the subject. After analyzing the material presented in [9, 16, 19], the following criteria were considered appropriate measures of success for an offset surface approximation method:

(1) the resulting approximation should be smooth (at least G^1),
(2) the resulting approximation should be accurate to some tolerance,
(3) the algorithm should generate a minimal amount of patches.

After examining the approximation methods presented in Section 2, it became apparent that such methods would be difficult to modify in order to develop a new method which satisfies the third measure of success. This is a result of the unwanted subdivision that occurs when the approximation methods of [9] and [16] are used in an adaptive algorithm as a result of the patch adjacency conditions which must be satisfied to achieve the first criterion. The approximation method of [19] also suffers from possibly unwarranted refinement when inserting new knots in either the u or v knot sequence of the progenitor surface to obtain a refined control polyhedron. In all three methods, the approximation may be refined in regions which do not warrant further definition.

To avoid the unnecessary refinement caused by patch adjacency requirements and the ripple effect caused by knot insertion, surfaces defined by meshes of triangular patches were examined and triangular interpolants were investigated. In particular, [21] developed a G^1 parametric interpolant to position and tangent data defined in a triangular mesh using triangular Bézier patches. This method was examined and then modified to produce an adaptive offset surface approximation algorithm designed to satisfy the three criteria for success. The remainder of this section is devoted to the description of this method and the modifications incorporated in order to develop an adaptive offset surface approximation method.

4.1. Bernstein-Bézier Notation

The triangular patches referenced throughout this section are of the Bernstein-Bézier form which are defined as:

$$\mathbf{b}(u, v, w) = \sum_{i+j+k=n} \mathbf{b}_{ijk} B^n_{ijk}(u, v, w),$$

where $\mathbf{b}_{ijk} \in \mathbb{R}^3$ are control points, u, v, and w are barycentric coordinates and

$$B^n_{ijk}(u, v, w) = \frac{n!}{i!\,j!\,k!} u^i v^j w^k, \quad i+j+k = n, \quad u+v+w = 1,$$

are the bivariate Bernstein polynomials of degree n. The abbreviations $\mathbf{i}$ for ijk, $|\mathbf{i}|$ for $i+j+k$ and $\mathbf{u}$ for (u, v, w) are used throughout this section to simplify some of the equations presented. Also, $\mathbf{e}_0, \mathbf{e}_1$ and $\mathbf{e}_2$ denote the vectors $[1,0,0]^T$, $[0,1,0]^T$ and $[0,0,1]^T$ respectively.

Boundary curves referenced throughout this section are Bernstein-Bézier curves which are of the form:

$$\mathbf{b}(t) = \sum_{i=0}^{n} \mathbf{b}_i B^n_i(t),$$

where $b_i \in \mathbb{R}^3$ are control points and $B^n_i(t)$ are the univariate Bernstein polynomials of degree n. Detailed development of curves and surfaces in Bernstein-Bézier representation and associated applications are found in [8].

4.2. Piper's Visually Smooth Triangular Interpolant

[21] employed a Clough-Tocher-like scheme where the mini-triangles in the split are degree elevated to quartics to facilitate generation of a visually smooth parametric interpolant to position and tangent information defined in a triangular mesh. Degree elevation was required as cubics were shown by counter-example not to have enough degrees of freedom to enable triangular Bézier patches to join in a visually smooth fashion in the parametric setting.

The basic method employed by [21] consists of generating a piecewise cubic C^0 surface which interpolates position and tangent plane data defined in a triangular mesh. Each patch in the triangular mesh is then split by joining each vertex to the centroid. This simple symmetric splitting technique is implemented in the Clough-

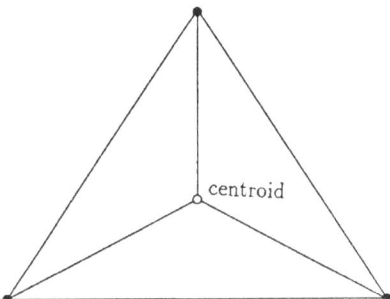

Figure 4. Splitting at the centroid

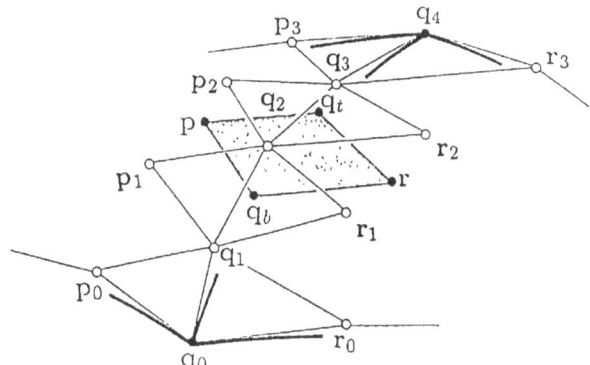

Figure 5. Candidate control points, sufficient G^1 condition

Tocher split triangle interpolant mentioned above and is discussed in [7]. Degree elevation is then performed on each sub-triangle. The unsplit triangle is referred to as the macro-triangle and the three sub-triangles resulting from the split are referred to as the mini-triangles. Geometrically this is illustrated in Fig. 4.

A sufficient G^1 condition is then used to adjust the control points $\mathbf{p}_1, \mathbf{p}_2, \mathbf{q}_2, \mathbf{r}_1$ and $\mathbf{r}_2$ shown in Fig. 5 such that neighboring patches join in a G^1 fashion. Since these adjustments lead to only continuous joins across interior edges formed by the Clough-Tocher split, the control points of the interior edges are then modified to maintain smoothness between patches according to [6]. The control points to be modified are illustrated in Fig. 6.

The control points labeled $\mathbf{a}, \mathbf{b}$ and $\mathbf{c}$ are computed as the centroids of the three that surround them. First, the three points labeled $\mathbf{a}$ are computed, then the three points labeled $\mathbf{b}$ are computed, and last the point labeled $\mathbf{c}$ can be computed.

4.3. Triangular Bézier Approximation

Piper's interpolant was modified to produce an adaptive offset surface approximation scheme. Since the cubic patches in the [21] interpolant are eventually degree

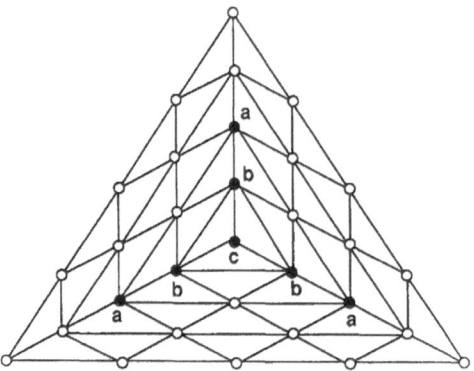

Figure 6. Interior edge adjustments

elevated to quartics, the modified interpolation scheme was designed to use quartics to generate the initial piecewise C^0 interpolant. The initial interpolant to position and tangent data in a triangular mesh was also modified to provide an adaptive interpolant to the position and tangent data of a parametric offset surface based on certain accuracy requirements. The adaptive interpolant was initiated with a simple triangulation of the offset surface domain (in the case of a rectangular domain, a square bisected along its diagonal). Triangular patches were then fit to this initial triangulation and a discrete error vector tolerance analysis performed on the boundary curves and interior of the approximating patches. Subdomains of triangular patches where accuracy requirements were not satisfied were subdivided further until a suitable piecewise C^0 interpolant was created.

The following sub-sections describe the components of the triangular offset surface approximation algorithm. The least squares techniques used to fit the boundary curves and interior points of the quartic triangular patches are described, adaptive refinement of the triangulated domain is discussed and evaluation of the resulting approximation is illustrated.

4.4. Boundary Curve Fitting

The subsection describes a method to fit the boundary curves of quartic triangular patches using least squares. The information we are given includes $u_a, v_a, \mathbf{p}, \mathbf{p}_u, \mathbf{p}_v$ and $u_b, v_b, \mathbf{q}, \mathbf{q}_u, \mathbf{q}_v$ where

$$\mathbf{p} = \mathbf{o}(u_a, v_a), \qquad \mathbf{q} = \mathbf{o}(u_b, v_b),$$
$$\mathbf{p}_u = \mathbf{o}_u(u_a, v_a), \quad \mathbf{q}_u = \mathbf{o}_u(u_b, v_b),$$
$$\mathbf{p}_v = \mathbf{o}_v(u_a, v_a), \quad \mathbf{q}_v = \mathbf{o}_v(u_b, v_b).$$

We want to find the quartic Bézier curve defined by $\mathbf{b}_i, i = 0, \ldots, 4$, such that $\mathbf{p}$ and $\mathbf{q}$ are interpolated at the endpoints and such that the tangent vectors $\mathbf{b}_1 - \mathbf{b}_0$ and $\mathbf{b}_3 - \mathbf{b}_4$ lie in the tangent planes defined by $\mathbf{p}, \mathbf{p}_u, \mathbf{p}_v$ and $\mathbf{q}, \mathbf{q}_u, \mathbf{q}_v$ respectively and such that $(\mathbf{b}(t_i) - \mathbf{o}(u_i, v_i))^2$ is minimized for $i = 1, \ldots, n$ where n is the number of fitting points used. For instance, if $n = 8, t_i, u_i$ and v_i can be computed as:

$$t_i = \frac{i}{9},$$

$$u_i = (1 - t_i)u_a + t_i u_b,$$

$$v_i = (1 - t_i)v_a + t_i v_b \quad \text{for} \quad i = 1, \ldots, 8.$$

We know $\mathbf{b}_1 - \mathbf{b}_0$ can be written as a linear combination of the partial derivatives at $\mathbf{p}$ since they span the tangent plane there. So we have:

$$\mathbf{b}_1 - \mathbf{b}_0 = \alpha_0 \mathbf{p}_u + \alpha_1 \mathbf{p}_v.$$

Similarly, we can write $\mathbf{b}_3 - \mathbf{b}_4$ as a linear combination of $\mathbf{q}_u$ and $\mathbf{q}_v$. So we also have:

$$\mathbf{b}_3 - \mathbf{b}_4 = \beta_0 \mathbf{q}_u + \beta_1 \mathbf{q}_v.$$

Since

$$\mathbf{b}(t_i) - \mathbf{o}(u_i, v_i) = \sum_{i=0}^{4} \mathbf{b}_i B_i^4(t_i) - \mathbf{o}(u_i, v_i)$$

$$= \mathbf{b}_0 B_0^4(t_i) + \mathbf{b}_1 B_1^4(t_i) + \mathbf{b}_2 B_2^4(t_i) + \mathbf{b}_3 B_3^4(t_i) + \mathbf{b}_4 B_4^4(t_i) - \mathbf{o}(u_i, v_i),$$

setting $\mathbf{b}(t_i) - \mathbf{o}(u_i, v_i) = 0$ yields:

$$\mathbf{b}_0 B_0^4(t_i) + (\mathbf{b}_0 + \alpha_0 \mathbf{p}_u + \alpha_1 \mathbf{p}_v)B_1^4(t_i) + \mathbf{b}_2 B_2^4(t_i)$$

$$+ (\mathbf{b}_4 + \beta_0 \mathbf{q}_u + \beta_1 \mathbf{q}_v)B_3^4(t_i) + \mathbf{b}_4 B_4^4(t_i) - \mathbf{o}(u_i, v_i) = 0.$$

Solving for the unknowns $\alpha_0, \alpha_1, \beta_0, \beta_1$ and $\mathbf{b}_2$ gives us:

$$\alpha_0 \mathbf{p}_u B_1^4(t_i) + \alpha_1 \mathbf{p}_v B_1^4(t_i) + \mathbf{b}_2 B_2^4(t_i) + \beta_0 \mathbf{q}_u B_3^4(t_i) + \beta_1 \mathbf{q}_v B_3^4(t_i)$$

$$= \mathbf{o}(u_i, v_i) - \mathbf{b}_0(B_0^4(t_i) + B_1^4(t_i)) - \mathbf{b}_4(B_3^4(t_i) + B_4^4(t_i)).$$

For $n = 8$, this gives us a system of 24 equations in 7 unknowns which can be solved using least squares. In matrix form we have:

$$\begin{pmatrix} \mathbf{p}_u B_1^4(t_1) & \mathbf{p}_v B_1^4(t_1) & \mathbf{e}_0 B_2^4(t_1) & \mathbf{e}_1 B_2^4(t_1) & \mathbf{e}_2 B_2^4(t_1) & \mathbf{q}_u B_3^4(t_1) & \mathbf{q}_v B_3^4(t_1) \\ \vdots & \vdots & \vdots & \vdots & \vdots & \vdots & \vdots \\ \mathbf{p}_u B_1^4(t_8) & \mathbf{p}_v B_1^4(t_8) & \mathbf{e}_0 B_2^4(t_8) & \mathbf{e}_1 B_2^4(t_8) & \mathbf{e}_2 B_2^4(t_8) & \mathbf{q}_u B_3^4(t_8) & \mathbf{q}_v B_3^4(t_8) \end{pmatrix}$$

$$\begin{pmatrix} \alpha_0 \\ \alpha_1 \\ \mathbf{b}_2 \\ \beta_0 \\ \beta_1 \end{pmatrix} = \begin{pmatrix} \mathbf{o}(u_1, v_1) - \mathbf{b}_0(B_0^4(t_1) + B_1^4(t_1)) - \mathbf{b}_4(B_3^4(t_1) + B_4^4(t_1)) \\ \vdots \\ \mathbf{o}(u_8, v_8) - \mathbf{b}_0(B_0^4(t_8) + B_1^4(t_8)) - \mathbf{b}_4(B_3^4(t_8) + B_4^4(t_8)) \end{pmatrix}.$$

4.5. *Interior Control Point Fitting*

This subsection describes a method to fit the interior control points of a quartic triangular patch using least squares. The information we are given includes $u_a, v_a, u_b, v_b, u_c, v_c$ and $\mathbf{b}_i$ where $u_a, v_a, u_b, v_b, u_c, v_c$ are parameter values of the offset surface such that:

$$\mathbf{b}_{300} = \mathbf{o}(u_a, v_a),$$
$$\mathbf{b}_{030} = \mathbf{o}(u_b, v_b),$$
$$\mathbf{b}_{003} = \mathbf{o}(u_c, v_c).$$

If $\mathbf{b}_i$ are the control points of a quartic triangular Bernstein-Bézier patch where all but the interior control points $\mathbf{b}_{112}, \mathbf{b}_{121}$ and $\mathbf{b}_{211}$ have been determined using some boundary curve fitting technique, we want to compute the interior control points such that $(\mathbf{b}(u_i, v_i, w_i) - \mathbf{o}(s_i, t_i))^2$ is minimized for $i = 0, \dots, n$ where n is the number of fitting points used. For instance, if $n = 10$, compute u_i, v_i, w_i and s_i, t_i as shown in the following **C** code segment:

```
i = 0;

for (j = 1; j <= 4; j++)
for (k = 1; k <= 5 - j; k++) {
    u[i] = i / 5.0;
    v[i] = j / 5.0;
    w[i] = 1.0 - u[i] - v[i];

    s[i] = u[i]*u_a + v[i]*u_b + w[i]*u_c;
    t[i] = u[i]*v_a + v[i]*v_b + w[i]*v_c;

    i = i + 1;
}
```

This generates a symmetric pattern of barycentric coordinates and desired corresponding surface parameter values. The symmetric pattern is illustrated in Fig. 7.

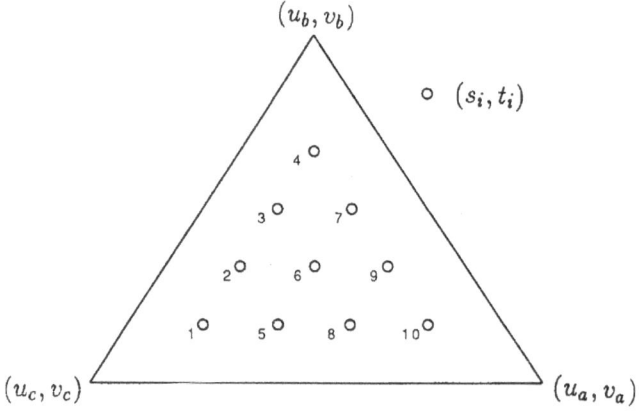

Figure 7. Parameter fitting pattern

Since

$$\mathbf{b}(\mathbf{u}_l) - \mathbf{o}(s_l, t_l) = \sum_{|i|=4} \mathbf{b}_i B_i^4(\mathbf{u}_l) - \mathbf{o}(s_l, t_l)$$

$$= \sum_{\substack{|i|=4 \\ i \neq 112,121,211}} \mathbf{b}_i B_i^4(\mathbf{u}_l)$$

$$+ \mathbf{b}_{112} B_{112}^4(\mathbf{u}_l) + \mathbf{b}_{121} B_{121}^4(\mathbf{u}_l) + \mathbf{b}_{211} B_{211}^4(\mathbf{u}_l) - \mathbf{o}(s_l, t_l),$$

setting $\mathbf{b}(\mathbf{u}_l) - \mathbf{o}(s_l, t_l) = 0$ and solving for $\mathbf{b}_{112}, \mathbf{b}_{121}$ and $\mathbf{b}_{211}$ yields:

$$\mathbf{b}_{112} B_{112}^4(\mathbf{u}_l) + \mathbf{b}_{121} B_{121}^4(\mathbf{u}_l) + \mathbf{b}_{211} B_{211}^4(\mathbf{u}_l) = \mathbf{o}(s_l, t_l) - \sum_{\substack{|i|=4 \\ i \neq 112,121,211}} \mathbf{b}_i (B_i^4(\mathbf{u}_l)).$$

For arbitrary n, this yields a system of $3n$ equations in 9 unknowns which can be solved using least squares. For $n = 10$, the result is the following system of equations:

$$\begin{pmatrix} B_{112}^4(\mathbf{u}_0) & B_{121}^4(\mathbf{u}_0) & B_{211}^4(\mathbf{u}_0) \\ \vdots & \vdots & \vdots \\ B_{112}^4(\mathbf{u}_9) & B_{121}^4(\mathbf{u}_9) & B_{211}^4(\mathbf{u}_9) \end{pmatrix} \begin{pmatrix} \mathbf{b}_{112} \\ \mathbf{b}_{121} \\ \mathbf{b}_{211} \end{pmatrix} = \begin{pmatrix} \mathbf{o}(s_0, t_0) - \sum_{\substack{|i|=4 \\ i \neq 112,121,211}} \mathbf{b}_i B_i^4(\mathbf{u}_0) \\ \vdots \\ \mathbf{o}(s_9, t_9) - \sum_{\substack{|i|=4 \\ i \neq 112,121,211}} \mathbf{b}_i B_i^4(\mathbf{u}_9) \end{pmatrix}$$

4.6. Error Vector Analysis

Once a boundary curve or interior point has been fit to a triangular patch, we must develop an estimate of how close the boundary curve or interior of the approximating patch is to a corresponding region of the offset surface. By computing a sequence of comparison parameter values and their desired corresponding parameter values of the exact offset surface, an iterative technique similar to that applied by [14] can be used to find a sequence of error vectors orthogonal to the exact offset surface. The technique applied is described in the remaining portion of this sub-section.

Given n, u_i, v_i and $\mathbf{f}_i, i = 0, \ldots, n-1$ where n is the number of comparison values u_i, v_i, and $\mathbf{f}_i$ are the corresponding points of the offset surface approximation, we want to find the maximum value of $\| \mathbf{f}_i - \mathbf{o}(\bar{u}_i, \bar{v}_i) \|$ where $\bar{u}_i, \bar{v}_i$ are adjusted values of u_i and v_i such that the error vectors are perpendicular to the exact offset surface.

The algorithm to adjust u_i and v_i until $\mathbf{f}_i - \mathbf{o}(u_i, v_i)$ is perpendicular to the offset surface is illustrated in Fig. 8. The following algorithmic notation describes the method for iterative parameter adjustment similar to that proposed by [14].

For $i = 0, \ldots, n-1$ perform the following actions:

(1) compute

$$\mathbf{o} = \mathbf{o}(u_i, v_i),$$
$$\mathbf{o}_u = \mathbf{o}_u(u_i, v_i),$$
$$\mathbf{o}_v = \mathbf{o}_v(u_i, v_i),$$
$$\mathbf{n} = \mathbf{n}(u_i, v_i) = \mathbf{o}_u \times \mathbf{o}_v / \| \mathbf{o}_u \times \mathbf{o}_v \|,$$
$$\mathbf{a} = \mathbf{f}_i - \mathbf{o}(u_i, v_i).$$

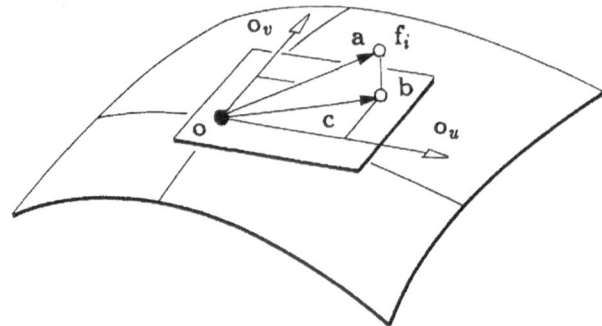

Figure 8. Error vector adjustment

(2) project $\mathbf{f}_i$ onto the tangent plane determined by $\mathbf{o}$ and $\mathbf{n}$, and call this point $\mathbf{b}$.

(3) compute $\mathbf{c} = \mathbf{b} - \mathbf{o}$.

(4) compute parameter adjustments du and dv to u_i and v_i as:

$$du = (\mathbf{c} \cdot \mathbf{o}_u)/\|\mathbf{o}_u\|^2,$$
$$dv = (\mathbf{c} \cdot \mathbf{o}_v)/\|\mathbf{o}_v\|^2.$$

(5) let

$$\bar{u} = u_i + du,$$
$$\bar{v} = v_i + dv.$$

(6) compute:

$$\bar{\mathbf{o}} = \mathbf{o}(\bar{u}, \bar{v}),$$
$$\bar{\mathbf{o}}_u = \mathbf{o}_u(\bar{u}, \bar{v}),$$
$$\bar{\mathbf{o}}_v = \mathbf{o}_v(\bar{u}, \bar{v}),$$
$$\bar{\mathbf{n}} = \mathbf{n}(\bar{u}, \bar{v}) = \bar{\mathbf{o}}_u \times \bar{\mathbf{o}}_v/\|\bar{\mathbf{o}}_u \times \bar{\mathbf{o}}_v\|,$$
$$\bar{\mathbf{a}} = \mathbf{f}_i - \bar{\mathbf{o}}(\bar{u}, \bar{v}).$$

if $\|\mathbf{a}\| > \|\bar{\mathbf{a}}\|$ then

(7) compute $\theta = arccos\left(\dfrac{\bar{\mathbf{a}}}{\|\bar{\mathbf{a}}\|} \cdot \bar{\mathbf{n}}\right)$

if $\theta < tolerance$ or $\pi - \theta < tolerance$ then

(8) stop iteration, parameter adjustment is complete.

else

(9) continue iteration after computing:

$$u_i = u_i + du,$$
$$v_i = v_i + dv.$$

else

(10) continue iteration after computing:

$$u_i = u_i + du/2,$$
$$v_i = v_i + dv/2.$$

4.7. Smoothing the C^0 Approximation

Smoothing the adjacent quartic patches along the edges of the macro-triangles of the Clough-Tocher split is accomplished using a sufficient G^1 condition presented in [8]. This is the same condition as presented by [21] but the development is simpler and the resulting equations are more tractable.

For the two adjacent quartic patches shown in 4.2, the sufficient G^1 condition for smoothness between these two patches can be stated as:

$$\frac{i}{4}\mathbf{d}_{i,4} + \left(1 - \frac{i}{4}\right)\mathbf{d}_{i,0} = \mathbf{0}, \quad i = 0, \ldots, 4$$

where

$$\mathbf{d}_{i,0} = [\alpha_0 \mathbf{p}_i + (1 - \alpha_0)\mathbf{r}_i] - [\beta_0 \mathbf{q}_i + (1 - \beta_0)\mathbf{q}_{i+1}]$$

and

$$\mathbf{d}_{i,4} = [\alpha_1 \mathbf{p}_{i-1} + (1 - \alpha_1)\mathbf{r}_{i-1}] - [\beta_1 \mathbf{q}_{i-1} + (1 - \beta_1)\mathbf{q}_i].$$

This yields 5 equations of the form:

$$\begin{aligned}
\mathbf{d}_{00} &= 0, \\
\tfrac{1}{4}\mathbf{d}_{14} + \tfrac{3}{4}\mathbf{d}_{10} &= 0, \\
\tfrac{1}{2}\mathbf{d}_{24} + \tfrac{1}{2}\mathbf{d}_{20} &= 0, \\
\tfrac{3}{4}\mathbf{d}_{34} + \tfrac{1}{4}\mathbf{d}_{30} &= 0, \\
\mathbf{d}_{44} &= 0,
\end{aligned} \tag{4.1}$$

from which the 1st and 5th equations of 4.1 can be used to determine $\alpha_0, \alpha_1, \beta_0$ and β_1 if $\mathbf{p}_0, \mathbf{r}_0, \mathbf{q}_0, \mathbf{q}_1$ and $\mathbf{p}_3, \mathbf{r}_3, \mathbf{q}_3, \mathbf{q}_4$ form respective coplanar quadrilaterals. This is precisely the situation occurring along boundaries of neighboring triangular patches in the C^0 triangular mesh in the offset surface approximation when the fitting techniques described in this section are followed.

In the adaptive offset surface approximation scheme, $\mathbf{p}_1, \mathbf{p}_2$, and $\mathbf{r}_1, \mathbf{r}_2$ are used as the points which are allowed to vary to satisfy the G^1 conditions. [21] used $\mathbf{q}_2$ as well to achieve the G^1 condition. In our scheme however, $\mathbf{q}_2$ was left fixed to avoid unnecessary modification of the boundary curves already determined to be appropriately adjusted to approximate the exact offset surface.

Solving the 2nd, 3rd and 4th equations in (4.1) for $\mathbf{p}_1, \mathbf{p}_2, \mathbf{r}_1$, and $\mathbf{r}_2$ yields the following underdetermined system of equations of the form $\mathbf{Ax} = \mathbf{b}$ for each of the coordinates of the candidate control points:

$$\begin{pmatrix} 3\alpha_0 & 3(1-\alpha_0) & 0 & 0 \\ 3\alpha_1 & 3(1-\alpha_1) & 3\alpha_0 & 3(1-\alpha_0) \\ 0 & 0 & 3\alpha_1 & 3(1-\alpha_1) \end{pmatrix} \begin{pmatrix} \mathbf{p}_1 \\ \mathbf{r}_1 \\ \mathbf{p}_2 \\ \mathbf{r}_2 \end{pmatrix}$$

$$= -\begin{pmatrix} \alpha_1 \mathbf{p}_0 + (1-\alpha_1)\mathbf{r}_0 - \beta_1 \mathbf{q}_0 - (3\beta_0 + (1-\beta_1))\mathbf{q}_1 - (3(1-\beta_0)\mathbf{q}_2 \\ -3(\beta_1\mathbf{q}_1 + (\beta_0 + (1-\beta_1))\mathbf{q}_2 + (1-\beta_0))\mathbf{q}_3) \\ \alpha_1 \mathbf{p}_3 + (1-\alpha_0)\mathbf{r}_3 - 3\beta_1 \mathbf{q}_2 - (\beta_0 + 3(1-\beta_1))\mathbf{q}_3 - (1-\beta_0)\mathbf{q}_4 \end{pmatrix},$$

where $\mathbf{A}$ is 3×4, $\mathbf{x}$ is 4×1 and $\mathbf{b}$ is 1×3.

The rows of A are linearly independent if $\alpha_0 \neq \alpha_1$. If $\alpha_0 = \alpha_1$, row 2 of the system is a linear combination of rows 1 and 3, and hence a 2×4 underdetermined system must be solved. As shown by [5] and presented here, there is a solution to the underdetermined system $Ax = b$ of the form $x = g + A^T t$ such that $\|x - g\|_2$ is minimized where g is an initial guess to the solution x.

Suppose we have a solution of the form $x - g = A^T t$ to the underdetermined system $Ax = b$ where A is an $m \times n$ matrix such that $m < n$. Consider any other solution $\bar{x} = x + d$ for some $d \neq 0$. Since $A\bar{x} = b$ we have:

$$A(x + d) = b,$$
$$A(g + A^T t + d) = b,$$
$$A(g + A^T t) + Ad = b,$$

and since $A(g + A^T t) = b$ this gives us $Ad = 0$.

Now,

$$\|\bar{x} - g\|_2^2 = [x - g]^T [x - g] + 2d^T [x - g] + d^T d$$
$$= [x - g]^T [x - g] + 2d^T A^T t + d^T d$$

and since $d^T A^T t = 0^T t = 0$, we have:

$$\|\bar{x} - g\|_2^2 = [x - g]^T [x - g] + d^T d$$
$$= \|x - g\|_2^2 + d^T d$$
$$> \|x - g\|_2^2 \quad \forall d \neq 0, \forall \bar{x} \neq x.$$

Using this result, the solution x such that $\|x - g\|_2$ is minimized may be determined assuming $x = g + A^T t$, so $Ax = b$ becomes:

$$A(g + A^T t) = b,$$
$$Ag + AA^T t = b,$$
$$AA^T t = b - Ag,$$

and solving for t yields

$$t = (AA^T)^{-1}(b - Ag),$$

so then x may be computed as $g + A^T t$.

The values of α_0 and α_1 when $\alpha_0 \approx \alpha_1$ gives us a matrix A which is near rank deficiency. In most practical cases within the context of the offset surface approximation scheme, α_0 and α_1 are similar in value. The decision of when to use the 2×4 system instead of the 3×4 system was based on the condition number of AA^T, $K(AA^T)$. For any matrix A, $K(A)$ can be written as

$$K(A) = \|A\| \|A^{-1}\|$$

and

$$A^{-1} = \frac{1}{\det(A)} \text{adj}(A),$$

where $\|\cdot\|$ is any matrix norm, det(**A**) is the determinant of **A** and adj(**A**) is the adjoint of **A** (the transpose of the matrix of cofactors of **A**). The Frobenius norm $\|\mathbf{A}\| = \sqrt{\sum_{i=0}^{m}\sum_{j=0}^{n} a_{ij}^2}$ will be used in the following development for the sake of the simplicity it provides.

The matrix $\mathbf{A}\mathbf{A}^T$ can be written entirely in terms of α_0, and α_1 and can be shown to be of the following form:

$$\mathbf{A}\mathbf{A}^T = \begin{pmatrix} a & d & 0 \\ d & b & d \\ 0 & d & c \end{pmatrix},$$

where

$$a = \alpha_0^2 + (1 - \alpha_0)^2,$$
$$b = \alpha_1^2 + (1 - \alpha_1)^2 + \alpha_0^2 + (1 - \alpha_0)^2,$$
$$c = \alpha_1^2 + (1 - \alpha_1)^2,$$
$$d = \alpha_0\alpha_1 + (1 - \alpha_0)(1 - \alpha_1).$$

Computing det($\mathbf{A}\mathbf{A}^T$) and adj($\mathbf{A}\mathbf{A}^T$) we get:

$$\det(\mathbf{A}\mathbf{A}^T) = \frac{1}{abc - d^2(a + c)}$$

and

$$\mathrm{adj}((\mathbf{A}\mathbf{A}^T)^{-1}) = \begin{pmatrix} bc - d^2 & -dc & d^2 \\ -dc & ac & -ad \\ d^2 & -ad & ab - d^2 \end{pmatrix},$$

so

$$(\mathbf{A}\mathbf{A}^T)^{-1} = \frac{1}{abc - d^2(a + c)} \begin{pmatrix} bc - d^2 & -dc & d^2 \\ -dc & ac & -ad \\ d^2 & -ad & ab - d^2 \end{pmatrix}$$

using Cramer's rule. Taking the norm of $\mathbf{A}\mathbf{A}^T$ and $(\mathbf{A}\mathbf{A}^T)^{-1}$ we get:

$$\|\mathbf{A}\mathbf{A}^T\| = \sqrt{a^2 + b^2 + c^2 + 4d^2},$$

and

$$\|(\mathbf{A}\mathbf{A}^T)^{-1}\|$$
$$= \frac{1}{abc - d^2(a + c)}\sqrt{(bc - d^2)^2 + (ac)^2 + (ab - d^2)^2 + 2(dc)^2 + 2(ad)^2 + 2(d^2)^2},$$

where a, b, c and d are known scalar quantities. $K(\mathbf{A}\mathbf{A}^T)$ can now be written as a function of α_0 and α_1.

Since double precision computations yielding an approximate accuracy of 16 decimal digits were used in the implementation of the approximation scheme, and since every factor of 10 in $K(\mathbf{A}\mathbf{A}^T)$ typically reduces the accuracy of the solution by one digit of precision [2], A condition number tolerance of 10^8 was considered a good indication that the solution of the 3×4 system would be inaccurate and the 2×4 system should be used. The 2×4 underdetermined system can be analyzed using the same techniques yielding:

$$K(\mathbf{A}\mathbf{A}^T) = \frac{1}{ac}(a^2 + c^2).$$

This shows that the matrix $\mathbf{A}\mathbf{A}^T$ is well conditioned for all values of α_0 and $\alpha_1 \in (0, 1)$.

4.8. Refining the Domain Triangulation

The adaptive refinement of the triangular mesh in the offset surface approximation scheme was accomplished using bisection along the longest edge of domain triangles. At any point in the fitting process of the C^0 approximation, a search is performed to determine the approximating patch in the triangular mesh with the worst domain triangle. The quality of the domain triangles was determined by computing the following normalized shape ratio [3].

$$nsr = 2r/R,$$

where r is the radius of the triangle's incircle and R is the radius of the triangle's circumcircle.

For a triangle with vertices $\mathbf{a}$, $\mathbf{b}$ and $\mathbf{c} \in \mathbb{R}^2$, let l_{ab}, l_{bc} and l_{ac} represent the lengths of the respective sides of triangle $\mathbf{abc}$. Let S be the semiperimeter and A be the area of triangle $\mathbf{abc}$. Then S and A are defined as:

$$S = 1/2(l_{ab} + l_{bc} + l_{ac})$$

and

$$A = 1/2 \begin{pmatrix} a_x & b_x & c_x \\ a_y & b_y & c_y \\ 1 & 1 & 1 \end{pmatrix}$$

and the inradius and circumradius are defined as:

$$r = A/S$$

and

$$R = (l_{ab}l_{bc}l_{ac})/4A.$$

These formulas are given by [10]. The normalized shape ratio equals 1 when the triangle being considered is equilateral and as the triangle becomes skinnier, the ratio approaches 0. With that in mind, a smaller ratio corresponds to a worse triangle. [4] point out that the normalized shape ratio favors skinny triangles with obtuse angles less than other skinny triangles.

This adaptive refinement strategy was used as a result of the research presented by [4, 10, 22, 25] which indicates that such a scheme will produce the minimal amount of refinement necessary with the best quality among the subtriangles. In addition, the irregular triangular subdivision techniques based on the location of the largest errors that occur on the boundary of individual triangles presented in [23], yielded poor results in comparison. Comparisons were based on both in the total number of triangles produced and the quality of individual triangles as measured with the *normalized shape ratio*.

4.9. Evaluation of the Triangular Approximation

Once the triangular offset surface approximation has been computed, the approximation must be evaluated for rendering and further analysis. Evaluation of the triangular offset surface approximation at an arbitrary domain point $\mathbf{p} = [u, v]^T$, requires the identification of the correct mini-triangle within correct macro-triangle among all the triangles defined by the triangular mesh. This may be accomplished using barycentric coordinates.

With an initial domain triangle **abc** and the evaluation point **p**, compute the barycentric coordinates of **p** with respect to triangle **abc**. The barycentric coordinates r, s, and t of **p** with respect to **abc** are:

$$r = \text{area}(\mathbf{p}, \mathbf{b}, \mathbf{c})/\text{area}(\mathbf{a}, \mathbf{b}, \mathbf{c}),$$
$$s = \text{area}(\mathbf{a}, \mathbf{p}, \mathbf{c})/\text{area}(\mathbf{a}, \mathbf{b}, \mathbf{c}),$$
$$t = \text{area}(\mathbf{a}, \mathbf{b}, \mathbf{p})/\text{area}(\mathbf{a}, \mathbf{b}, \mathbf{c}),$$

where

$$\text{area}(\mathbf{a}, \mathbf{b}, \mathbf{c}) = 1/2 \begin{pmatrix} a_x & b_x & c_x \\ a_y & b_y & c_y \\ 1 & 1 & 1 \end{pmatrix}.$$

These formulas can be found in [3].

If r, s and t are all greater than or equal to zero, then **p** lies within triangle **abc**. If r, s or t is less than zero then the point **p** lies within another domain triangle located opposite to one the edges of triangle **abc** as indicated in Fig. 9.

If a negative barycentric coordinate is computed, the appropriate neighbor of triangle **abc** is tested. For instance, if the barycentric coordinate u is negative, the neighbor of triangle **abc** along edge **bc** is determined as the next search candidate. This technique is repeated until **p** is found to be within one of the domain triangles of the triangular mesh. Once the macro-triangle has been correctly identified, the correct mini-triangle can be identified similarly. The barycentric coordinates used in the identification of the correct mini-triangle are the coordinates used in the evaluation of the quartic patch defined for this subtriangle. This resulting point is the point in the approximate offset surface which corresponds to the point **p**.

5. Approximation Implementations

Implementation specifics are described for a rectangular based and a triangular based parametric offset surface approximation scheme using the C programming

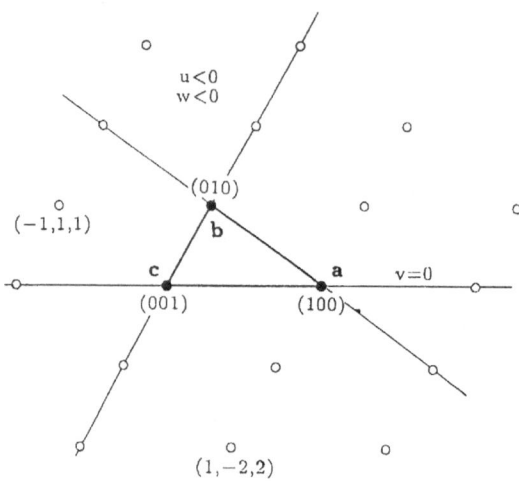

Figure 9. Barycentric coordinates wrt triangle **abc**

language. Details include several data structures and utility functions used for efficient storage and manipulation. Finally, some examples of the rectangular approximation and the triangular approximation are compared in terms of the total number of patches required to generate the approximation and the effects of the parameters used in the triangular scheme.

5.1. Rectangular Offset Approximation Implementation

The ideas of [9] described in Section 3, were used in a rectangle based offset surface approximation scheme. The implementation used adaptive subdivision to develop a quadtree representation of the approximate offset surface where each leaf on the quadtree corresponds to a bicubic Hermite approximating patch defined on a subdomain of the offset surface. In this adaptive scheme, uniform n-fold subdivision was not enforced. In such an implementation, the one-to-one patch adjacency requirements are not maintained and hence the surface is not continuous along boundaries of subpatches which do not correspond one-to-one with their neighbors.

Despite the discontinuity in the approximation, this rectangle based approximation scheme provided an initial approximation scheme to build upon. A reasonable user interface for an interactive offset surface approximation program was established based on the experience gained from this initial implementation. The differential analysis proposed by [9] to determine the maximum error attained by the approximate offset surface was not implemented. Instead, the accuracy achieved by each subpatch in the approximation was determined to be accurate enough through a discrete error tolerance comparison. In these comparisons, the error vectors were adjusted until they were perpendicular to the offset surface as suggested by [16, 19].

The data structure used to implement the rectangular approximation method is illustrated in the following C code statements:

```
typedef struct rnode RNODE;

struct rnode {
  int accepted;
  int level;
  double umin, vmin;
  double umax, vmax;
  double h[4] [4] [3];
  RNODE *quad[4];
};
```

In the RNODE structure, accepted indicates whether a particular subpatch has been accepted as an accurate approximation, level indicates the level of subdivision of the subpatch, umin, vmin, umax and vmax are the values of the corners of the subpatch domain, h stores the coefficients of the bicubic Hermite approximating patch defined over the subpatch domain and quad stores pointers to subpatches at the next level of subdivision.

5.2. Triangular Offset Approximation Implementation

A triangular based offset surface approximation method was implemented as described in Section 4. The data structure used to implement the triangular approximation method is illustrated in the following C code statements:

```
typedef struct tnode TNODE;

struct tnode {
  int accepted;
  int processed;
  int smoothed[3];
  int d[3];
  double ratio;
  double b[15] [3];
  double B[3] [15] [3];
  TNODE *nbr[3];
};
```

In the TNODE structure, accepted indicates whether a particular macro-triangle has been accepted as an accurate approximation, processed indicates whether the macro triangle has been split into mini-triangles and each mini-triangle has been degree elevated to quartic degree, smoothed indicates whether a particular mini-triangle and its neighbor have been smoothed, d stores indices of a global domain point array corresponding to the macro-triangle's vertices, ratio stores the *normalized shape ratio* of the domain triangle, b stores the control net of the macro-triangle, B stores the control net of the mini-triangles and nbr stores pointers to the neighbors of the macro-triangle.

As shown in the TNODE structure, control nets of macro-triangles and mini-triangles in the triangular mesh are stored in linear arrays of points. This storage technique requires routines to compute the total number of points in the control net of a nth degree triangular patch and to map the ijk indices of the b_{ijk}'s to the i indices of the b[i]'s. The following macros provide a consistent method of coding the various functions required without the expense of a function call:

```
#define TRINUM(n) (((n+1)*(n+2))>>1)
#define NMAP(n) ((n<<1)+3)
#define TPIMAP(base,i,j) (((i)*base-(i))>>1)+(j)).
```

The macro TRINUM() computes the triangles numbers $(n + 1)(n + 2)/2$ and provides the number of points in the control net of an nth degree triangular patch. The macro NMAP() computes the number of control points along the boundary of an nth degree triangular patch. The macro TPIMAP() computes a triangular patch control point index using the number of control points along the boundary of an nth degree triangular patch (base) and the i and j components of an ijk index. The formula for the triangle numbers was taken from [8]. The formulas for the macros NMAP and TPIMAP were developed from material presented in [1] and [24]. Figure 10 illustrates the associated mapping these functions provide for quartic patches and Fig. 11 shows how the mini-triangles are stored in the B[i]'s.

5.3. Observation and Comparison

The results produced by the rectangle based approximation and the triangle based approximation were observed and comparisons made based on the criteria for a successful approximation method as stated in Section 4.

The criterion for smoothness is not statisfied by the rectangular approximation as one-to-one patch adjacency is not maintained. C^1 continuity could be attained, however, if uniform subdivision were performed to the highest level of subdivision attained as a post process after the adaptive rectangular approximation is complete. For a maximum level of subdivision, $l_{max} > 0$, this results in a total of $2^{2l_{max}}$ patches. The criterion for smoothness is satisfied by the triangular approximation method as the resulting approximation is G^1.

The accuracy criterion is satisfied in a similar fashion in both implementations as error tolerance analysis is performed at a discrete number of points in both approximation methods.

Further comparison of the two approximation methods was accomplished by examining the total number of surface patches required for approximation and the aesthetics of the surfaces produced. Four different rational bicubic Bézier progenitor surfaces were used as test cases for the offset surface approximation schemes. All test cases were contained in the unit cube. An error tolerance of 10^{-3} was used in all approximations. The results are shown in Figs. 12–19.

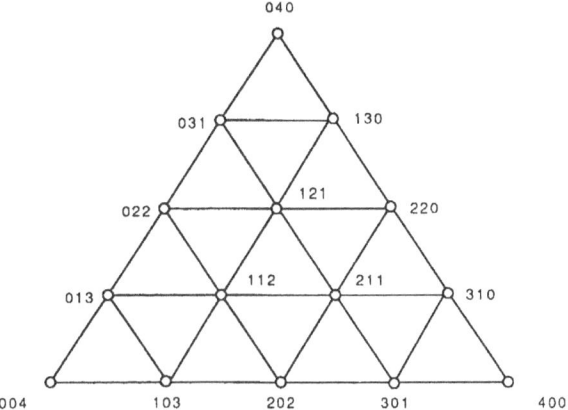

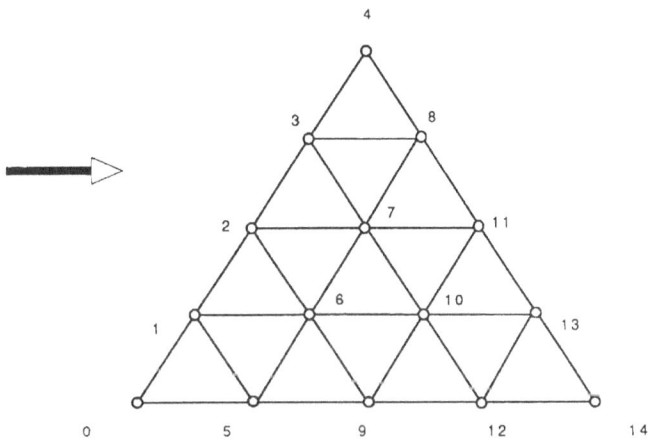

Figure 10. Quartic triangular patch index mapping

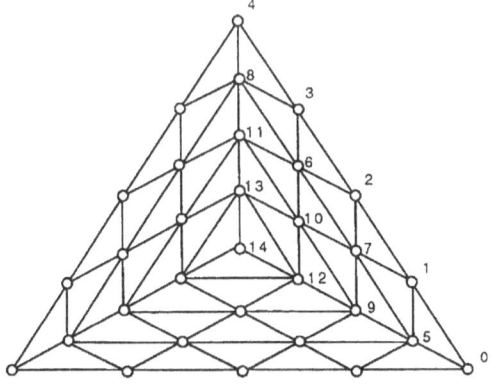

Figure 11. Mini-triangle storage

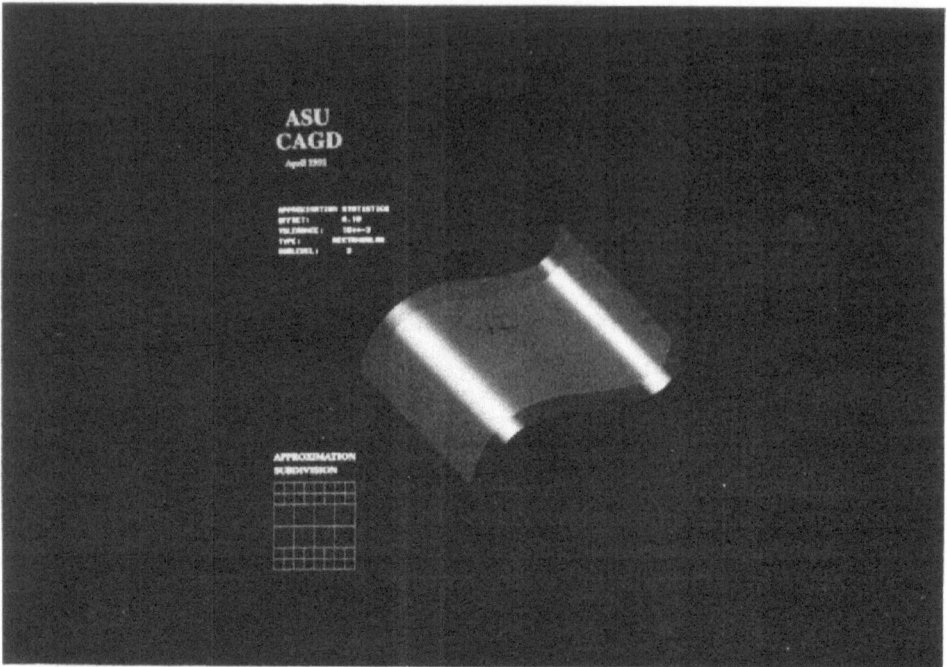

Figure 12. Adaptive rectangular offset surface approximation. Rational cubic Bézier progenitor surface, test case 1

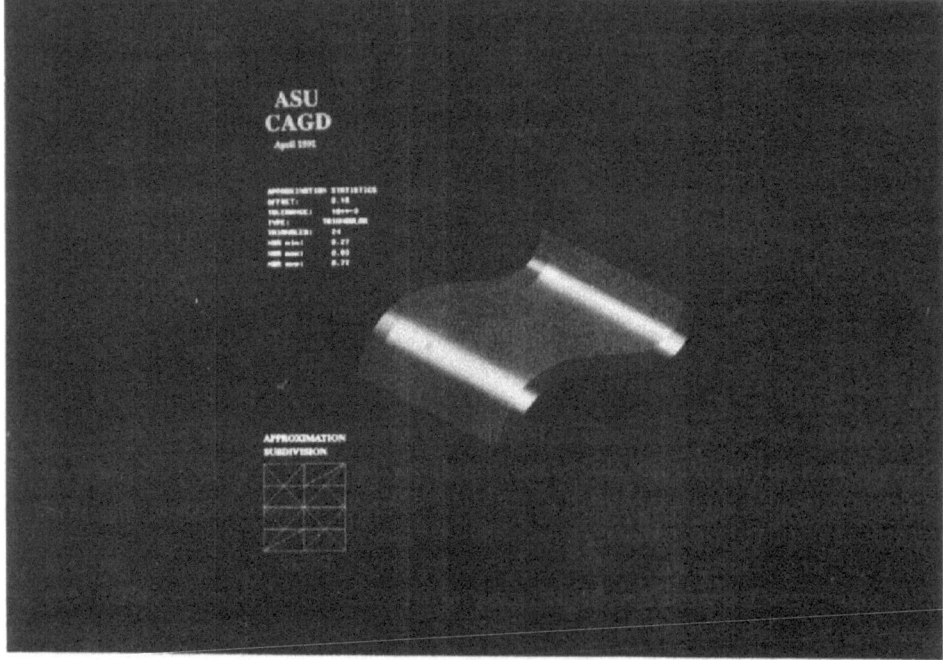

Figure 13. Adaptive triangular offset surface approximation. Rational cubic Bézier progenitor surface, test case 1

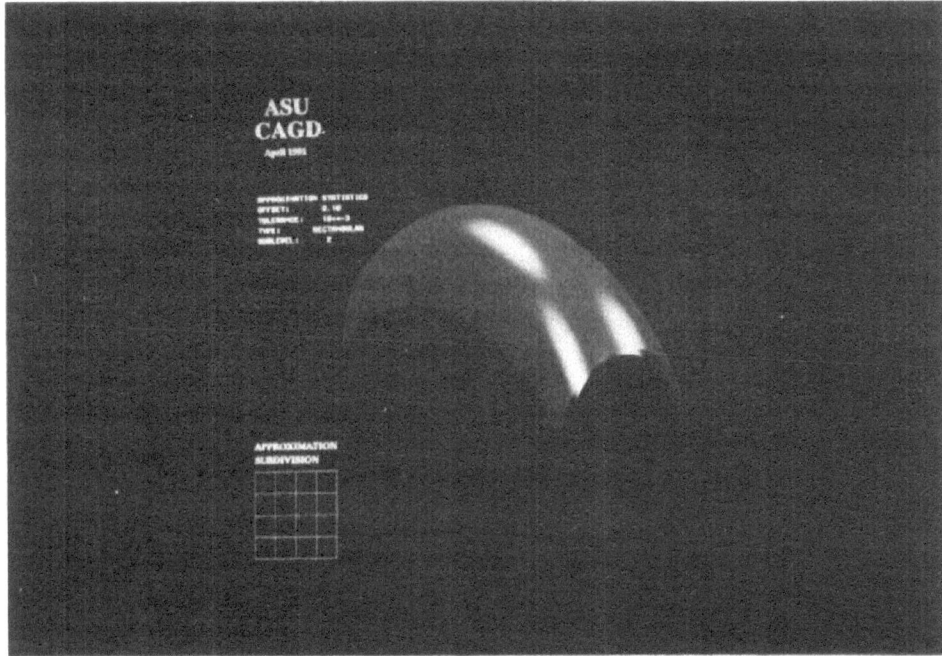

Figure 14. Adaptive rectangular offset surface approximation. Rational cubic Bézier progenitor surface, test case 2

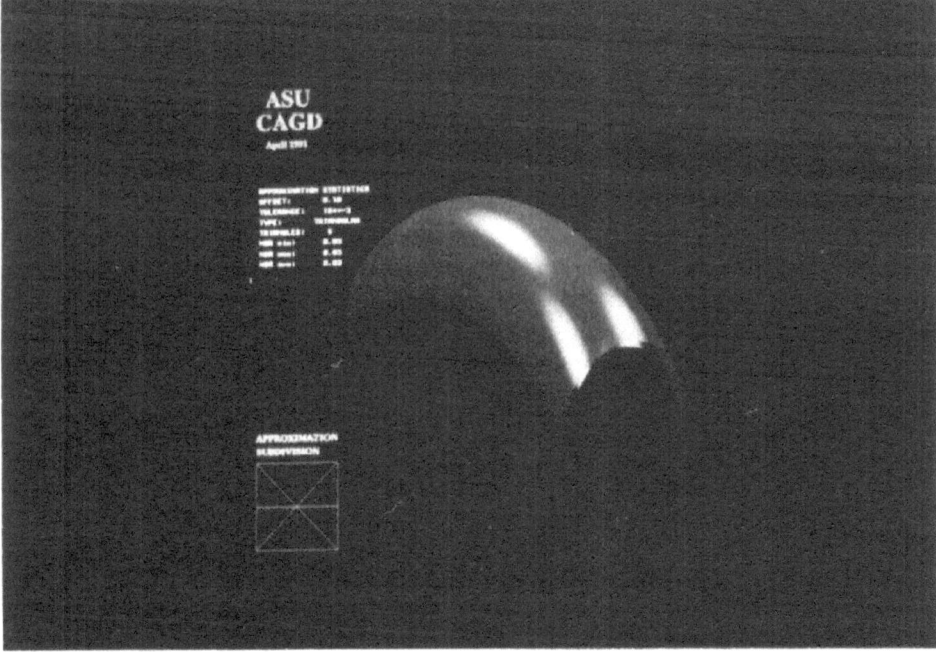

Figure 15. Adaptive triangular offset surface approximation. Rational cubic Bézier progenitor surface, test case 2

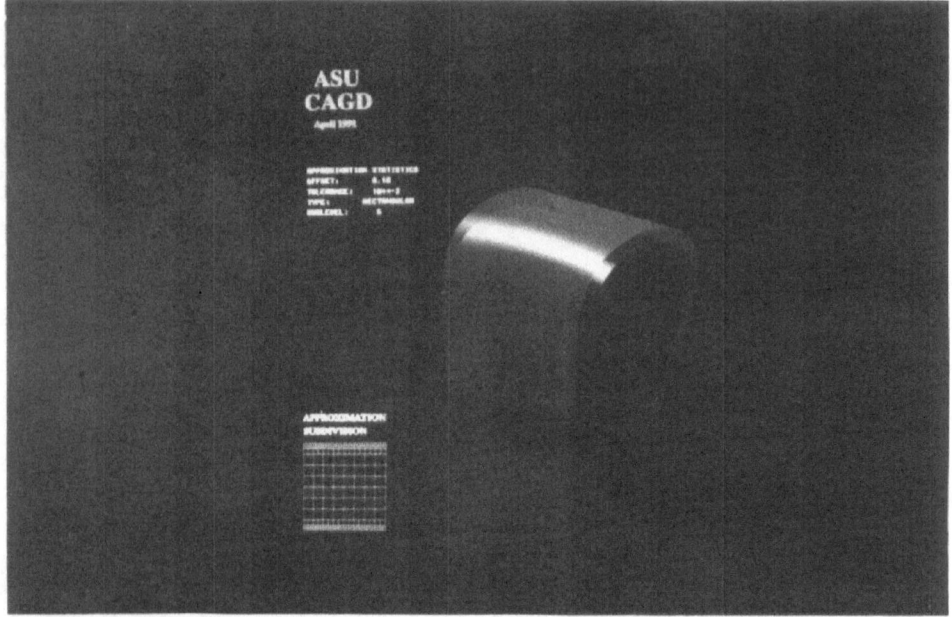

Figure 16. Adaptive rectangular offset surface approximation. Rational cubic Bézier progenitor surface, test case 3

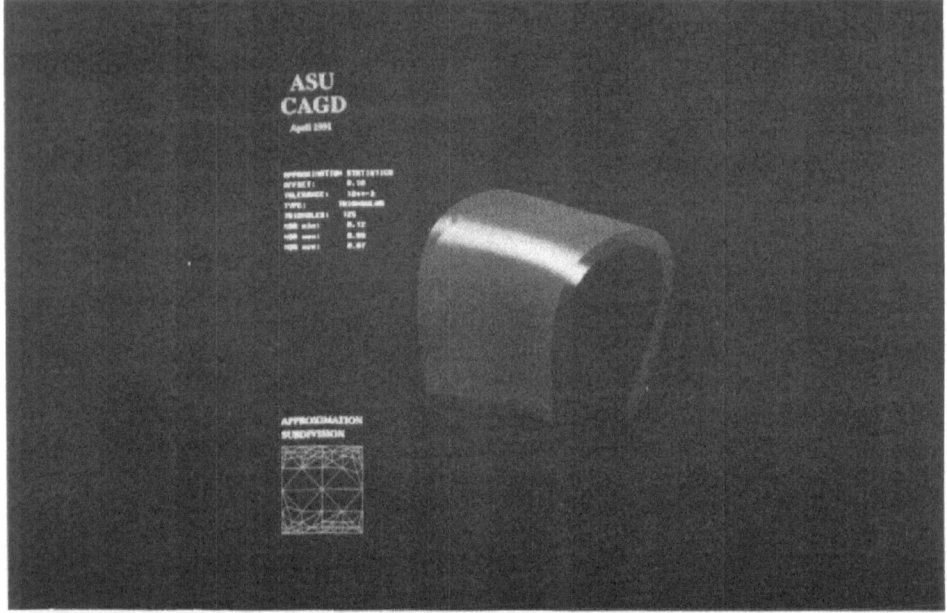

Figure 17. Adaptive triangular offset surface approximation. Rational cubic Bézier progenitor surface, test case 3

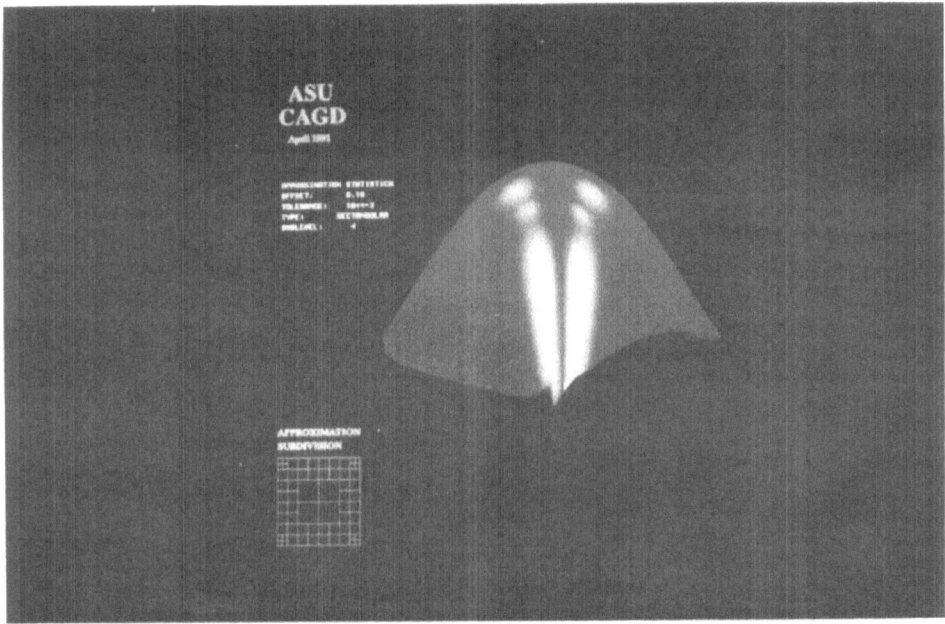

Figure 18. Adaptive rectangular offset surface approximation. Rational cubic Bézier progenitor surface, test case 4

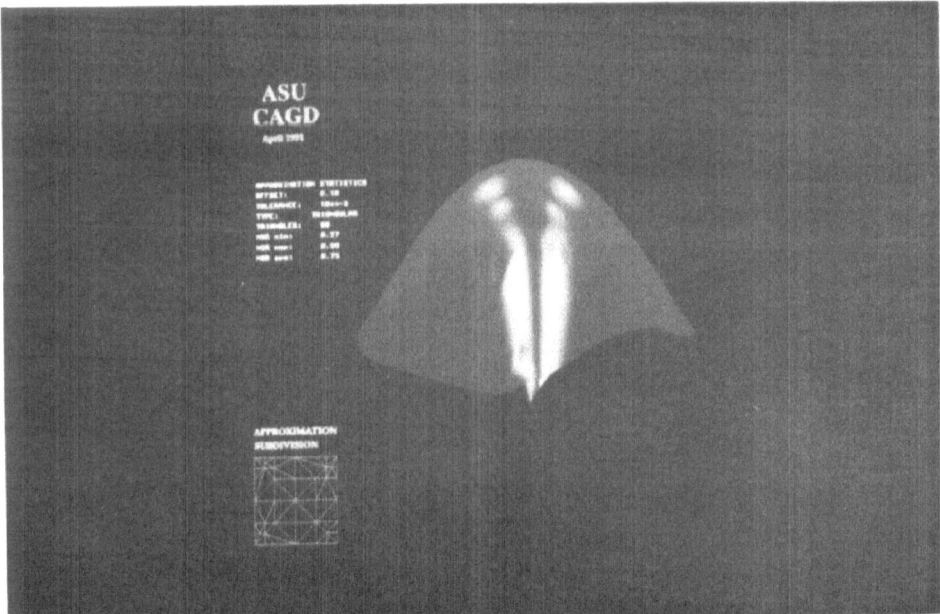

Figure 19. Adaptive triangular offset surface approximation. Rational cubic Bézier progenitor surface, test case 4

In all examples of the triangular approximation shown, least squares techniques were employed to fit quartic triangular patches in the C^0 approximation. In all test cases, the total number of patches produced by the triangular approximation is comparable or considerably less than the total number of patches produced by the rectangular approximation. The total number of patches considered for the rectangular approximation is that required for C^1 approximation ($2^{2l_{max}}$). The total number of patches considered for the triangular approximation is that required for G^1 approximation (three times the number of triangular patches required for the C^0 approximation). The number of patches produced by the two methods of approximation can be determined by examination of the subdivision map labeled *APPROXIMATION SUBDIVISION* in the photos. The illustrations of these test cases are shown in Figs. 12–19. The test cases provide a transition from surfaces which are fairly easy to approximate to surfaces which require much subdivision to generate a reasonable approximation. In all illustrations, the progenitor surface is shaded red (opaque) and the offset surface approximation is shaded blue (translucent).

The triangular approximation method was implemented such that the smoothing process was accomplished after the initial fitting process had been completed. By examining the triangular approximation before smoothing was performed, it was observed that in many cases, the initial piecewise C^0 surface appeared to be smoother than the G^1 surface. It appeared that in some test cases, oscillations resulted from the application of the smoothing process. It was also noted that if oscillation occurred in a test case, changing the fitting technique employed did not make any difference. This suggests that oscillations are introduced in some approximations as the result of a lack of flexibility in the sufficient G^1 condition used in the smoothing process.

6. Conclusion and Further Research

One of the goals of this research was to develop an adaptive offset surface approximation method for producing an accurate, smooth result with a minimal number of approximating patches. Triangular Bernstein-Bézier patches were found to posses the necessary properties to accomplish this goal. Triangular patches lend themselves to a variety of boundary curve fitting and interior point fitting techniques which can be coded efficiently. The triangular offset surface approximation method proposed was partially successful in that the piecewise C^0 approximate satisfied the criteria for an adaptive algorithm producing a minimal amount of accurate approximating patches. However, the oscillations produced when generating the G^1 surface using a sufficient G^1 condition indicate that the smoothing process requires refinement and perhaps the sufficient G^1 condition needs to be reformulated.

Further research needs to be done to develop a more flexible G^1 condition so that oscillations are not introduced into the triangular approximation. Once this problem is solved, various extensions to the triangular approximation should be considered. These extensions include using higher degree triangular patches, rational triangular patches and the development of more accurate methods of fitting boundary curves and interior points. Methods incorporating these ideas should be tuned to satisfy the smoothness, accuracy and minimality criteria within the context of an adaptive offset surface approximation scheme.

Acknowledgements

The authors would like to thank Gerald Farin and Thomas Foley for their contributions in the development of the triangular approximation algorithm. We would also like to thank Wayne Woodland for the photography used to illustrate this research. This research was supported by NSF grant DMC-8807747, NSF/AFOSR grant DMS-9116930 and DOE contract DEFG0287ER25041.

References

[1] Alfeld, P.: A trivariate Clough-Toucher scheme for tetrahedral data. Comput. Aided Geom. Des. *1*, 169-181 (1984).

[2] Atkinson, K. E.: An introduction to numerical analysis. New York: J. Wiley 1978.

[3] Barnhill, R. E.: Representation and approximation of surfaces. In: Mathematical software III. (Rice, J. R., ed.), pp. 69-129. San Diego: Academic Press 1977.

[4] Barnhill, R. E., Farin, G., Hansford, D.: A subdivision process for triangular mesh refinement. Technical Report TR 93/005, Computer Science Department, Arizona State University 1993.

[5] Boehm, W., Gose, G., Kahmann, J.: it Methoden der Numerischen Mathematik. Braunschweig: Vieweg 1985.

[6] Farin, G. E.: Smooth interpolation to scattered 3D data. In: Surfaces in CAGD (Barnhill, R. E., Boehm, W., eds.), pp. 43-63. Amsterdam: North-Holland 1983.

[7] Farin, G. E.: Triangular Bernstein-Bézier patches. Comput. Aided Geom. Des. *3*, 83-127 (1986).

[8] Farin, G. E.: Curves and surfaces for computer aided geometric design, 2nd edn. San Diego: Academic Press 1990.

[9] Farouki, R. T.: The approximation of non-degenerate offset surfaces. Comput. Aided Geom. Des. *3*, 15-43 (1986).

[10] Frey, W. H.: Selective refinement: a new strategy for automatic node placement in graded triangular meshes. General Motors Research Publication GMR-5432, Mathematics Department, 1986.

[11] Frost, T. M.: Parametric offset surfaces. M.S. Thesis, Department of Computer Science and Engineering, Arizona State University, Tempe, 1991.

[12] Gregory, J. A.: Geometric continuity. In: Mathematical methods in computer aided geometric design (Lyche, T., Schumaker, L., eds.), pp. 353-372. New York: Academic Press 1989.

[13] Hoschek, J.: Offset curves in the plane. Comput. Aided Des. *17*, 77-82 (1985).

[14] Hoschek, J.: Intrinsic parametrization for approximation. Comput. Aided Geom. Des. *5*, 27-31 (1988).

[15] Hoschek, J., Wissel, N.: Optimal approximate conversion of spline curves and spline approximation of offset curves. Comput. Aided Des. *20*, 475-483 (1988).

[16] Hoschek, J., Schneider, F. J., Wassum, P.: Optimal approximate conversion of spline surfaces. Comput. Aided Geom. Des. *6*, 293-306 (1989).

[17] Klass, R.: An offset spline approximation for plane cubic splines. Comput. Aided Des. *15*, 297-299 (1983).

[18] Lyche, T., Cohen, E., Morken E.: Knot refinement algorithms for tensor product B-spline surfaces. Comput. Aided Geom. Des. *2*, 133-139 (1985).

[19] Patrikalakis, N. M., Prakash, P. V.: Free-form plate modeling using offset surfaces. In: Computers in offshore and arctic engineering—1987 (Chung, J. S., Angelides, D., eds.), pp. 37-44.

[20] Pham, B.: Offset approximation of uniform B-splines. Comput. Aided Des. *20*, 471-474 (1988).

[21] Piper, B. R.: Visually smooth interpolation with triangular Bézier patches. In: Geometric modeling: algorithms and new trends (Farin, G., ed.), pp. 221-233. Philadelphia: SIAM 1987.

[22] Rosenberg, I. G., Stenger, F.: A lower bound on the angles of triangles constructed by bisecting the longest side. Comput. *29*, 390-395 (1975).

[23] Scarlatos, L., Pavlidis, T.: Hierarchical triangulation using terrain features. In: Visualization '90. (Kaufman, A., ed.), pp. 168–185. IEEE Computer Society, Los Alamitos, 1990.
[24] Schumaker, L. L., Volk, W.: Efficient evaluation of multivariate polynomials. Comput. Aided Geom. Des. *3*, 149–154 (1986).
[25] Stynes, M.: On faster convergence of the bisection method for all triangles. Math. Comput. *35*, 1195–1201 (1980).
[26] Tiller, W., Hanson, E. G.: Offsets of two-dimensional profiles. IEEE Comput. Graphics Appl. *4*, 36–46 (1984).

Dr. R. E. Barnhill
Dr. T. M. Frost
Computer Science Department
Arizona State University
Tempe, AZ 85287-5406
U.S.A.

Computing Suppl. 10, 35–41 (1995)

Unimodal Properties of Generalized Ball Bases

P. J. Barry, Minneapolis, and R. N. Goldman, Houston

Abstract. A necessary and sufficient condition is derived for a sequence of functions $b_0(t), \ldots, b_n(t)$ to be unimodal in an interval $[a, b]$. This condition is applied to show that the Generalized Ball basis of degree n is always unimodal whenever n is odd and is never unimodal whenever n is even except for the cases $n = 2, 4$. A new proof of the unimodality of the Bernstein basis is also provided.

Key words: Ball basis, Bernstein basis, unimodality.

1. Introduction

Ball's basis shares many important properties with the Bernstein basis. Curves represented using Ball's basis as blending functions are affine invariant, lie in the convex hull of their control vertices, interpolate their first and last control points, are nondegenerate, and satisfy the variation diminishing property [5, 6, 8]. Moreover, there is a recursive evaluation algorithm for curves written in terms of Ball's basis which is more efficient than de Casteljau's recursive evaluation algorithm for curves of the same degree represented in Bezier form [5, 8].

Ball first introduced his basis for cubic polynomials [1–3]. Later this basis was generalized to polynomials of arbitrary odd degrees by Goodman and Said [6, 8], who also noted that these generalized basis functions are closely related to two point Hermite interpolation. The purpose of this paper is to show that these generalized Ball bases of odd degree are unimodal.

A sequence of functions $b_0(t), \ldots, b_n(t)$ is said to be *unimodal* in the interval $[a, b]$ if and only if for any parameter $t^* \in [a, b]$ and any index i such that $b_i(t^*) \geq b_j(t^*)$ for all j

$$\text{if } k < j < i \text{ or } i < j < k, \text{ then } b_j(t^*) \geq b_k(t^*).$$

Thus a sequence of functions $b_0(t), \ldots, b_n(t)$ is said to be unimodal in an interval $[a, b]$ if and only if the sequence of numbers $b_0(t^*), \ldots, b_n(t^*)$ has only one local maximum for each $t^* \in [a, b]$. The significance of this unimodality property for curve design is discussed in Barry et al. [4], where it is proved that the Bernstein bases of arbitrary degree, as well as certain other spline bases common in computer aided design, are unimodal.

In Section 2 we derive a necessary and sufficient condition for a sequence of functions $b_0(t), \ldots, b_n(t)$ to be unimodal in an interval $[a, b]$, and in Section 3 we apply these criteria to show that the generalized Ball bases of odd degree are unimodal in the interval $[0, 1]$. We also provide a plausible extension of Ball's basis to even degrees, and we show that these even degree basis functions are never unimodal for any degree greater than four. Finally we use the unimodality criterion developed in Section 2 to furnish a new proof that the Bernstein bases of arbitrary degree are unimodal in the interval $[0, 1]$.

2. A Necessary and Sufficient Condition for Unimodality

The following Lemma provides necessary and sufficient conditions for the sequence of functions $b_0(t), \ldots, b_n(t)$ to be unimodal in the interval $[a, b]$.

Lemma. *Suppose that $b_0(t), \ldots, b_n(t)$ is a sequence of continuous functions defined on the interval $[a, b]$ which satisfy the following two conditions:*

(i) *Each difference $b_k(t) - b_{k-1}(t)$ has exactly one (counting multiplicities) real root r_k in the interval (a, b) for $k = 1, \ldots, n$. This means that if $u \in (a, r_k)$ and $v \in (r_k, b)$, then $\text{sign}\{b_k(u) - b_{k-1}(u)\} = -\text{sign}\{b_k(v) - b_{k-1}(v)\}$ for $k = 1, \ldots, n$.*

(ii) *Let $r_0 = a$ and $r_{n+1} = b$. Then there is an index $0 \le i \le n$ and a parameter t^* for which*

    a. $r_i < t^* < r_{i+1}$

    b. $b_i(t^*) > b_j(t^*)$ *for all j and the sequence of numbers $b_0(t^*), \ldots, b_n(t^*)$ has only one local maximum.*

Then the sequence of functions $b_0(t), \ldots, b_n(t)$ is unimodal in the interval $[a, b]$ if and only if

(iii) $a = r_0 < r_1 \le r_2 \le \cdots \le r_{n-1} \le r_n < r_{n+1} = b$.

Moreover in this case, $b_k(t) \ge b_j(t) \ge (t)$ for all $t \in [r_k, r_{k+1}]$ and all $0 \le j, k \le n$.

Proof: First suppose that Conditions (i), (ii), and (iii) are satisfied; we shall show that the sequence of functions $b_0(t), \ldots, b_n(t)$ is unimodal in the interval $[a, b]$ and that $b_k(t)$ is maximal for all $t \in [r_k, r_{k+1}]$. We begin by examining the interval $[r_i, r_{i+1}]$. Condition (ii) says that the sequence of numbers $b_0(t^*), \ldots, b_n(t^*)$ has only one local maximum. Since the functions $b_0(t), \ldots, b_n(t)$ are continuous, two functions with adjacent indices cannot change relative size until they attain equality. Therefore by Conditions (ii) and (iii), the sequence of functions $b_0(t), \ldots, b_n(t)$ must be unimodal over the entire interval $[r_i, r_{i+1}]$ and $b_i(t) \ge b_j(t)$ for all $t \in [r_i, r_{i+1}]$.

Now we proceed by induction on k to show that the sequence of functions $b_0(t), \ldots, b_n(t)$ is unimodal in $[r_k, r_{k+1}]$ for $i \le k \le n$ and that $b_k(t)$ is maximal for all $t \in [r_k, r_{k+1}]$. By the inductive hypothesis if $k > i$, then the sequence of functions $b_0(t), \ldots, b_n(t)$ is unimodal in $[r_{k-1}, r_k]$ and $b_{k-1}(t)$ is maximal for all t in this interval. Furthermore $b_k(r_k) - b_{k-1}(r_k) = 0$, and by assumption r_k is a root of multiplicity one; moreover r_k is the only parameter in (a, b) where $b_k(t) = b_{k-1}(t)$. Therefore for all $t \in [r_k, r_{k+1}]$, $b_{k-1}(t) \le b_k(t)$. Since, in addition, none of the other functions with adjacent indices can change their relative sizes until they attain equality, it follows from Condition (iii) that their relative sizes remain invariant in $[r_k, r_{k+1}]$. Thus the sequence of functions $b_0(t), \ldots, b_n(t)$ is unimodal in $[r_k, r_{k+1}]$ and $b_k(t)$ is maximal for all $t \in [r_k, r_{k+1}]$. (If $r_{k+1} > r_k = r_{k-1} = \cdots = r_{k-j} \ge r_{i+1}$, then the sequence of functions $b_0(t), \ldots, b_n(t)$ must still be unimodal in $[r_k, r_{k+1}]$ and $b_k(t)$ must still be maximal for all $t \in [r_k, r_{k+1}]$ because $b_{k-j-1}(t) < b_{k-j}(t) < \cdots < b_{k-1}(t)$ once $t > r_k$ since, by assumption, each root r_p has multiplicity 1.) Now it follows by induction on k that this result is true for all $i \leq k \leq n$. A similar symmetric argument works for all $0 \leq k \leq i$. Hence we conclude that the sequence of functions $b_0(t), \ldots, b_n(t)$ is unimodal in $[a, b]$.

Conversely, suppose that Conditions (i) and (ii) are satisfied, but that Condition (iii) fails to hold; we shall show that the sequence of functions $b_0(t), \ldots, b_n(t)$ cannot be unimodal in the interval $[a, b]$. Consider the two sets of indices

$$L = \{k \mid 0 < k < i \text{ and } r_k > r_{k+1}\}$$
$$H = \{k \mid i+1 < k < n+1 \text{ and } r_k < r_{k-1}\}.$$

Since Condition (iii) is not satisfied, at least one of these sets is nonempty. We shall treat only the case where H is nonempty; the argument for the case when L is nonempty is entirely symmetrical.

Let h be the smallest index in H. Since $h - 1 \geq i + 1$ and $h - 1$ is not in H, it follows that $r_{h-1} \geq r_{i+1} > t^*$. Therefore by Condition (ii)

$$b_{h-2}(t^*) > b_{h-1}(t^*) > b_h(t^*). \tag{*}$$

(Notice that the second inequality actually becomes equality if $r_h = t^*$; we shall treat this special case separately below.) Now the main idea of the proof is that as t moves from t^* in the direction of r_h, $b_{h-2}(t)$ remains larger than $b_{h-1}(t)$ but $b_{h-1}(t)$ becomes smaller than $b_h(t)$ as soon as t moves past r_h, thus violating unimodality.

To simplify the remainder of the analysis, recall that $r_{h-1} > t^*$. Therefore, there are only three cases to consider depending upon whether r_h is greater than, less than, or equal to t^*.

Case 1: $t^* < r_h < r_{h-1}$

By Condition (i) $b_{h-1}(t)$ and $b_h(t)$ must change relative size at $t = r_h$, but $b_{h-1}(t)$ and $b_{h-2}(t)$ cannot change relative size until $t = r_{h-1}$. Therefore it follows from (*) that for all $s \in (r_h, r_{h-1})$

$$b_{h-1}(s) < b_{h-2}(s)$$
$$b_{h-1}(s) < b_h(s)$$

which violates unimodality.

Case 2: $r_h < t^* < r_{h-1}$

Again by Condition (i) $b_{h-1}(t)$ and $b_h(t)$ must change relative size at $t = r_h$, but $b_{h-1}(t)$ and $b_{h-2}(t)$ cannot change relative size until $t = r_{h-1}$. Therefore it follows from (*) that for all $s \in (r_0, r_h)$

$$b_{h-1}(s) < b_{h-2}(s)$$
$$b_{h-1}(s) < b_h(s)$$

which again violates unimodality.

Case 3: $r_h = t^* < r_{h-1}$

By Condition (i) $b_{h-1}(t)$ and $b_h(t)$ change relative size at $t^* = r_h$, but $b_{h-1}(t)$ and $b_{h-2}(t)$ do not change relative size until $t = r_{h-1}$. Therefore it follows from the first inequality of (∗) that either for all $s \in (r_h, r_{h-1})$ or for all $s \in (r_0, r_h)$, we have

$$b_{h-2}(s) > b_{h-1}(s)$$
$$b_{h-1}(s) < b_h(s)$$

which violates unimodality.   □

Condition (i) of the Lemma—that for $k = 1, \ldots, n, b_k(t) - b_{k-1}(t)$ has exactly one (counting multiplicities) real root r_k in the interval (a, b)—may seem, at first, to be a rather stringent constraint. However, for sequences of functions such as the Bernstein and Ball bases, which are totally positive in the interval (a, b) [5, 6, 7], each difference $b_k(t) - b_{k-1}(t)$ can have at most one real root in (a, b). Moreover, if $b_k(t) - b_{k-1}(t)$ has no real root in (a, b), we can simply let $r_k = b$ and the Lemma and its proof remain valid. In any event, we shall verify by explicit computation in Section 3 that for the Bernstein and Ball bases, the functions $b_k(t) - b_{k-1}(t)$ do indeed have exactly one real root in the interval of interest.

Before proceeding with our examples we need a quick and easy way to verify that Condition (ii) of the Lemma is satisfied. The following observation supplies such a fast and simple test.

Suppose that $b_0(t), \ldots, b_n(t)$ is a sequence of continuous functions defined on the interval $[a, b]$ such that

a. $b_k(t) \geq 0 \quad a \leq t \leq b \quad 0 \leq k \leq n$

b. $0 \leq \underset{t \to a}{\text{Lim}}\, b_k(t)/b_{k-1}(t) < 1 \quad 1 \leq k \leq n.$

Then clearly there is some neighborhood of $t = a$ such that the sequence of functions $b_0(t), \ldots, b_n(t)$ is unimodal and $b_0(t)$ is maximal in this neighborhood. Thus when the two inequalities (a) and (b) hold, Condition (ii) of the Lemma is automatically satisfied.

We close this section with one final remark to help with the computation of the roots needed to test Condition (iii) of the Lemma. Notice that

$$b_k(r_k) - b_{k-1}(r_k) = 0 \Leftrightarrow b_k(r_k)/b_{k-1}(r_k) = 1 \quad \text{provided that } b_{k-1}(r_k) \neq 0.$$

We make use of this observation below to simplify the calculation of the roots r_k, $k = 1, \ldots, n$, in each of the examples in Section 3.

3. Examples of Unimodal Bases

By applying the Lemma, we shall now provide examples of some bases that are unimodal and other bases that are not unimodal.

Example 1. *Bernstein Basis–Degree n*

The Bernstein basis of degree n is defined by

$$B_k(t) = \binom{n}{k} t^k (1-t)^{n-k} \quad k = 0, 1, \ldots, n$$

$$\binom{n}{k} = n!/k!(n-k)! \qquad k = 0, 1, \ldots, n.$$

Taking ratios of functions with adjacent indices, we find that

$$B_k(t)/B_{k-1}(t) = (n+1-k)t/k(1-t) \quad k = 1, 2, \ldots, n.$$

To apply the Lemma, observe that

$$B_k(t) \geq 0 \quad 0 \leq t \leq 1 \quad k = 0, 1, \ldots, n$$

$$\lim_{t \to 0} B_k(t)/B_{k-1}(t) = 0 \quad k = 1, 2, \ldots, n$$

$$r_k = k/(n+1) \qquad k = 1, 2, \ldots, n.$$

Therefore by the Lemma, the Bernstein bases of arbitrary degree are unimodal in the interval $[0, 1]$.

Example 2. *Ball's Basis—Degree 2m + 1*

The generalized Ball basis of degree $2m+1$ is defined in [8] by setting

$$B_k(t) = \binom{m+k}{k} t^k (1-t)^{m+1} \qquad k = 0, 1, \ldots, m$$

$$B_{2m+1-k}(t) = \binom{m+k}{k} t^{m+1}(1-t)^k \quad k = 0, 1, \ldots, m.$$

Again taking ratios of functions with adjacent indices, we find that

$$B_k(t)/B_{k-1}(t) = (m+k)t/k \qquad\qquad k = 1, \ldots, m$$
$$B_{m+1}(t)/B_m(t) = t/(1-t) \qquad\qquad k = m+1$$
$$B_{2m+2-k}(t)/B_{2m+1-k}(t) = k/(m+k)(1-t) \quad k = 1, \ldots, m.$$

To apply the Lemma, observe that

$$B_k(t) \geq 0 \quad 0 \leq t \leq 1 \qquad\qquad\qquad k = 0, 1, \ldots, 2m+1$$

$$\lim_{t \to 0} B_k(t)/B_{k-1}(t) = 0 \qquad\qquad\qquad k = 1, \ldots, m+1$$

$$\lim_{t \to 0} B_{2m+2-k}(t)/B_{2m+1-k}(t) = k/(m+k) < 1 \quad k = 1, \ldots, m$$

$$r_k = k/(m+k) \qquad\qquad\qquad\qquad k = 1, \ldots, m$$

$$= 1/2 \qquad\qquad\qquad\qquad\qquad k = m+1$$

$$= m/(3m+2-k) \qquad\qquad\qquad k = m+2, \ldots, 2m+1.$$

Therefore by the Lemma, the generalized Ball bases of odd degree are unimodal in the interval $[0, 1]$.

Example 3. *Ball's Basis—Degree 2m*

In Ball's basis of degree $2m + 1$ only the two basis functions $B_m(t)$, $B_{m+1}(t)$ are of total degree $2m + 1$; the remainder of the basis functions have degree less than or equal to $2m$. Therefore one way to construct a generalized Ball basis of degree $2m$ is to replace the two basis functions $B_m(t)$, $B_{m+1}(t)$ by their sum $B_m(t) + B_{m+1}(t)$, which is a single polynomial of degree $2m$, and leave the remaining basis functions unchanged. Thus a generalized Ball basis of degree $2m$ can defined by setting

$$B_k(t) = \binom{m+k}{k} t^k (1 - t)^{m+1} \qquad k = 0, \ldots, m - 1$$

$$B_k(t) = \binom{2m}{m} t^m (1 - t)^m \qquad\qquad k = m$$

$$B_{2m-k}(t) = \binom{m+k}{k} t^{m+1} (1 - t)^k \quad k = 0, \ldots, m - 1.$$

Once again taking ratios of functions with adjacent indices, we find that

$$\begin{aligned}
B_k(t)/B_{k-1}(t) &= (m+k)t/k & k &= 1, \ldots, m - 1 \\
B_m(t)/B_{m-1}(t) &= 2t/(1 - t) & k &= m \\
B_{m+1}(t)/B_m(t) &= t/2(1 - t) & k &= m + 1 \\
B_{2m+1-k}(t)/B_{2m-k}(t) &= k/(m+k)(1 - t) & k &= 1, \ldots, m - 1.
\end{aligned}$$

To apply the Lemma, observe that

$$\begin{aligned}
B_k(t) &\geq 0 \quad 0 \leq t \leq 1 & k &= 0, 1, \ldots, 2m \\[4pt]
\underset{t\to 0}{\text{Lim}}\, B_k(t)/B_{k-1}(t) &= 0 & k &= 1, \ldots, m + 1 \\[4pt]
\underset{t\to 0}{\text{Lim}}\, B_{2m+1-k}(t)/B_{2m-k}(t) &= k/(m+k) < 1 & k &= 1, \ldots, m - 1 \\[4pt]
r_k &= k/(m+k) & k &= 1, \ldots, m - 1 \\[4pt]
&= 1/3 & k &= m \\[4pt]
&= 2/3 & k &= m + 1 \\[4pt]
&= m/(3m + 1 - k) & k &= m + 2, \ldots, 2m.
\end{aligned}$$

Therefore Ball's basis for even degree is unimodal if and only if $m = 1, 2$, that is, for degree $= 2, 4$. Hence Ball's bases for even degree are *not* unimodal in the interval $[0, 1]$ for any degree ≥ 6.

One final observation: Barry et al. [5] describe a technique for modifying the de Casteljau algorithm by reducing the number of recursive computations; this technique always results in curve schemes that are variation diminishing in the interval $[0, 1]$. They call such curve schemes *hedges*. Generalized Ball bases of both even and odd degree are instances of hedges. Thus our last example illustrates that Ball's bases of even degree ≥ 6 represent a collection of hedges which are not unimodal in the interval $[0, 1]$.

Acknowledgement

This work was supported in part by National Science Foundation Grant CCR-9113239.

References

[1] Ball, A. A.: CONSURF Part1: introduction to conic lofting tile. CAD 6, 243–249 (1974).
[2] Ball, A. A.: CONSURF Part 2: description of the algorithms. CAD 7, 237–242 (1975).
[3] Ball, A. A., CONSURF Part 3: how the program is used. CAD 9, 9–12 (1977).
[4] Barry, P. J., Beatty, J. C., Goldman, R. N.: Unimodal properties of B-spline and Bernstein basis functions. CAD 24, 627–636 (1992).
[5] Barry, P. J., DeRose, T. D., Goldman, R. N.: Pruned Bezier curves. Proceeding of Graphics Interface 90, 229–238 (1990).
[6] Goodman, T. N. T., Said, H. B.: Shape preserving properties of the generalised Ball basis. CAGD 8, 115–121 (1991).
[7] Polya, G., Schoenberg, I. J.: Remarks on the De La Vallee Poussin means and convex conformal maps of the circle. Pacific J. Math. 8, 296–234 (1958).
[8] Said, H. B.: A generalised Ball curve and its recursive algorithm. ACM Trans. Graphics 8, 360–371 (1989).

Phillip J. Barry
Department of Computer Science
University of Minnesota
4-192 EE/CSci Building
200 Union Street SE
Minneapolis, MN 55455
U.S.A.

Ronald N. Goldman
Department of Computer Science
Rice University
P.O. Box 1892
Houston, TX 77251
U.S.A.

Computing Suppl. 10, 43–60 (1995)

© Springer-Verlag 1995

Nef Polyhedra: A Brief Introduction

H. Bieri, Bern

Abstract. A Nef polyhedron is any set in $\mathbb{R}^d$ ($d \in \mathbb{N}_0$) which can be obtained by applying a finite number of Boolean set operations *cpl* and $\bigcap$ to (open) linear halfspaces. This paper summarizes the fundamentals of the theory of Nef polyhedra and illustrates them by means of simple 2D examples. The notions of Nef polyhedron, locally adjoined pyramid and face of a Nef polyhedron are carefully explained. Data structures for representing Nef polyhedra are discussed, and an implemented test-system is presented with the aid of a small application.

Key words: Polyhedra, Nef polyhedra, geometric modeling, solid modeling, constructive solid geometry.

1. Introduction

17 years ago, Walter Nef published his book *Beiträge zur Theorie der Polyeder—mit Anwendungen in der Computergraphik* [23], in which he founded a mathematically sound theory of a new kind of polyhedra which should be at the same time general, simple and applicable—and not too far away from a user's conception of polyhedra, especially in CAD and Computer Graphics. He succeeded surprisingly well in meeting simultaneously all these not wholly compatible requirements, and at a time when *Solid Modeling* was still in its pioneering phase [27]. Nef introduces a class of d-dimensional polyhedra which is closed with regard to the ordinary set operations and to the topological operations closure and interior (and therefore also to regularization), he gives an original new definition of the faces (vertex, edge, etc.) of a polyhedron, and he presents algorithms for the set operations *complement* and *intersection*. These are just a few of the achievements in his book, and all of them refer to topics which continue to be of research interest [28, 29] Nef's book was acclaimed from the beginning, but already some of its first reviewers feared that its language, i.e. German, and its "small local publisher" were not very good conditions to make it widely known, which unfortunately turned out to be true. Subsequent publications became better known, but they deal primarily with new applications [3, 7–9]. [10] resumes the very subject of Nef's book, but it is a rather condensed paper and hard to understand without knowing [23]. Since 1988 some applications in picture analysis continuing [3, 8] have been published [4, 6, 13], but these papers do not extend the theory itself. However groundwork, too, progressed, and Nef's theory became continuously more mature. Especially better algorithms for the operations with polyhedra than in [23] have been developed and are now implemented and ready for publication[1]. Not a bad moment, it seems to us, to summarize the basics of

---

[1] [5], which has been written shortly after the present paper, introduces a new intersection algorithm.

Nef's theory of polyhedra from the actual point of view- and to introduce the notion "*Nef polyhedron*" to clearly distinguish it from the several different notions of "general polyhedron". More than [23] and [10], this brief introduction addresses itself to readers in the field of Geometric Modeling where, in spite of a considerable amount of research, *d*-dimensional polyhedra and the application of set and topological operations to them represent still a topical domain for further investigations [28]. Most researchers in the field work with different kinds of combinatorial or topological polyhedra [1, 11, 15, 18–20, 22, 25, 26, 30, 31, 33]. They take notice of our results ([10] is often quoted) as we do of theirs, but active cooperation is very limited so far (cf., however, [12]). We hope that papers like the present one can help to improve this situation.

The following 4 sections introduce Nef polyhedra, locally adjoined pyramids, faces of Nef polyhedra and geometric models of Nef polyhedra. Section 6 presents a provisional implemented modeling system. The paper gives definitions, basic properties, comments and examples, but no proofs. Almost all proofs can be found in Nef's book, and we will refer to them using brackets {,}. (By {1; 3.1}, e.g., we refer to Theorem 3.1 in Chapter 1 of [23].) Nef polyhedra are not only meant to be a mathematical subject, they shall also provide a basis for ambitious solid modeling in practice. It is because of lack of space, therefore, and not of material or relevance, that we only include one very small application of Nef polyhedra in this brief introduction.

2. Nef Polyhedra

We start from the d-dimensional real *Euclidean space* $\mathbb{R}^d$ ($d \geq 0$). Let L be a *linear subspace* in $\mathbb{R}^d$ of dimension $m \leq d$, then for every *point* $x \in \mathbb{R}^d$ the set $M = L + x$ is an *affine subspace* or *plane*, respectively, of dimension m. Affine subspaces of dimension $d - 1$ are called (linear) *hyperplanes*. For every point set $A \subseteq \mathbb{R}^d$ there exists a smallest affine subspace *aff* A containing it, i.e. the intersection of all hyperplanes containing A. *aff* A is called the *affine hull* of A. The empty set, too, is by definition an affine subspace (of dimension -1). Hence *aff* $\varnothing = \varnothing$ and *aff* $\mathbb{R}^d = \mathbb{R}^d$. By *cpl* A we denote the *complement* of $A \subseteq \mathbb{R}^d$, by *int* A its interior, by *clos* A its *closure*, by *bnd* A its *boundary*, and by *ext* A its *exterior*: *bnd* $A = clos\ A \cap clos(cpl\ A)$, *ext* $A = cpl(clos\ A)$. A is called *relatively open* if A is open with respect to the *relative topology* induced by the natural topology of $\mathbb{R}^d$ on *aff* A. $x \in A$ is called a *relatively interior point* of A if x is an interior point of A with respect to this relative topology, and the set of all relatively interior points $\in A$ is called *relative interior* of A and will be denoted by *relint* A. $A^* = clos(int\ A)$ is called the *regularization* of A. Because of $(A^*)^* = A^*$, A^* is a (closed) *regular set*. In classical *Constructive Solid Geometry* (CSG) regularization is a very important operation [27, 22, 31]. Every (linear) hyperplane $H^0 \subset \mathbb{R}^d$ can be defined by means of a *linear function* $h: \mathbb{R}^d \to \mathbb{R}, h \neq 0$: $H^0 = h^{-1}(0)$. The *positive* and *negative open* (linear) *halfspaces* corresponding to H^0 can be defined analogously by $H^+ = h^{-1}(\mathbb{R}^+) = \{x \in \mathbb{R}^d; h(x) > 0\}$ and $H^- = h^{-1}(\mathbb{R}^-) = \{x \in \mathbb{R}^d; h(x) < 0\}$. $\bar{H}^+ = H^+ \cup H^0$ and $\bar{H}^- = H^- \cup H^0$ are the positive and negative *closed* (linear) *halfspaces* corresponding to H^0. (In Computer Graphics it would be more usual to give equivalent definitions by means of a *normal vector $n \neq 0$* and *scalar products* $\langle , \rangle$, e.g. $H^0 = \{x \in \mathbb{R}^d; \langle x, n \rangle + \alpha = 0\}$.) Sometimes it will be practical to write just H instead of H^0 and to assume without loss of generality a given halfspace to be positive. H^0 is the common boundary of H^+ and H^- (and of $\bar{H}^+$ and $\bar{H}^-$). H, H^+ and H^- are relatively open and form together a *partition* of $\mathbb{R}^d$.

After these preparations, we can define *Nef polyhedra*. Actually we will give 4 equivalent definitions of them in order to stress several of their various properties right from the beginning.

Definition 1. A *Nef polyhedron* in $\mathbb{R}^d$ is a set $P \subseteq \mathbb{R}^d$ that can be obtained by applying a finite number of set operations *cpl* and $\cap$ (*intersections*) to a finite number of given open (linear) halfspaces.

Observations:

- We could include the set operations $\cup$ (*union*) and $\setminus$ (*difference*) in Definition 1. This is not necessary, however, because of $P_1 \cup P_2 = cpl(cpl\, P_1 \cap cpl\, P_2)$ and $P_1 \setminus P_2 = P_1 \cap cpl\, P_2$.
- Because of $H^+ = cpl\, \bar{H}^-$, Definition 1 would not need to assume the halfspaces to be open.
- It follows immediately from Definition 1 that the class of Nef polyhedra in $\mathbb{R}^d$ is closed with respect to the set operations.
- The class of Nef polyhedra in $\mathbb{R}^d$ is also closed with respect to the topological operations *interior* and *closure* {6; 1.1, 3}.
- The set of initially given open halfspaces and their corresponding hyperplanes (i.e. their boundaries) is never enlarged as a result of the application of set or topological operations.
- It is easily seen that e.g. affine subspaces and any finite point set in $\mathbb{R}^d$ as well as their complements are Nef polyhedra. Hence, Nef polyhedra are point sets $\subseteq \mathbb{R}^d$ which do not have to be nonempty, bounded, closed, connected or regular. This implies that Nef polyhedra need not be *solids* (according to the not yet wholly established definition of "solid" in Geometric Modeling [22]).
- A *convex polytope* is normally defined as the convex hull of a nonempty finite set of points $\subset \mathbb{R}^d$ [17]. Since such a convex hull is the intersection of a finite number of closed halfspaces, all convex polytopes are Nef polyhedra.
- An *elementary polyhedron* is normally defined as the union of a finite number of convex polytopes [19]. Therefore, all elementary polyhedra are Nef polyhedra.
- A *polyhedral set* is normally defined as the intersection of a finite number of closed halfspaces [17]. Therefore, all polyhedral sets are Nef polyhedra.
- The set of all points belonging to the *simplices* of an *m*-dimensional *simplicial complex* in $\mathbb{R}^d (0 \leq m \leq d)$ is normally called a *linear d-polyhedron* [25]. Hence, all linear *d*-polyhedra are Nef polyhedra.

With regard to Geometric Modeling it is probably most natural to introduce Nef polyhedra within the context of Constructive Solid Geometry (CSG). Here solids are defined by means of *binary construction trees* (*CSG trees*). In Backus-Naur form, a CSG is normally defined recursively, as follows [22, 27]:

$$\langle\text{CSG tree}\rangle ::= \langle\text{primitive}\rangle |$$
$$\langle\text{CSG tree}\rangle\langle\text{set operation}\rangle\langle\text{CSG tree}\rangle |$$
$$\langle\text{CSG tree}\rangle\langle\text{transformation}\rangle$$

Of course, the construction of a CSG tree is assumed to be finite. In practice, a common set of *primitives* is the set {cube, sphere, cylinder, cone}. Typical transformations are *translation*, *rotation* and *scaling*. More controversial is the choice of the set operations, most common are the regularized union, intersection and difference [27]. *Halfspace models* use primitives of the form $\{x \in \mathbb{R}^d; f(x) \geq 0\}$, where f is not necessarily a linear function [22]. In general, halfspace models use the ordinary set operations, not the regularized ones [22, 24]. A Nef polyhedron can be understood as a special halfspace model:

Definition 2. A Nef polyhedron in $\mathbb{R}^d$ is a set $P \subseteq \mathbb{R}^d$ that can be represented by a CSG tree with (closed) linear halfspaces as primitives and the ordinary union, intersection and difference as set operations.

The restriction to linear halfspaces makes Nef polyhedra a natural generalization of the elementary polyhedra. But more important, it allows a sound and comprehensive theory of d-dimensional polyhedra as well as effective procedures for their construction and manipulations.

Which properties does a set $P \subseteq \mathbb{R}^d$ have to possess to be a Nef polyhedron? The following characterization defines Nef polyhedra rather by description than by construction [2]:

Definition 3. A set $P \subseteq \mathbb{R}^d$ is a Nef polyhedron iff there exist finitely many relatively open sets $A_1, \ldots, A_l$ and $B_1, \ldots, B_m$ in $\mathbb{R}^d$ such that $P = \bigcup_{i=1}^{l} A_i$ and $cpl(P) = \bigcup_{j=1}^{m} B_j$.

Example 1. The closed quadrangle P in Fig. 1a is a 2-dimensional nonconvex elementary polyhedron. Figure 1b shows that P is the union of 9 (pairwise disjoint) relatively open sets. $cpl\,P$ is open, hence also relatively open.

Let $\mathbf{H} = \{H_1, \ldots, H_r\}$ $(r \geq 0)$ be a family of (linear) hyperplanes in $\mathbb{R}^d$. $\mathbf{H}$ defines an *arrangement* $A(\mathbf{H})$ consisting of all nonempty intersections $C = \bigcap_{i=1}^{r} H_i^{s_i}$, where

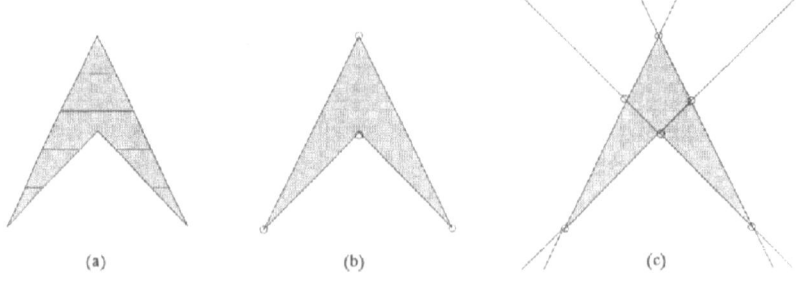

(a)   (b)   (c)

Figure 1

$s_i \in \{-, 0, +\}$. All *cells* $C \in A(\mathbf{H})$ are relatively open convex sets of dimension $\in \{0, \ldots, d\}$, and together they form a (finite) partition of $\mathbb{R}^d$ (cf. [14], where cells are called "faces"). Using $A(\mathbf{H})$, we give a last definition of Nef polyhedra which shows that the relatively open sets $A_1, \ldots, A_l$ and $B_1, \ldots, B_m$ in Definition 3 may be required to be pairwise disjoint:

Definition 4. A point set $P \subseteq \mathbb{R}^d$ is a Nef polyhedron if there exists a finite family $\mathbf{H}$ of hyperplanes in $\mathbb{R}^d$ such that P is the union of certain cells of the corresponding arrangement $A(\mathbf{H})$ $\{2; 6\}$.

$\mathbf{H}$ as well as the corresponding arrangement $A(\mathbf{H})$ will be called P-compatible. $\mathbf{H}$ is not uniquely defined by P.

Figure 1c shows that the polyhedron $P \subset \mathbb{R}^2$ in Example 1 may be obtained as the union of three 2-dimensional, eight 1-dimensional and six 0-dimensional cells belonging to the arrangement defined by 4 straight lines (i.e. hyperplanes).

In general, an application will deal with several (but finitely many) Nef polyhedra. A *point-in-polyhedron test*, e.g., may include a one-point polyhedron $\{x\}$. By *refinement* it will always be possible to construct $\mathbf{H}$ and $A(\mathbf{H})$ to be compatible with all Nef polyhedra which occur.

Working with Nef polyhedra means working with relatively open sets and relatively interior points. The often awkward distinctions between sets which are open, or closed or none of both, which are typical when working with other kinds of polyhedra, can thus be avoided.

3. Locally Adjoined Pyramids

When working with Nef polyhedra the most important tools are the so-called *locally adjoined pyramids*, which is fortunate because pyramids are very simple Nef polyhedra.

Definition 5. A point set $Q \subseteq \mathbb{R}^d$ is called a *cone with apex* 0 if $Q = \lambda Q$ for all $\lambda > 0$. Q is called a *cone* if there exists a point $x \in \mathbb{R}^d$ such that $Q - x$ is a cone with apex 0. x is then called an *apex* of Q.

(This definition of a cone is slightly different from the one in [17] where an apex of Q always belongs to Q.) A cone Q may be empty or a one-point set or it consists of rays (halflines) with a common endpoint being an apex of Q. The set $N(Q)$ of all apices of Q is an affine subspace of $\mathbb{R}^d$ $\{1; 2\}$, $\{1; 3.1\}$. Either $N(Q) \subseteq Q$ or $N(Q) \cap Q = \emptyset$. $\mathbb{R}^d$ and $\emptyset$ are cones, and $N(\mathbb{R}^d) = N(\emptyset) = \mathbb{R}^d$. $N(Q) = Q$ iff Q is a nonempty affine subspace.

Example 2. Let H_1^+ and H_2^+ be open halfspaces in $\mathbb{R}^3$ whose boundaries (i.e. hyperplanes) intersect in a straight line L. Let $Q = H_1^+ \cup H_2^+$. Q is a nonconvex open cone with apex set $N(Q) = L$. $N(Q) \cap Q = \emptyset$.

Definition 6. A point set $Q \subseteq \mathbb{R}^d$ is called a *pyramid* if Q is a cone and also a Nef polyhedron.

(This definition is different from the classic one which requires a pyramid to be closed and bounded [17].) The cone of Example 2 is a pyramid. According to Definition 4, every Nef polyhedron can be represented as a union of cells belonging to an arrangement $A(\mathbf{H})$. In case of a pyramid Q there exists always a Q-compatible family $\mathbf{H}$ of hyperplanes for which $\bigcap_{H\in\mathbf{H}} H = N(Q)$ $\{2;9\}$. In this case every cell C of $A(\mathbf{H})$ is a (relatively open) convex pyramid with apex set $N(Q)$. Hence Q is a union of such convex pyramids. In most practical cases it even holds that $\mathbf{H}$ consists of $\leq d$ hyperplanes intersecting in one point, which means that the number of cells $\in A(\mathbf{H})$ is $\leq 3^d$.

Now let P be a Nef polyhedron, $A(\mathbf{H})$ a P-compatible arrangement and x any point in $\mathbb{R}^d$. Among all neighborhoods of x consisting of cells $\in A(\mathbf{H})$ there exists a smallest one. We denote it by $U_0(x)$ and define $P^x = x + \mathbb{R}^+(P\cap U_0(x) - x) = \{z = x + \lambda(y - x);\ y\in P\cap U_0(x),\ \lambda > 0\}$. It follows that $P^x = x + \mathbb{R}^+(P\cap U(x) - x)$ for every neighborhood $U(x)$ of x which is $\subset U_0(x)$, i.e. which is "sufficiently small" $\{3;8\}$. That is, P^x expresses the "local properties" of P "around" x.

- For every $x\in\mathbb{R}^d$ P^x is a pyramid with apex x $\{3;3\}$.
- For all P-compatible families $\mathbf{H}$ of hyperplanes we get the same P^x, i.e. P^x depends only on P and x $\{3;9\}$.

Definition 7. For $P\subseteq\mathbb{R}^d$ a Nef polyhedron and $x\in\mathbb{R}^d$ we call P^x *the pyramid locally adjoined to P in x.*

Often we will call P^x less precisely just "*locally adjoined pyramid (of x)*". x is not uniquely determined by P^x, in general, i.e. the same pyramid may be locally adjoined to P in more than one point. In the next section we will make use of this fact to define the faces of a Nef polyhedron. In case P is the open positive (or negative) halfspace belonging to a hyperplane H^0, there exists just three pyramids locally adjoined to $P = H^+$:

- $P^x = \mathbb{R}^d$ for every $x\in H^+$.
- $P^x = H^+$ for every $x\in H^0$.
- $P^x = \phi$ for every $x\in H^-$.

This together with the following formulaes guarantees that for any Nef polyhedra P given according to Definitions 1 and 2, and for any point $x\in\mathbb{R}^d$, P^x can easily be constructed:

- $(\operatorname{cpl} P)^x = \operatorname{cpl} P^x$ $\{3;15\}$.
- $(P_1\cap P_2)^x = P_1^x\cap P_2^x$ $\{3;13\}$.
- $(P_1\cup P_2)^x = P_1^x\cup P_2^x$ $\{3;14\}$.
- $(P_1\setminus P_2)^x = P_1^x\setminus P_2^x$.

One further reason for the usefulness of locally adjoined pyramids results from the following facts which provide e.g. a convenient basis for a number of important *set membership classification* algorithms [34].

- $x\in P$ iff $x\in P^x$ $\{3;7\}$.
- $x\in \operatorname{clos} P$ iff $P^x \neq \varnothing$ $\{3;19\}$.

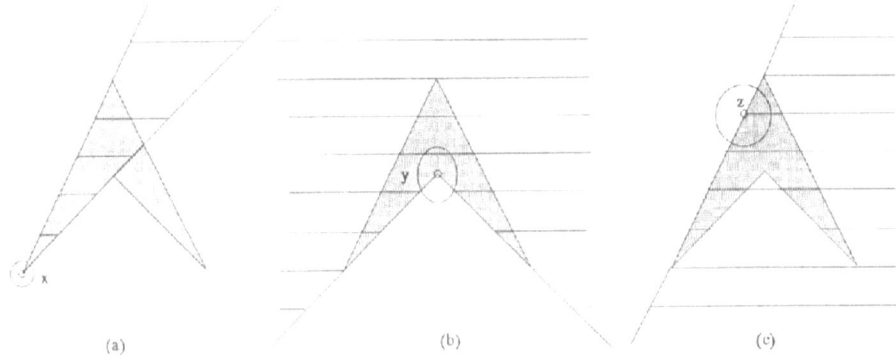

Figure 2

- $x \in relint\ P$ iff $P^x = aff\ P$ $\{3; 19\}$.
- $x \in int\ P$ iff $P^x = \mathbb{R}^d$ $\{3; 19\}$.
- $x \in ext\ P$ iff $P^x = \varnothing$.
- $x \in bnd\ P$ iff $P^x \neq \varnothing$, $P^x \neq \mathbb{R}^d$.

Figure 2 shows again the simple (Nef-) polyhedron of Example 1, together with 3 points which we denote by x, y and z. Each of these points is surrounded by a "sufficiently small" neighborhood. It is easily seen that the pyramids locally adjoined to P in these points can be described as follows:

(a): $Q = P^x$ is closed and convex, $N(Q) = \{x\}$.
(b): $Q = P^y$ is closed and nonconvex, $N(Q) = \{y\}$.
(c): $Q = P^z$ is a closed halfspace $\bar{H}^+$, $z \in N(Q) = H^0$.

4. Faces of Nef Polyhedra

Traditionally, solids in $\mathbb{R}^3$ have often been described or even defined by their boundaries, and the *boundary representation* (B-rep) is probably still the most popular technique to implement solids in CAD-systems. In case of a convex polytope P of any dimension d, it is well possible to describe *bnd P*, and hence P itself, by means of the proper faces of P which are by definition themselves convex polytopes (of lower dimensions) [17]. In the special case where P is a simplex, the faces of P are again simplices. In case of a *linear d-polyhedron P F* is called an *m-face* of P if F is an m-dimensional simplex belonging to a simplicial complex associated with P ($m \leq d$) [25]. Although (nonconvex) *elementary polyhedra* can be understood as linear d-polyhedra, the naive conception of "face" does hardly coincide with the notion of m-face, in general, as the faces of a polyhedron should have some maximality property. There still does not seem to exist a definition for the faces of d-dimensional elementary polyhedra which is at the same time "natural", precise and handy. Nef's definition of the face of a polyhedron is certainly one of the very central ingredients of his theory. It is precise, general and has proven to be very handy. And we hope to make clear that it is also surprisingly "natural":

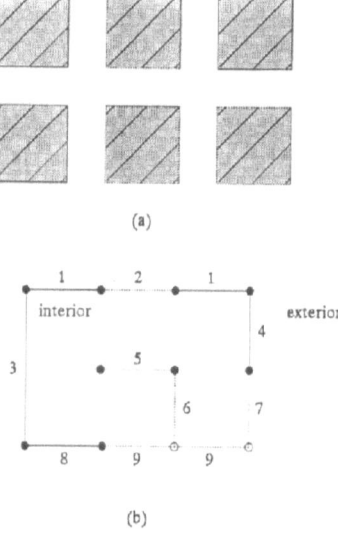

(a)

(b)

Figure 3

Let $P \subseteq \mathbb{R}^d$ be a Nef polyhedron. By $x \sim y$ iff $P^x = P^y$, an equivalence relation $\sim$ is defined on $\mathbb{R}^d$.

Definition 8. The *faces* of a Nef polyhedron $P \subseteq \mathbb{R}^d$ are the equivalence classes of the relation $\sim$.

That is, all points in $\mathbb{R}^d$ with the same locally adjoined pyramid form together a face F of P. From now on we will rather write P^F instead of P^x to indicate all points in which the pyramid is locally adjoined to P. By $\mathbf{F}(P)$ we denote the set of all faces of P.

It is easily seen that for the (elementary) polyhedron P of Fig. 1a the 4 vertices (one-point sets), the 4 edges (without endpoints), *int* P and *ext* P are the faces of P. That is, the 10 relatively open sets shown in Figure 1(b) are precisely the faces of P.

Example 3. In order to illustrate more thoroughly Nef's definition of the faces of polyhedra we choose the slightly artificial example of Fig. 3. Figure 3a shows a scene" of 6 Nef polyhedra $\subset \mathbb{R}^2$, i.e. 6 squares of the same size, 3 of which are closed and 3 open. Each such square has 10 faces: 4 vertices, 4 edges, interior and exterior. Now the 6 squares are "put together" to form one new Nef polyhedron P, i.e. a "rectangle" which is neither closed nor open. Figure 3b shows that $\mathbf{F}(P)$ consists of 11 vertices, 9 edges, and 2 two-dimensional faces, i.e. *int* P and *ext* P. Edges 1 and 9 are not connected. (Vertices contained in P are represented by filled dots.) Of course, P can be understood as the result of applying union operations to the 6 squares properly positioned.

Not everybody likes faces of polyhedra which are not connected. It is not difficult, of course, to handle such a face as a sequence of connected components. But when we designed our algorithms, this variant turned out to be simply not necessary.

In order to illustrate further the "naturalness" of Nef's definition of the face of a polyhedron, let us imagine P to be an illuminated space embedded in an opaque ("misty") surrounding. Let us interpret P^x to be that part of the visual cone of an observer at position x which contains P. We conclude that P looks to 2 observers "the same" iff their positions belong to the same face of P.

The following basic properties of the faces of a Nef polyhedron $P \subseteq \mathbb{R}^d$ are already illustrated by Examples 1 and 3:

- The number of faces of P is finite and ≥ 1 {6; 1.1}.
- The faces of P are pairwise disjoint. Their union is $\mathbb{R}^d$ {6; 3}.
- Every face F of P is nonempty and relatively open {6; 2}.
- For every face F of P we have either $F \subseteq P$ or $F \cap P = \varnothing$ {6; 4}.
- Every face F of P is a Nef polyhedron {6; 1.1}.
- If $int\, P \neq \varnothing$ then $int\, P \in F(P)$. More generally: if $relint\, P \neq \varnothing$ then $relint\, P \in F(P)$.
- If $ext\, P \neq \varnothing$ then $ext\, P \in F(P)$.
- Let $F \in F(P)$. Every P-compatible set of hyperplanes is F-compatible (but not vice versa, in general) {6; 1.1}.
- $aff\, F = N(P^F)$ for every face F of P {6; 2}.

[23] assumes P^F to be nonempty, and $ext\, P$ becomes a face of P only in [10]. This simplifies Definition 8 and a number of statements, but occasionally it is still necessary to consider only *proper faces* F of P, i.e. faces for which $F \subseteq clos\, P$ holds. The union of all proper faces is $clos\, P$.

Let $P \subseteq \mathbb{R}^d$ a Nef polyhedron and $F_1, F_2 \in F(P)$. It holds that $F_1 \cap clos\, F_2 = \varnothing$ or $F_1 \subset clos\, F_2$ {6; 10} (cf. Fig. 3). In the latter case we say that F_1 is *incident* to F_2 and arrive at a *partial ordering relation* $\prec$ on $F(P)$: $F_1 \prec F_2$ iff F_1 is incident to F_2. We call $\prec$ *incidence relation* on $F(P)$. $\prec$ represents an important tool to implement Nef polyhedra efficiently (cf. Section 5).

Let $P \subseteq \mathbb{R}^d$ be a convex polytope. P is also a Nef polyhedron, and it holds that for every proper face $F \in F(P)$ its closure is a "classical" face of P, and, conversely, every (nonempty) "classical" face of P is the closure of a face $F \in F(P)$. This again shows that faces defined by Definition 5 are not "too far away" from the "intuitive" conception of "faces" of polyhedra.

As every face F of a Nef polyhedron P is relatively open, it is near at hand to define its *dimension* by setting $dim\, F := dim(aff\, F)$. According to common practice, we call 0-dimensional faces *vertices* and 1-dimensional faces *edges*. (We have already used these terms "naively" in our examples, which now is justified.) Since faces of Nef polyhedra are Nef polyhedra it might be reasonable to define the dimension of any Nef polyhedra P analogously, i.e. by setting $dim\, P := dim(aff\, P)$. In [23] there appears no such definition simply because it does not prove necessary. Another- and probably more appropriate-way to define the dimension of a Nef polyhedron might consist in setting $dim\, P$ to the maximum of the dimensions of the cells $\subseteq P$, according to Definition 4. This definition would at least be analogous to the one for linear p-polyhedra.

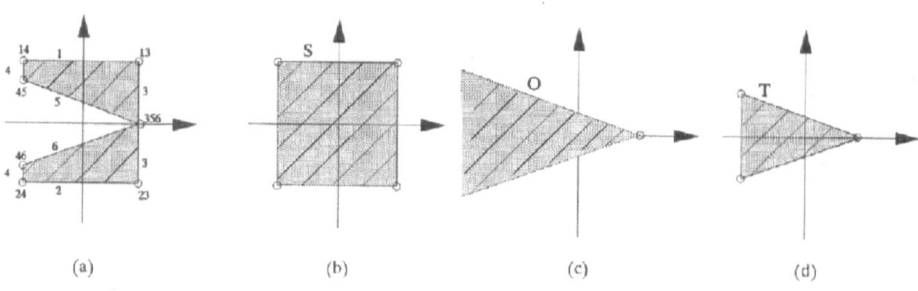

Figure 4

With regard to the elementary set operations the following properties are essential:

- P and $cpl\,P$ have the same faces.
- Every face of $P_1 \cap P_2$ is the union of certain intersections $F_1 \cap F_2$ where $F_1 \in F(P_1)$, $F_2 \in F(P_2)$. (The statement "every face of $P_1 \cap P_2$ is an intersection $F_1 \cap F_2$, where $F_1 \in F(P_1)$, $F_2 \in F(P_2)$" needs not be true (cf. in [23] the counterexample following $\{6;16\}$.) In particular: Every vertex of $P_1 \cap P_2$ is a 0-dimensional intersection $F_1 \cap F_2$, where $F_1 \in F(P_1)$, $F_2 \in F(P_2)$ $\{6;16\}$.

Example 4. Before we discuss representations for Nef polyhedra, another small example shall illustrate Nef's approach compared to the classical CSG-approach [22]. Figure 4a shows a very simple CAD-operation: a triangle is to be cut out of a square S (Fig. 4b). We do this by means of a difference operation $D = S \backslash O$ where O is an open pyramid (Fig. 4c). O can be defined and constructed very easily as will be demonstrated in Section 6. D results to be closed and bounded, it can be considered an elementary (2-dimensional) polyhedron. Figure 4d illustrates the "classic" approach. Instead of O a closed triangle T is used, and the regularized difference $S \backslash^* T$ is performed in order to arrive at an "acceptable" result which happens to coincide with D. Of course, this procedure is also practicable with Nef polyhedra, although less natural and efficient than Nef's approach. (With some geometric modeling systems D would still not be "acceptable" because D is a "*non-manifold solid*" [20, 31].)

5. Geometric Models of Nef Polyhedra

In Geometric Modeling, the following *three-level view* has become quite popular [27, 22]: A *physical object* is modeled by a *mathematical object* which consists of a point set $A \subseteq \mathbb{R}^d$ and, possibly, nongeometric information. In practice, A is often assumed to be a *solid*, i.e. (among other things) a set that is infinite, closed and bounded. To the mathematical object, more precisely to A, a *finite representation* is assigned which is called a *geometric model*. In case of solid objects, the geometric model is called *solid model*. The most prominent solid models are the *decomposition model*, the *constructive model* and the *boundary model* [22]. From the point of view of implementation, geometric models will often be conceived as *abstract data types* (cf. [10], where geometric models are called "information structures"). These abstract data types have finally to be *implemented* by means of *data structures* and *procedures*, which can be considered the fourth level of abstraction of the modeling process [16]. Nef polyhedra are mathematical objects of course, but they need not be solids. In spite of this, when looking for appropriate geometric models to represent Nef polyhedra, it seems reasonable to consider the three popular solid models mentioned above.

According to Definition 4, every Nef polyhedron $P \subseteq \mathbb{R}^d$ is a union of cells with respect to any P-compatible arrangement $A(H)$. Therefore, representing P by such cells, i.e. relatively open, convex polyhedra of dimension $\in \{0, \ldots, d\}$ we immediately arrive at a (nonregular) *cell decomposition* [22]. It is not hard to see, however, that this representation is inefficient, in general, because the number of cells involved may become exceedingly large. But in the important special case of pyramids, this cell decomposition model proves useful.

Our first actual implementations of Nef polyhedra (cf. [7,9]) were in APL and used a simple kind of *constructive model*: Let $P \subseteq \mathbb{R}^d$ a Nef polyhedron, H a P-compatible family of hyperplanes and $H^0 \in H$. Let c^0, c^+, and c^- be the *characteristic functions* belonging to H^0, H^+ and H^-, i.e. defined by $c^s(x) = 1$ iff $x \in H^s (s \in \{0, +, -\})$. If H^0 is given by a corresponding affine function or normal vector, respectively, the evaluation of c^s for any $x \in \mathbb{R}^d$ is trivial. Now P can be defined uniquely by an *expression tree* T where every leaf represents a characteristic function c^s and every nonleaf node one of the Boolean functions $\neg$ (not), $\wedge$ (and) and $\vee$ (or). (For reasons of simplicity, we implemented T as a *character string* to be interpreted for any given $x \in \mathbb{R}^d$ by the APL-operator *execute*.) This constructive model offers a number of advantages. Point-in-polyhedron tests, for instance, are almost trivial. It is especially well suited to represent closed and relatively open convex Nef polyhedra because of:

- Every closed convex Nef polyhedron $P \subseteq \mathbb{R}^d$ is the intersection of a finite number of closed halfspaces $\{7; 1\}$. (Hence P is a polyhedral set.)
- Every relatively open Nef polyhedron $P \subseteq \mathbb{R}^d$ is the intersection of a finite number of hyperplanes and open halfspaces $\{7; 13\}$.

Of course, when working with this constructive model there appear the well-known *redundancy problems* (e.g. *null object detection*) [32], and for large examples our old implementation is certainly quite inefficient.

At the 4th Computational Geometry Workshop in Würzburg in 1988, a less obvious geometric model for Nef polyhedra has been presented which, when interpreted appropriately, can be said to be a *boundary model* [10] and which we sometimes call "*Würzburg structure*" and denote by "*W-structure*". Let $P \subseteq \mathbb{R}^d$ be again a Nef polyhedron. As each face $F \in F(P)$ is uniquely determined by its locally adjoined pyramid P^F, and P is the disjoint union of certian of its faces, it is not too surprising that the set $Q(P)$ of all pyramids $Q \subseteq \mathbb{R}^d$ locally adjoined to P defines P uniquely (the proof of this statement has not yet been published). Actually P is already defined uniquely by those $Q = P^F \in Q(P)$ for which F is a *minimal element* with respect to the incidence relation $\prec$. Since $int(P)$ and $ext(P)$ (if not empty) are the only faces of P which are not contained in $bnd P$, it seems justified to call the W-structure a boundary model of P. Making use of $Q(P)$ it is very easy, for instance, to determine the corresponding representations of $clos\,P$ and $int\,P$ [10]:

- $Q(clos\,P) = \{clos\,Q; Q \in Q\}$.
- $Q(int\,P) = \{cpl(clos(cpl\,Q)); Q \in Q\}$;

i.e. we basically only have to find the complement and the closure of pyramids which can easily be done by means of their representations as unions of cells. The intersection algorithm in [10] makes use of this boundary model.

In the meantime it has proved useful to extend the Würzburg structure. Especially the incidence relation $\prec$ on $F(P)$ has been added, i.e. the set of all pairs (F_1, F_2) of faces of P for which $F_1 \subset clos\,F_2$ holds. This "*extended Würzburg structure*" ("*EW-structure*") of P will normally be determined from the W-structure of P by pre-processing.

The decomposition model, the constructive model and the boundary model as described above are *geometrically complete models*, i.e. they define $P \subseteq \mathbb{R}^d$ uniquely with respect to any geometric construction involving P [27].

6. An Implemented Test-System

There already exists a working implementation POLYHEDRA based on the EW-structure and implemented in Turbo Pascal which allows modeling with Nef polyhedra. Due to limitations of space, we only demonstrate it by means of the very simple Example 4 which discusses the construction of the difference polyhedron D of a closed square S and an open pyramid O (Fig. 4a). We assume Cartesian coordinates ξ_1, ξ_2 and define $S := \{x \in \mathbb{R}^2; -1 \le \xi_1, \xi_2 \le 1\}$ (Fig. 4b) and $O := H_1^- \cap H_2^-$, where H_1^- and H_2^- denote the negative open halfspaces defined by the linear functions $h_1(x) := 4\xi_1 + 1\xi_2 - 1$ and $h_2(x) := -4\xi_1 + 1\xi_2 - 1$ (Fig. 4c).

The EW-structure as implemented in POLYHEDRA separates properly *metric* and *topological* information [22] and leans on [10]: For each Nef polyhedron $P \subseteq \mathbb{R}^d$, its metric information consists of a list of linear functions defining a P-compatible family $\mathbf{H}$ of hyperplanes. Each linear function is specified by a (unique) number as its identifier and by $d + 1$ coefficients with respect to the assumed coordinate system. The topological information of P consists of a list of all faces of P, where each $F \in F(P)$ is specified by a number, its locally adjoined pyramid P^F, a list of all faces $F' \in F(P)$ being "down-incident" to F, i.e. satisfying $F' \prec F$, and a list of all faces $F'' \in F(P)$ being "up-incident" to F, i.e. satisfying $F \prec F''$, Locally adjoined pyramids are implemented by means of cell decomposition (cf. Section 5): Let $Q \in Q(P)$ and $\mathbf{H}(Q)$ a subset of $\mathbf{H}$ which is Q-compatible. $\mathbf{H}(Q)$ can be chosen much smaller than $\mathbf{H}$, in general. (In most practical cases, $\mathbf{H}(Q)$ will consist of $\le d$ hyperplanes intersecting in one point.) Let $\mathbf{H}(Q) = \{H_1, \ldots, H_r\}$ and $A(\mathbf{H}, Q)$ the arrangement defined by $\mathbf{H}(Q)$. Each cell $C \in A(\mathbf{H}, Q)$ is implemented as a sign-tuple $(s_1, \ldots, s_r)$, where $s_i \in \{-, 0, +\}$, signifying $C \subset H^{s_i}(i = 1, \ldots, r)$. $A(\mathbf{H}, Q)$ itself is implemented as a list of all these sign-tuples, and the locally adjoined pyramid Q as a Boolean list of the same length, with values TRUE indicating those cells $\in A(\mathbf{H}, Q)$ which are contained in Q. As an

```
number of face =   1
  numbers of linear functions:
       1
  adjoined pyramid:
     0    0    FALSE
     1    -    TRUE
     2    +    FALSE

number of face = -1
  numbers of linear functions:

  adjoined pyramid:
     0         FALSE


number of face =   0
  numbers of linear functions:

  adjoined pyramid:
     0         TRUE
```

Figure 5. W-structure of an open-negative halfspace $\subset \mathbb{R}^2$ (topological part)

```
begin
{01} polin('hspopn.pfl',hsp1);
{02} linlstin('lin1.lfl',linlst1);
{03} inslinlst(linlst1,hsp1);
{04} prepro(hsp1);

{05} polin('hspopn.pfl',hsp2);
{06} linlstin('lin2.lfl',linlst2);
{07} inslinlst(linlst2,hsp2);
{08} prepro(hsp2);

{09} polcut(hsp1,hsp2,pyramid);
{10} polsave(pyramid,'pyramid.pfl');
{11} linlst:=pyramid^.linlst;
{12} linlstsave(linlst,'pyramid.lfl');
end.
```

Explanations

{01} reading the first halfspace (topological part, "op" for "open", "n" for "negative").
{02} reading its list of linear functions (containing in this case just one function).
{03} this list is combined with the topological part.
{04} determination of some further data, e.g. the incidences between the faces.

{05 to 08} the same procedure for the second halfspace.

{09} intersection of the 2 halfspaces, thus getting the open pyramid O.
{10} save the topological part of the pyramid under the name "pyramid.pfl".
{11} extraction of the list of linear functions.
{12} saving this list under the name "pyramid.lfl".

Figure 6. Program for computing and storing the EW-structure of the open pyramid O

```
          number of face = 12
            numbers of linear functions:
                 1 2
            adjoined pyramid:
(a)         0    0 0     FALSE
            1    0 -     FALSE
            2    0 +     FALSE
            3    - 0     FALSE
            4    - -     TRUE
            5    - +     FALSE
            6    + 0     FALSE
            7    + -     FALSE
            8    + +     FALSE
            down-incident faces nr:
            up-  incident faces nr:     1    2   -1    0

          number of face =  1
            numbers of linear functions:
                 1
            adjoined pyramid:
            0    0     FALSE
            1    -     TRUE
            2    +     FALSE
            down-incident faces nr:    12
            up-  incident faces nr:    -1    0

          number of face =  2
            numbers of linear functions:
                 2
            adjoined pyramid:
            0    0     FALSE
            1    -     TRUE
            2    +     FALSE
            down-incident faces nr:    12
            up-  incident faces nr:    -1    0

          number of face = -1
            numbers of linear functions:

            adjoined pyramid:
            0        FALSE
            down-incident faces nr:    12    1    2
            up-  incident faces nr:

          number of face =  0
            numbers of linear functions:

            adjoined pyramid:
            0        TRUE
            down-incident faces nr:    12    1    2
            up-  incident faces nr:


(b)             1     4.00     1.00    -1.00
                2    -4.00     1.00    -1.00
```

Figure 7. EW-structure of the open pyramid O: **a** topological part, **b** list of linear functions

```
begin
  {01} polin('square.pfl',square);
  {02} linlstin('square.lfl',linlst1);
  {03} inslinlst(linlst1,square);
  {04} prepro(square);
  {aa} rotate(square,10,square1);

  {05} polin('pyramid.pfl',pyramid);
  {06} linlstin('pyramid.lfl',linlst2);
  {07} inslinlst(linlst2,pyramid);
  {08} prepro(pyramid);
  {aa} rotate(pyramid,10,pyramid1);

  {09} poldiff(square1,pyramid1,diff1);
  {aa} rotate(diff1,-10,diff);
  {10} polprint(diff);
  {11} linlst:=diff^.linlst;
  {12} linlstprint(linlst);
end.
```

Explanations

{01} to {04}: reading and preprocessing square

{05} to {08}: reading and preprocessing pyramid

{09} to {12}: building and printing the difference polyhedron

{aa}   square and pyramid are rotated by 10 degrees because the procedure
                  "poldiff" assumes the polyhedra concerned to be in "general position" with
                  respect to the coordinate system.

Figure 8. Program for computing and printing the EW-structure of the difference polyhedron
$$D = S \setminus O$$

example, Fig. 9 lists the implementation of the locally adjoined pyramid Q belonging to vertex 356 in Fig. 4a. Obviously Q contains a cell 5, e.g, which is the set $H_3^0 \cap H_5^- \cap H_6^+$.

The only primitives of POLYHEDRA are (linear) open halfspaces. As the topological part of the EW-structure is identical for all positive (negative) open halfspaces, it is prestored. Figure 5 shows it for any $H^- \subset \mathbb{R}^2 : F(H^-)$ consists of $H^0 = bnd\ H^-$, $H^+ = ext\ H^-$ and $H^- = int\ H^-$. Accordingly, $Q(H^-)$ consists of $H^-, \varnothing$ and $\mathbb{R}_2$. In case of positive open halfspaces, H^- and H^+ simply reverse their roles. The metric part of the EW-structure of each open halfspace consists of the coefficients of a suitable linear function. Thus each primitive is defined by specifying $d + 1$ coefficients $\in \mathbb{R}$ and associating them with the desired topological part (cf. Fig. 6).

All other objects (i.e. Nef polyhedra) are constructed by performing set operations. Figure 6 shows how the open pyramid O is found by intersecting H_1^- and H_2^+ with the aid of the routine *"polcut"*. Figure 7 shows the resulting EW-structure. By means of Figure 4(b) it should be easy to identify the cells of the arrangement defined by

<pre>
       number of face = 356
       numbers of linear functions:
             3 5 6
       adjoined pyramid:
(a)   0   0 0 0   TRUE
       5   0 - +   TRUE
       7   0 + -   TRUE
      10   - 0 -   TRUE
      12   - - 0   TRUE
      13   - - -   FALSE
      14   - - +   TRUE
      16   - + -   TRUE
      20   + 0 +   FALSE
      23   + - +   FALSE
      24   + + 0   FALSE
      25   + + -   FALSE
      26   + + +   FALSE
       down-incident faces nr:
       up- incident faces nr: 3   5   6   -1   0

       number of face =   3
       numbers of linear functions:
             3
       adjoined pyramid:
       0   0   TRUE
       1   -   TRUE
       2   +   FALSE
       down-incident faces nr: 13 23 356
       up- incident faces nr: -1   0

       number of face =   4
       numbers of linear functions:
             4
       adjoined pyramid:
       0   0   TRUE
       1   -   TRUE
       2   +   FALSE
       down-incident faces nr: 14 24 45 46
       up- incident faces nr: -1   0

(b)   1   1.00   -0.00   -1.00
             2   -1.00   0.00   -1.00
             3   -0.00   1.00   -1.00
             4   0.00   -1.00   -1.00
             5   4.00   1.00   -1.00
             6   -4.00   1.00   -1.00
</pre>

Figure 9. Except of the EW-structure of the difference polyhedron $D = S/O$: **a** topological part (for three faces only), **b** list of linear functions

H_1^- and H_2^+, the faces of O and their locally adjoined pyramids, and the incidence relation $<$ on $\mathbf{F}(O)$.

Objects can be stored and reactivated in POLYHEDRA, and a number of often used elementary polyhedra (cube, tetrahedron, and wedge, e.g.) will normally be made available in a library. We assume the closed square S (which could éasily be constructed starting from 4 linear functions) to be prestored (Fig. 8).

Figure 8 shows how the desired polyhedron D can be found by computing the set difference of S and O with the aid of the routine "poldiff". Finally, Fig. 9 shows an excerpt of the (already somewhat voluminous) EW-structure of D. Taking into account that POLYHEDRA automatically renumbered the linear functions involved, i.e. S is now defined by means of $h_1, \ldots, h_4$ and O by means of h_5 and h_6, it should again be easy to interpret the listing by means of Fig. 4a.

References

[1] Abramowski, S., Müller, H.: Geometrisches Modellieren. Wissenschaftsverlag 1991.
[2] Bieri, H.: Eine Charakterisierung der Polyeder. Elem. Math. *35*, 143–144 (1980).
[3] Bieri, H.: Computing the Euler characteristic and related additive functionals of digital objects from their bintree representation. Comput. Vis. Graphics, Image Proc. *40*, 115–126 (1987).
[4] Bieri, H.: Hyperimages—an alternative to the conventional digital images. In: Eurographics '90 (Vandoni, C. E., Duce, D. A., eds.), pp. 341–352. Amsterdam: North-Holland 1990.
[5] Bieri, H.: Boolean and topological operations for Nef polyhedra. CSG 94 Set-theoretic solid modelling: techniques and applications, pp. 35–53. Information Geometers 1994.
[6] Bieri, H., Mayor, D.: A ternary tree representation of generalized digital images. In: Virtual worlds and multimedia (Magnenat Thalmann, N., Thalmann, D., eds.), pp. 23–35. New York: J. Wiley 1993.
[7] Bieri, H., Nef, W.: A sweep-plane algorithm for computing the volume of polyhedra represented in Boolean form. Lin. Alg. Appl. *52/53*, 69–97 (1983).
[8] Bieri, H., Nef, W.: Algorithms for the Euler characteristic and related additive functionals of digital objects. Comput. Vis. Graphics, Image Proc. *28*, 166–175 (1984).
[9] Bieri, H., Nef, W.: A sweep-plane algorithm for computing the Euler characteristic of polyhedra represented in Boolean form. Computing *34*, 287–302 (1985).
[10] Bieri, H., Nef, W.: Elementary set operations with d-dimensional polyhedra. In: Computational geometry and its applications (Noltemeier, H., ed.), pp. 97–112. Berlin Heidelberg New York Tokyo: Springer 1988 (Lecture Notes in Computer Science vol. 333).
[11] Brisson, E.: Representing geometric structures in d dimensions: topology and order. Proceedings of the 5th Annual ACM Symposium on Computational Geometry, pp. 218–227. New York: ACM Press 1989.
[12] Dobrindt, K., Melhorn, K., Yvinec, M.: A complete and efficient algorithm for the intersection of a general and a convex polyhedron. In: Algorithms and data structures. (Dehne, F., Sack J.-R., Santoro N., Whitesides S., eds.), pp. 314–324. Berlin Heidelberg New York Tokyo: Springer 1993 (Lecture Notes in Computer Science 709).
[13] Dürst, M. J., Bieri, H., Kunii, T. L.: Two linear time algorithms for the Euler number of hierarchically represented digital pictures. Department of Information Science, University of Tokyo, Technical Report 90–007 (1990).
[14] Edelsbrunner, H.: Algorithms in combinatorial geometry. Berlin Heidelberg New York Tokyo: Springer 1987.
[15] Ferrucci, V., Paoluzzi, A.: Extrusion and boundary evaluation for multidimensional polyhedra. Comput. Aided Des. *23*, 40–50 (1991).
[16] Gomes, J., Velho, L.: Abstraction paradigms for computer graphics. To appear in: Vis. Comput. (1995).
[17] Grünbaum, B.: Convex polytopes. New York: Wiley 1967.
[18] Hansen, H. O., Christensen, N. J.: A model for n-dimensional boundary topology. In: Second symposium on solid modeling and applications (Rossignac, J., Turner, J., Allen, G., eds.), pp. 65–73. ACM Press 1993.
[19] Hadwiger, H.: Vorlesungen über Inhalt, Oberfläche und Isoperimetrie. Berlin Heidelberg: Springer 1957.
[20] Hoffmann, C. M.: Geometric and solid modeling—an introduction. Morgan Kaufmann 1989.
[21] Lienhardt, P.: Topological models for boundary representation: a comparison with n-dimensional generalized maps. Comput. Aided Des. *23*, 59–82 (1991).
[22] Mäntylä, M.: An introduction to solid modeling. Computer Science Press 1988.
[23] Nef, W.: Beiträge zur Theorie der Polyeder—mit Anwendungen in der Computergraphik (Contributions to the theory of polyhedra—with applications in computer graphics). Herbert Lang 1978.

[24] Okino, N., Kakazu, Y., Kubo, H.: TIPS-1: technical information processing system for computer-aided design, drawing and manufacturing. In: Computer languages for numerical control (Hatvany, J., ed.), pp. 141–150. Amsterdam: North-Holland 1973.

[25] Paoluzzi, A., Bernardini, F., Cattani, C., Ferrucci, V.: Dimension-independent modeling with simplicial complexes. ACM Trans. Graphics *12*, 56–102 (1993).

[26] Rappoport, A.: Then n-dimensional extended convex difference tree (ECDT) for representing polyhedra. Proceedings of the Symposium on Solid Modeling Foundations and CAD/CAM Applications, pp. 139–147. New York: ACM Press 1991.

[27] Requicha, A. A. G.: Representations for rigid solids: theory, methods, and systems. ACM Comput. Surv. *12*, 437–464 (1980).

[28] Requicha, A. A. G., Rossignac, J. R.: Solid modeling and beyond. IEEE Comput. Graphics Appl. *12*, 31–44 (1992).

[29] Rossignac, J. R.: Through the cracks of the solid modeling milestone. Eurographics '91 State of The Art Report on Solid Modeling. In: From object modelling to advanced visualization (Coquillart, S., Strasser, W., Stucki, P., eds.), pp. 1–75. Berlin Heidelberg New York Tokyo: Springer 1994.

[30] Rossignac, J. R., O'Connor, M. A.: SGC: a dimension-independent model for pointsets with internal structure and incomplete boundaries. In: Geometric modeling for product engineering (Wozny, M. J., Turner, J. U., Preiss, K., eds.), pp. 145–180. Amsterdam: North-Holland 1989.

[31] Rossignac, J. R., Requicha, A. A. G.: Constructive non-regularized geometry. Comput. Aided Des. *23*, 21–32 (1991).

[32] Rossignac, J. R., Voelcker, H. B.: Active zones in CSG for accelerating boundary evaluation, redundancy elimination, interference detection, and shading algorithms. ACM Trans. Graphics *8*, 51–87 (1989).

[33] Sobhanpanath, C.: Extension of a boundary representation technique for the description of n-dimensional polytopes. Comput. Graphics *13*, 17–23 (1989).

[34] Tilove, R. B.: Set membership classification: a unified approach to geometric intersection problems. IEEE Trans. Comput. *C-29*, 847–883 (1980).

Hanspeter Bieri
Institut für Informatik und
angewandte Mathematik
Neubrüggstrasse 10
CH-3012 Bern
Switzerland

Computing Suppl. 10, 61–77 (1995)

© Springer-Verlag 1995

Complex PDE Surface Generation for Analysis and Manufacture

M. I. G. Bloor and **M. J. Wilson**, Leeds

Abstract. An outline of the way in which the PDE method can be used to generate complex surfaces from individual surface patches is given, with particular emphasis on the influence of the boundary conditions and the effects of parametrisation. The constraints imposed by the requirements of analysis and typical rapid prototyping techniques on the representation of the surface geometry are addressed. These ideas are then employed to obtain the generic design of an engine inlet port of the type which induces swirl, and the results of an analysis of the flow using a powerful automatic mesh generator and analysis package (VECTIS) are presented.

Key words: PDE method, design for function, surface design.

1. Introduction

An approach to designing and generating complex surfaces by regarding their definition as the solution to a suitably posed boundary value problem is examined in this paper. The method was introduced in the area of blend generation where the formulation of the problem falls directly into a form appropriate to this procedure.

The problem of blend generation in computer-aided-design is essentially that of being able to generate a smooth surface to act as a bridging 'secondary' transition between neighbouring 'primary' surfaces [1, 2]. Mathematically we may view the calculation of a blending surface as a boundary value problem in that we require a function $\underline{X}$ on a domain Ω with boundary $\partial\Omega$, on which is specified boundary data [3]. The boundary data will typically be in the form of $\underline{X}$ and a number of its parametric derivatives on $\partial\Omega$. The number of derivatives specified will depend on the required degree of continuity between the blend and the surfaces which meets. We view u & v as coordinates of a point in $\Omega(\subset R^2)$, and $\underline{X}$ as a mapping from that point in Ω to a point in 3-space: $R^2(\Omega) \rightarrow E^3$. To satisfy these requirements we regard $\underline{X}$ as the solution of a partial differential equation

$$L_{u,v}^m(\underline{X}) = \underline{F}(u, v) \tag{1}$$

where $L_{u,v}^m()$ is an elliptic partial differential operator of order m in the independent variables u & v, while $\underline{F}$ is a vector-valued function of u & v. There is a wide variety of methods for finding the solution of elliptic PDEs, ranging from elementary separation of variables, through Green's Functions, to numerical techniques. For more information the reader is referred to Williams [4] and Smith [5]. The actual PDEs we have solved in previous papers have been based on the biharmonic equation ($\nabla^4 \phi = 0$), and for the moment we will restrict our discussion to surfaces that are the solutions of the following fourth-order elliptic PDE:

$$\left(\frac{\partial^2}{\partial u^2} + \mathbf{a}^2 \frac{\partial^2}{\partial v^2}\right)^2 \underline{X} = 0 \tag{2}$$

This requires boundary conditions on the function value and its normal parametric derivatives on $\partial\Omega$; in other words, in terms of a blend problem, tangent plane continuity can be achieved with this order of differential equation. In an earlier paper [6], we illustrated the potential of the method by the use of examples of practical significance which were obtained from either analytical or numerical solutions of a suitably chosen PDE.

The important point is to understand the nature of the control of the surface which is afforded by the boundary conditions and the parameters in the problem. The normal derivative boundary conditions control the direction and speed of approach of the surfaces to the trimlines, and the smoothing parameter $\mathbf{a}$ controls the relative smoothing of the dependent variables between the u and v directions. For large $\mathbf{a}$, changes in the u direction occur over a relatively short length scale, namely $1/\mathbf{a}$ times the length scale in the v direction over which similar changes take place. Thus by adjusting the value of $\mathbf{a}$ one can change the properties of the blending surface, either propagating the effects of the boundary curves over substantial parts of the blend, or confining their effects of regions near the edges of the blend. These points can best be illustrated by reference to simple examples produced earlier by the authors [7].

It is common practice to represent complex surfaces in terms of simple polynomial functions of two parameters, e.g. Bezier surfaces [8, 9] B-spline surfaces [10, 11], Rational B-spline surfaces [12]. Surface design using such representations has been discussed by a number of authors. Tiller [12], for instance, describes a method of using rational B-spline curves to produce a variety of different surface shapes. He describes this method as 'skinning', in which a surface is defined by a series of plane cross-sectional profiles through which the surface must pass, which are arranged along a 'path profile' which acts to orient the cross sections and order them in space. Woodward [10], also, describes a variety of techniques for producing free-form surfaces which can be collectively labelled with the term 'cross-sectional design'. In a later paper [13], Woodward goes on to describe in some detail interactive skinning techniques using B-spline surface interpolation which, by defining an object in 'outline' by a collection of projection and section curves, enables him to produce a wide variety of shapes.

Here, we will indicate how the PDE method, introduced for blends, can be used to generate complex free-form surfaces. Unlike earlier work, where the emphasis was on designing a shape using a single surface patch, we will describe how multiple patches can be used to design even more complex surfaces. The surfaces generated by this method are expressed parametrically, often in terms of transcendental functions of the surface parameters rather than simple polynomial expressions, and hence the resulting surfaces tend to be smooth. To use this method to generate a free-form surface, one seeks a solution to an equation of the type already discussed, e.g. Eq. 2.

For complex surfaces, it will generally be necessary to construct the surface from a number of parametric patches, and we need to be conscious of the fact that each patch must be bounded by specified curves, with adjacent surfaces having the required degree of continuity across their common boundary. However, when constructing a free-form surface from a number of PDE patches, one still has considerable freedom with which to choose the boundary conditions, despite the continuity requirements, in order to achieve the desired shape. The solution remains sensitive to the choice of boundary conditions, and this fact is turned to the designer's advantage as it is the boundary conditions that provide him or her with a powerful tool for surface manipulation.

A convenient way to demonstrate the relationship between parameters and surface changes is through the series of examples given in earlier papers [6, 14, 15] to which the reader is referred rather than repeating here.

For the purposes of this paper, we shall be concerned with what may be called periodic (or closed band) solutions. Thus, Eq. 2 is to be solved over the region $(u, v: 0 \le u \le 1; 0 \le v \le 2\pi)$ since we require the solution to be periodic in v. In this case, we may write the general solution in the form,

$$\underline{X}(u, v) = \underline{A}_0(u) + \sum_{n=1}^{\infty} (\underline{A}_n(u) \cos nv + \underline{B}_n(u) \sin nv) \tag{3}$$

where

$$\underline{A}_0(u) = \underline{a}_{00} + \underline{a}_{01} u + \underline{a}_{02} u^2 + \underline{a}_{03} u^3$$
$$\underline{A}_n(u) = \underline{a}_{n1} \exp(anu) + \underline{a}_{n2} u \exp(anu) + \underline{a}_{n3} \exp-(anu) + \underline{a}_{n4} u \exp-(anu)$$
$$\underline{B}_n(u) = \underline{b}_{n1} \exp(anu) + \underline{b}_{n2} u \exp(anu) + \underline{b}_{n3} \exp-(anu) + \underline{b}_{n4} u \exp-(anu) \tag{4}$$

and $\underline{a}_{n1}, \underline{a}_{n2}, \underline{a}_{n3}, \underline{a}_{n4}$ and $\underline{b}_{n1}, \underline{b}_{n2}, \underline{b}_{n3}, \underline{b}_{n4}$ are vector-valued constants, determined by the boundary conditions imposed on the isoparametric lines $u = 0$ and $u = 1$, which form the edges of the patch.

The boundary conditions imposed on the solution are of the form

$$\underline{X}(0, v) = \underline{G}_0(v)$$
$$\underline{X}(1, v) = \underline{G}_1(v)$$
$$\underline{X}_u(0, v) = \underline{s}_0(v)$$
$$\underline{X}_u(1, v) = \underline{s}_1(v); \tag{5}$$

where the subscript u denotes a partial derivative with respect to u.

2. Generic Inlet Port Design

Let us now consider the problem of creating a surface representation of the 'interior' surface of a typical, modern inlet port of an internal combustion engine. As an example which combines most of the likely complications which may arise in single valve systems, we consider the inlet geometry for a diesel engine shown in Fig. 1.

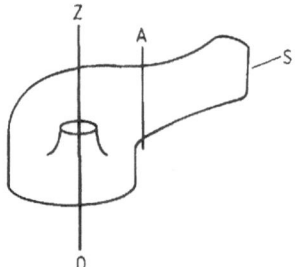

Figure 1. Inlet geometry for a diesel engine

Notice that this is a mould of the space within the inlet port through which the gas flows before entering the combustion chamber. For this reason the volume forms part of the domain over which a fluid dynamical analysis needs to be carried out and we shall concentrate at this stage in trying to represent this type of geometry using a boundary value approach. It follows that once this has been done, the definition of the mechanical part surrounding this space requires further, though generally less critical, surface generation to complete the definition of a solid object.

Before deciding on the way this problem can most conveniently be tackled, it is worth taking note of the functional aspects of this design so that these essential-features are preserved in the geometry to be generated. The gas enters this section of the inlet port over the surface marked S and is accelerated in its passage to the combustion chamber by a slight reduction in cross-sectional area of the duct through which it flows. Furthermore, when the gas reaches station A, it has a component of angular momentum about O in the direction OZ. This ensures that a swirling flow is created within the combustion chamber characterised by a 'swirl coefficient' specified in the design. Another feature that is of significance is the geometry which forms the valve guide. This boss-like structure with a central hole must be located symmetrically on the line OZ.

In order to create this type of geometry we shall use a fourth-order partial differential equation so that the boundary conditions can be chosen to ensure tangent continuity between adjacent patches. Bearing in mind the interpretation of the mathematical form of the surfaces created as outlined in the previous section, i.e. periodic, two distinct surface structures can be identified in the required geometry, providing the complication of the valve guide feature is ignored for the present. In other words, the type of surface to be created initially is as shown in Fig. 2. The surface labelled S_1, which we shall refer to as the swirl chamber, attaches with tangent continuity to the surface S_2, which we shall refer to as the inlet duct, and each of these can be formed from a single periodic patch. At this stage it is instructive to restrict ourselves to an analytic solution where the influence of design parameters and boundary conditions is readily apparent. We shall demonstrate that quite realistic results can be obtained even within the framework of closed form solutions. We consider the surface obtained by solving the system of Eqs. 2 of partial differential equations: where the smoothing parameter **a** is a constant for each component: we shall see that the freedom on the choice of **a** is a very powerful design

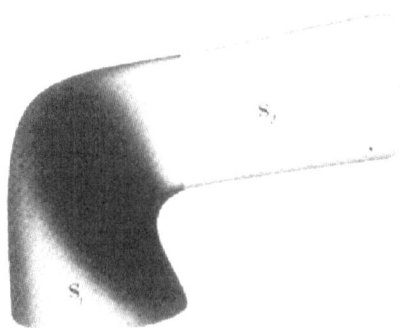

Figure 2. Inlet duct and swirl chamber (without valve guide)

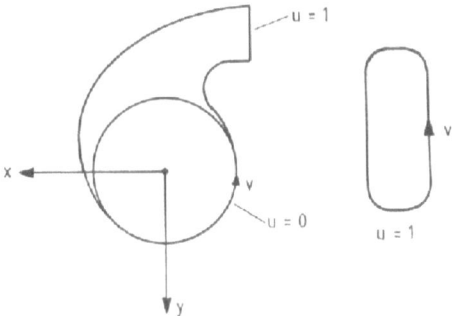

Figure 3. Coordinate system for swirl chamber

tool in the system allowing various features or influences from the boundary conditions to be either propagated or smoothed out through the surface.

We need to set up the problem as a boundary value problem in (u, v) space with boundary conditions specified along curves in the (u, v) plane which correspond to closed curves in E^3. A Cartesian coordinate system is set up as shown in Fig. 3. First of all, considering the surface S_1, we take one of the boundary curves to be the cross-sectional outline of the swirl chamber at the inlet. On this curve, u is taken to be unity and the shape is given parametrically in terms of v. The other boundary curve is taken to be $u = 0$, and is the exit from the swirl chamber at its junction with the combustion chamber. Again the curve is given in a similar way in terms of v. Knowing that separable solutions to Eq. 2 are of the form sinusoidal function times exponential function, to find such a solution the choice of boundary conditions must reflect this. Thus, in the present problem for $0 \leq v \leq 2\pi$ the curve at $u = 1$ is defined by

$$x = xd$$
$$y = yd + y1 \sin v + y3 \sin 3v$$
$$z = zd + z1 \cos v + z3 \cos 3v \qquad (6)$$

and at $u = 0$ by

$$x = \cos v$$
$$y = \sin v$$
$$z = 0 \tag{7}$$

Notice that the correspondence between points on the two curves in space having the same value of v, has to be considered carefully with regard to position and order around the curves. For this type of geometry, the parametrisation of the boundary conditions can be used to good effect to influence the final shape. This is an aspect which will be examined in a later study, alongside other design parameters. With the boundary conditions on x, y, z now determined, it remains to specify the boundary conditions on the derivatives. In the present case, the specification of the derivative boundary conditions is a major part of the design process. The way that appropriate values and distributions can be determined is intuitively straightforward and will be best illustrated by reference to the final shape produced. For the moment, we take

$$\underline{X}_u(0, v) = \begin{pmatrix} 0 \\ 0 \\ sz_0(1 + szv \cos v) \end{pmatrix} \tag{8}$$

$$\underline{X}_u(0, v) = \begin{pmatrix} sx_1(1 + sxv \cos v) \\ 0 \\ sz_1 \end{pmatrix} \tag{9}$$

where the suffix notation has been used for partial differentiation, and sx_1, sz_0, sxv and szv are positive constants. The boundary conditions on this problem indicate that a solution can be expressed in the form:

$$\underline{X}(u, v) = \underline{A}_0(u) + \sum_{n=1}^{3} (\underline{A}_n(u) \cos nv + \underline{B}_n(u) \sin nv) \tag{10}$$

where the $\underline{A}$'s and $\underline{B}$'s are of the form indicated in Eq. 4. The choice of derivative conditions, in particular the variation with v around the boundary curves can be understood by reference to Fig. 1 where the requirement for a variation of the speed of approach of the surface to the boundary curves is apparent. It is clear that neither sxv nor szv should exceed unity otherwise the direction of approach of the surface to the boundary is reversed over a part of the boundary. If further design features needed to be introduced through the derivative conditions, the means by which this could be achieved is apparent from the results obtained in this rather limited solution. The solution shown in Fig. 4 corresponds to parameter values as follows: $xd = -1$, $yd = -2$, $zd = 2$, $sz_0 = 1$, $sx_1 = 5$, $szv = 0.8$, $sxv = 0.5$, $y1 = 0.3$, $y3 = 0.04$, $z_1 = 1$, $z_3 = -0.08$, $a = 0.5$.

The influence of the choice of the parameter **a** can be deduced by reference to the authors' earlier paper [7] where it was shown that the influence of the boundary data associated with a particular mode extends only over a length scale of $O(1/a)$ in u.

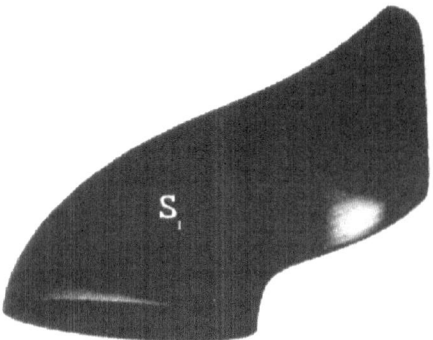

Figure 4. Swirl chamber

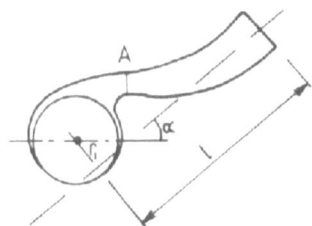

Figure 5. Geometry of inlet duct

If we now turn our attention to S_2, we see that the boundary conditions at the junction curve with S_1 are, in part, determined by the requirements of positional and tangent plane continuity. Although not strictly necessary, we will take the same parametrisation of this junction curve in terms of u and v as was used for S_1. Hence we have

$$\underline{X}(1, v) = \begin{pmatrix} xd \\ yd + y_1 \sin v + y_3 \sin 3v \\ zd + z_1 \cos v + z_3 \cos 3v \end{pmatrix}. \tag{11}$$

Referring to Fig. 5 which shows the general shape of the inlet duct and the some of the parameters used to define the overall geometry, we see that a possible requirement for this type of geometry is a bend in the duct for the avoidance of neighbouring parts of the cylinder head. For the same reason, the offset of the duct from the origin, denoted by r_1 is also allowed. Taking the cross section of the duct at its entrance to be of a similar form to that at its junction with the swirl chamber, we can take the boundary condition on position at $u = 2$, say, to be

$$\underline{X}(2, v) = \begin{pmatrix} -r_1 \sin \alpha - l \cos \alpha - y_s \sin \alpha \\ r_1 \cos \alpha - l \sin \alpha + y_s \cos \alpha \\ zd1 + ze_1 \cos v + ze_3 \cos 3v \end{pmatrix}, \tag{12}$$

where

$$y_s = ye_1 \sin v + ye_3 \sin 3v \tag{13}$$

Here, we have simply two rectangular regions of parameter space joined along the line $u = 1$. Although the condition on the tangent plane at $u = 1$ restricts the choice of the derivative condition there, it does not fix it, and the actual scale of the distribution chosen depends on the value of the derivative at $u = 2$ and the location required for the bend in the duct. Intuitively, it can be seen that increasing the magnitude of the derivative at an end pushes the bend further from that end. Bearing in mind these constraints, the derivative conditions are taken as:

$$\underline{X}_u(1,v) = sc_1 \begin{pmatrix} sx_1(1 + sxv\cos v) \\ 0 \\ sz_1 \end{pmatrix} \tag{14}$$

and

$$\underline{X}_u(2,v) = \begin{pmatrix} sl_1\cos\alpha - sa_1 y_s\sin\alpha \\ sl_1\sin\alpha + sa_1 y_s\cos\alpha \\ sz_1(ze_1\cos v + ze_3\cos 3v) \end{pmatrix}. \tag{15}$$

The solution of Eq. 2 satisfying these conditions is of the form of Eq. 3 and the resulting shape is shown in Fig. 6 for parameter values as follows: $sc_1 = 0.2$, $ye_1 = 0.5$, $ye_3 = 0.04$, $ze_1 = 0.5$, $ze_3 = -0.1$, $r_1 = 0.5$, $1 = 5$, $zd_2 = 2.5$, $sl_1 = 3$, $sa_1 = 0.2$, $a = 0.3$.

To obtain a very gradual transition from the conditions imposed at the entrance to those imposed at the swirl chamber junction, we need to choose a small value of $\mathbf{a}$.

These results serve to illustrate the control on the surface shape which is exercised by the boundary conditions and smoothing parameter. It should be borne in mind that this has been demonstrated within the very restrictive framework of a closed form solution and a constant value of $\mathbf{a}$ over the various harmonics. When these restrictions no longer apply, so that for instance $\mathbf{a}$ could be allowed to vary according to position and length scale of the solution, it can be appreciated that the method has great versatility as a design tool (although a solution would have to be sought numerically), and a study of this has been made by Cheng [16, 17].

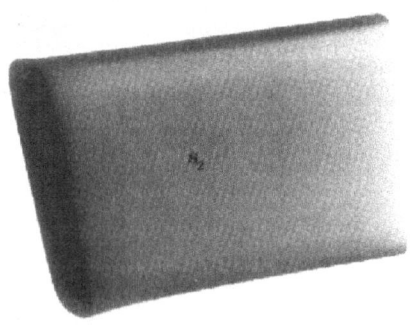

Figure 6. Inlet duct

We now return to the surface S_1 which requires modification in order to accommodate the valve guide structure. It is evident from Figs. 1 and 9 that the surface in the neighbourhood of the z axis needs replacing by an approximately cylindrical surface, which we will call S_3, blended with tangent plane continuity to the swirl chamber. To achieve this without dealing with an intersection problem, a part of the surface S_1, sufficiently large to contain the required surface, is removed surrounding the z axis. The boundary of this hole is formed from isoparametric lines of the original surface, namely $u = u_0$, $u = u_1$, $v = v_0$ and $v = v_1$. This closed boundary now provides the boundary conditions on one closed boundary contour of a new periodic PDE patch (S_3), since all the properties of the surface S_1, are known and thus position and derivative conditions can be found. For this new patch S_3 a new set of parameters is required which we denote by s and t, the latter being the periodic variable.

If we take the hole in S_1 to be $s = 0$, we need to specify $\underline{X}$ and its normal derivative in the new parameter space at $s = 0$ in terms of t for $0 \leq t \leq 2\pi$. For convenience, we take the parametrisation in t to match the existing parametrisation, in u on the isoparametric v lines, and in v on the isoparametric u lines, with a suitable scaling so that t covers the range 0 to 2π. The point $t = 0$ on $s = 0$ corresponds to the point (u_0, v_0) on S_1. The other closed contour bounding the patch to be created is $s = 1$, and this is taken as a plane circle concentric with the z axis. The derivative boundary conditions on $s = 1$ are straightforward, being derived from the curves $u = u_0$, $u = u_1$, $v = v_0$, and $v = v_1$ on S_1, but those at $s = 0$ require some care. Notice that here, the 'overall' parameter space has become somewhat more complicated, in that the (s, t) parameter space grows out of a hole in the (u, v) space.

If we denote $\underline{X}$ on the surface S_1 by $\underline{X}_1(u, v)$, the boundary conditions for the new boundary value problem in the (s, t) plane are given by:

for $u_0 < u < u_1$ and $0 < t < t_1$

$$\underline{X}(0, t) = \underline{X}_1(u, v_0)$$
$$\underline{X}_s(0, t) = sp\underline{X}_{1v}(u, v_0) \tag{16}$$

for $v_0 < v < v_1$ and $t_1 < t < t_2$

$$\underline{X}(0, t) = \underline{X}_1(u_1, v)$$
$$\underline{X}_s(0, t) = -sp\underline{X}_{1u}(u_1, v) \tag{17}$$

for $u_1 > u > u_0$ and $t_2 < t < t_3$

$$\underline{X}(0, t) = \underline{X}_1(u, v_1)$$
$$\underline{X}_s(0, t) = -sp\underline{X}_{1v}(u, v_1) \tag{18}$$

for $v_1 > v > v_0$ and $t_3 < t < 2\pi$

$$\underline{X}(0, t) = \underline{X}_1(u_0, v)$$
$$\underline{X}_s(0, t) = sp\underline{X}_{1u}(u_0, v) \tag{19}$$

where sp is a constant scaling parameter. Notice that the ordering of the variables in the inequalities indicates the correspondence of the old u, v parameters on the boundary to the new t parameter. At the corners of the patch, the u derivatives are not defined, in other words we have singularities at these points, and these will be discussed shortly as they have decisive effect on the shape of the surface in the corner.

At $s = 1$ we have:

$$\underline{X}(1, t) = \begin{pmatrix} rg\cos(t - \theta) \\ rg\sin(t - \theta) \\ zg \end{pmatrix} \tag{20}$$

and

$$\underline{X}(1, t) = \begin{pmatrix} 0 \\ 0 \\ sg \end{pmatrix} \tag{21}$$

where rg is the 'guide' radius and sg is a constant slope parameter. The parameter θ is chosen to avoid excessive twisting between $s = 0$ and $s = 1$ of the isoparametric s lines.

The shape of surface resulting from the solution of this boundary value problem is shown in Fig. 7 and the parameter values are $rg = 0.3$, $zg = 1.7$, $sg = 1$ and $\theta = 0.5$.

All that is required to complete the description is an axially symmetric surface which represents the outer surface of the valve guide and meets the surface just created as $s = 1$. This again is a periodic patch which has a boundary at $s = 2$ consisting of a cirle centred on the z axis and in the plane $z = zg_2$. The additional boundary conditions are

$$\underline{X}(2, t) = \begin{pmatrix} rh\cos(t - \theta) \\ -rh\sin(t - \theta) \\ zg_2 \end{pmatrix} \tag{22}$$

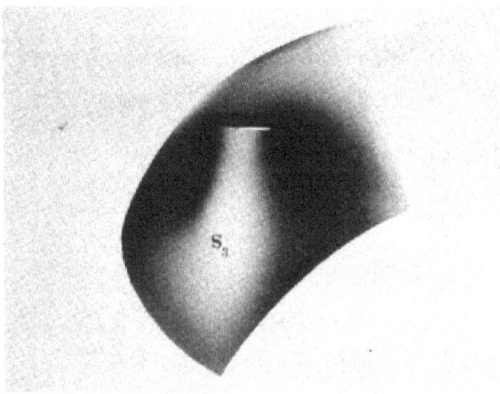

Figure 7. Valve guide

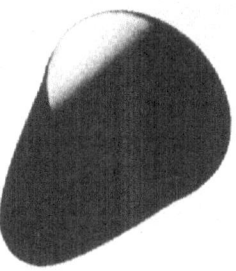

Figure 8. Outer surface of valve guide

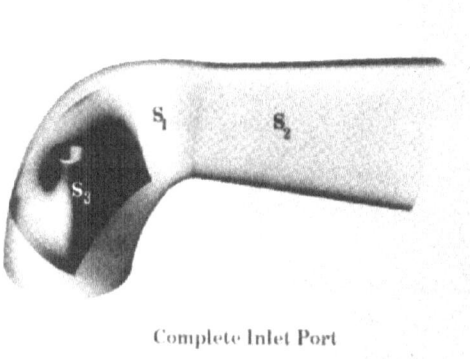

Complete Inlet Port

Figure 9. Complete inlet port

and

$$\underline{X}_s(1, t) = \begin{pmatrix} 0 \\ 0 \\ sg_2 \end{pmatrix} \tag{23}$$

and the resulting shape is shown in Fig. 8 for $rh_2 = 0.1$, $zg_2 = 0.5$ and $sg = -0.5$.

Collecting these four patches together produces the complete surface shown in Fig. 9 which is of the general form of the inlet port geometry that we set out to create (compare with Fig. 1).

3. Corner Singularities

The boundary conditions 16, 17, 18, 19 used in the solution of Eq. 2 ensure that there will be singularities on the $s = 1$ boundary of S_3, corresponding to the four vertices of the (u, v) rectangle in the S_2 parameter plane. It is important that these are analysed, as they determine completely the shape of the surface in the neighbourhood of the corners. We will simplify the analysis by choosing the smoothing parameter **a** and sp to be unity. The extension to other constant values of the smoothing parameters simply represents a rescaling of the coordinate v.

We need only consider one corner, say (u_1, v_0) corresponding to $s = 0$, $t = t_1$. Choosing polar coordinates (r, θ) defined by

$$r \cos \theta = t - t_1$$
$$r \sin \theta = s \tag{24}$$

the equation satisfied by X is

$$\left(\frac{\partial^2}{\partial r^2} + \frac{1}{r} \frac{\partial}{\partial r} + \frac{1}{r^2} \frac{\partial^2}{\partial \theta^2} \right)^2 X = 0 \tag{25}$$

while the boundary conditions, obtained from Eqs. 16, 17, 18, 19 can be written in the following form.

On $\theta = 0$:

$$\underline{X} = \underline{\alpha} r$$
$$\underline{X}_\theta = -\underline{\beta} r \tag{26}$$

where terms of $O(r^2)$ have been nelgected, $\underline{\alpha} = \underline{X}_v(u_1, v_0)$ and $\underline{\beta} = \underline{X}_u(u_1, v_0)$.

On $\theta = \pi$:

$$\underline{X} = -\underline{\beta} r$$
$$\underline{X}_\theta = -\underline{\alpha} r \tag{27}$$

where, again, only leading order terms in r have been retained. It has been assumed that at the corner of the hole, the original surface S_2 had a well defined normal so that $\underline{\alpha}$ and $\underline{\beta}$ were not both zero.

The form of the boundary conditions suggests that we look for a local solution of the form

$$\underline{X} = r\underline{F}(\theta) \tag{28}$$

Substitution of Eq. 28 in Eq. 25 yields an ordinary differential equation for $\underline{F}$ which we can solve, the arbitrary constants of integration being determined by the boundary conditions. The solution is

$$\underline{F}(\theta) = \{\underline{\alpha}[(\pi - \theta)\cos\theta + (1 + \theta)\sin\theta] + \underline{\beta}[(\theta - 1 - \pi)\sin\theta + \theta\cos\theta]\}/\pi \tag{29}$$

Note that the solution in the vicinity of the singularity is entirely determined by the nature of the singularity itself, not the boundary conditions on distant parts of the patch. We also notice that Eq. 29 indicates that $\underline{X}$ lies in the plane of α and β, the tangent plane at the corner on the original surface. Furthermore, if we take the ratio of the two expressions in square brackets, we have a monotonic function of θ, for θ in the range 0 to 2π. Thus the surface in the corner is asymptotically planar without any double points and is a physically acceptable surface.

4. Application to Physical Analysis and Manufacture

In obtaining the surface as a combination of PDE patches, each of which is a solution of a boundary value problem, there is no problem in ensuring that adjacent patches meet perfectly with the desired degree of continuity. In other words, there are no holes in the final surface, which, if need be, can be represented in discrete form by a set of quadrilateral surface patches [22, 23]. This is an extremely important point, both from the point of view of analysis and rapid prototyping through such manufacturing techniques as stereo-lithography (SLA) methods. For instance, in CFD analysis, conservation laws are discretised and fluxes of physical quantities across cells are calculated. Clearly if there are holes in the cells, inaccuracies can arise. More importantly, holes in the surface cause difficulties for inside/outside testing which may be needed for automatic mesh generation. Also, in SLA fabrication, holes in the surface cause a leakage of resin with consequent flaws in the final product.

In term of rapid prototyping techniques more generally, one of the 'standards' for passing surface information about a model to rapid prototyping machines is the STL format. This entails producing a triangular faceted representation of the bounding surface, which is easily obtained from the closed quadrilateral representation produced by the present method. To demonstrate this point, Fig. 10 shows the isoparametric lines over the complete surface.

As it happens, the CFD code (VECTIS) which we use to demonstrate the physical analysis of our model, requires a closed surface of triangular facets bounding the solution domain before the automatic mesh generation and CFD analysis can be carried out. Thus, this code provides us with an ideal test both for analysis and rapid fabrication. The geometry for the inlet port, which we have just created, was

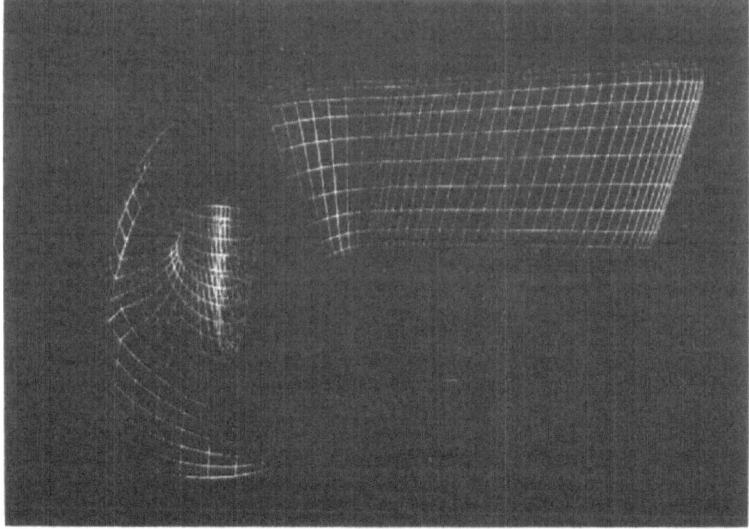

Figure 10. Isoparametric lines of individual PDE surface patches making up the complete port

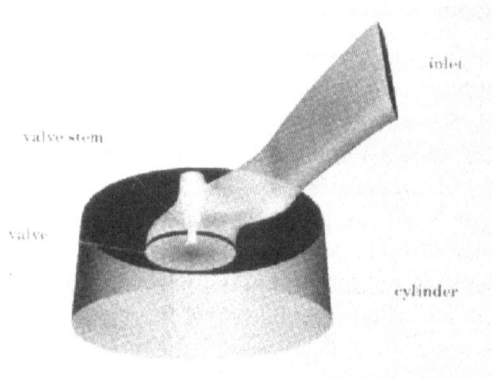

Figure 11. Inlet port, valve and cylinder, ready for CFD calculation. Note surfaces have been rendered as transparent

supplemented by additional geometry of the combustion chamber and a valve (created using the PDE method) to make a physically realistic model for analysis (see Fig. 11). The surface data were put into STL format and were immediately ready for the automatic mesh generation of VECTIS, without the necessity for any surface 'stiching', i.e. the closing of surface holes. Examples of the solution are shown in Figs. 12 to 14 which are taken directly from the VECTIS post processor.

Bearing in mind that the geometry exhibits perfect connectivity between all the patches, owing to the boundary value nature of the solution, this combination of PDE geometry with a sophisticated code for automatic mesh generation, analysis, and presentation of results represents a powerful tool for design for function.

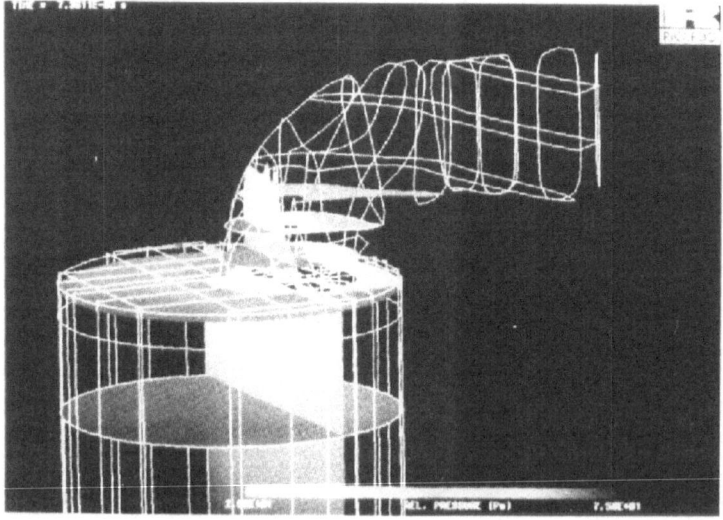

Figure 12. Pressure surfaces on selected coordinate surfaces, calculated using the flow code VECTIS

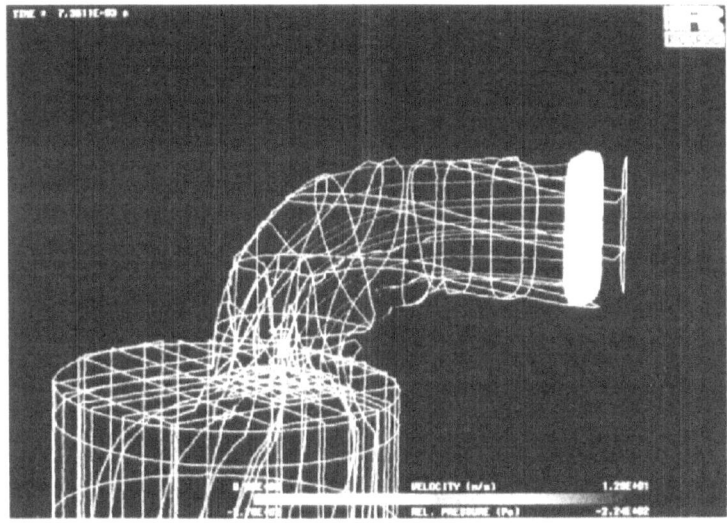

Figure 13. Selected particle tracks through inlet port and cylinder, calculated using the flow code
VECTIS

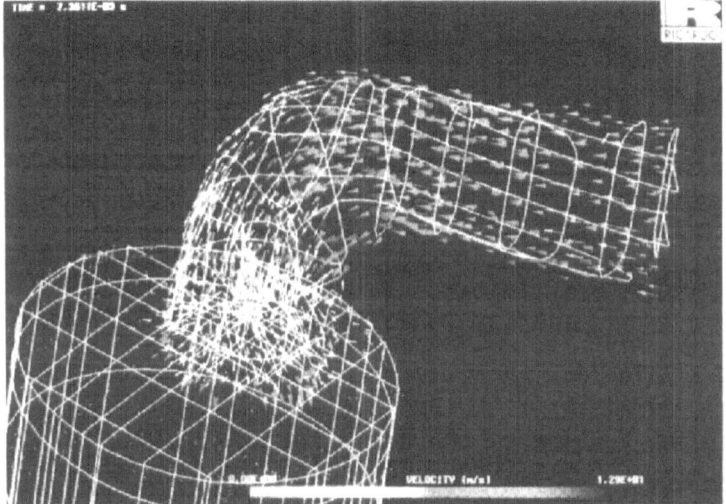

Figure 14. Selected velocity vectors, calculated using the flow code VECTIS

5. Conclusions

We have shown that the PDE method is readily applied to free-form surface design
of complex objects. By defining the boundary curves of a patch of surface, and
varying the derivative conditions imposed upon them and the smoothing parameter
in the PDE, it is possible to create a wide variety of shapes. We have illustrated the

way in which a surface of a realistic object can be created from a small number of patches. For more flexibility, a sixth order equation can be solved, in which case curvature conditions are required at the boundary. However, in many cases of practical interest, it is not necessary to go this far, and perfectly adequate results can be obtained from a fourth order system.

A very important aspect of this approach to the geometry definition, is the fact that, owing to the boundary value nature of the problem, no trimmed patches are used, and the resulting surface is free from holes. This is an extremely important feature when it comes to subsequent physical analysis and fabrication, because there is no difficulty with the mesh generation and no leakage of conserved quantities in the first case, and no leakage of material in the second.

The low level of parametrisation of the complex surface shape, which is a feature of this method, opens up the possibility of automatic optimisation of design. Simple examples of this have been carried out [18–21], but for problems of the analytic-complexity illustrated in this paper this has not yet been done. However, the principles have been established, and it remains to examine ways in which this extremely desirable goal can, in some measure, be achieved by seeking efficient strategies within the constraints imposed by computational power at this time.

Acknowledgements

The authors would like to acknowledge the support of the DTI for EUREKA project EU776, *Integration of CAD, CAE Tools & Fast Free-form Fabrication* (CARP, Computer-Aided Rapid Prototyping), and to the SERC for provision of computing equipment. Additionally, they would like to thank Ricardo Engineering Ltd. for the provision of VECTIS as part of this project, in particular Richard Johns and Karl John; and they would also like to thank Henry Bensler of VW for his support of and interest in this work. Many thanks are also due to Chris Dekanski for producing the VECTIS results, and to Joanna Brown for help with the data conversion.

References

[1] Woordwark, J. R.: Blends in geometric modelling. In: The mathematics of surfaces II (Martin. R. R., ed.), pp. 255–297. Oxford: Oxford University Press 1987.
[2] Hoffmann, C., Hopcroft, J.: The potential method for blending surfaces. In: Geometric modelling: algorithms and new trends (Farin, G., ed.), pp. 347–364. Philadelphia: SIAM 1987.
[3] Bloor, M. I. G., Wilson, M. J.: Generating blend surfaces using partial differential equations. CAD *21*, 165–171 (1989).
[4] Williams, W. E.: Partial differential equations. Oxford: Oxford University Press 1980.
[5] Smith, G. D.: Numerical solution of partial differential equations. Oxford: Oxford University Press 1987.
[6] Bloor, M. I. G., Wilson, M. J.: Using partial differential equations to generate free-form surfaces. Comput. Aided Des. *22*, 202–212 (1990).
[7] Bloor, M. I. G., Wilson, M. J.: Blend design as a boundary-value problem. In: Theory and practise of geometric modelling (Straber, W., Seidel H.-P., eds.), pp. 221–234. Berlin Heidelberg New York Tokyo: Springer 1989.
[8] Bezier, P.: Example of an existing system in the motor industry: the UNISURF system. Proc. Roy. Soc. London Ser. *A321*, 207–218 (1971).
[9] Bezier, P.: The mathematical basis of the UNISURF CAD system. London: Butterworth 1986.
[10] Woodward, C. W.: Cross-sectional design of B-spline surfaces. Comput. Graphics *11*, 193–201 (1987).

[11] Piegl, L., Tiller, W.: Curve and surface constructions using rational B-splines. CAD *19*, 485–498 (1987).

[12] Tiller, W.: Rational B-splines for curve and surface representation. IEEE Comput. Graph. Applic., 61–69 (1983).

[13] Woodward, C. D.: Skinning techniques for interactive B-spline surface interpolation. CAD *20*, 441–451 (1988).

[14] Bloor, M. I. G., Wilson, M. J.: Geometric design of Hull forms using partial differential equations. In: CFD and CAD in ship design (van Oortmeressen, G., ed.), pp. 65–73. Amsterdam: Elsevier 1990.

[15] Bloor, M. I. G., Wilson, M. J.: Design of free-form surfaces using partial differential equations. In: Curves and surface modelling (Hagen, H., ed.), pp. 173–190. Philadelphia: SIAM 1992.

[16] Cheng, S. Y., Bloor, M. I. G., Saia, A., Wilson, M. J.: Blending between quadric solids using partial differential equations. Adv. Des. Automation *1*, 257–263 (1990).

[17] Cheng, S. Y., Bloor, M. I. G., Saia, A., Wilson, M. J.: Blending parametric surfaces using partial differential equations. The Mathematics of Surfaces IV, IMA (to appear).

[18] Lowe, T. W., Bloor, M. I. G., Wilson, M. J.: Functionality in surface design. Adv. Des. Automation *1*, 43–50 (1990).

[19] Lowe, T., Bloor, M. I. G., Wilson, M. J.: Functionality in blend design. CAD *22*, 655–665 (1990).

[20] Doan, N., Bloor, M. I. G., Wilson, M. J.: A strategy for the automated design of mechanical parts. Second Symposium of Solid Modelling and Applications (Rossignac, J., Turner, J., Allen, G. A., eds.), pp. 15–21. New York: ACM Press 1993.

[21] Wilson, D. R., Bloor, M. I. G., Wilson, M. J.: An automated method for the incorporation of functionality in the geometric design of a shell. Second Symposium of Solid Modelling and Applications (Rossignac, J., Turner, J., Allen, G. A., eds.), pp. 253–259. New York: ACM Press 1993.

[22] Brown, J. M., Bloor, M. I. G., Bloor, M. S., Wilson, M. J.: Generation and modification of non-uniform B-spline surface approximations to PDE surfaces using the finite-element method. Adv. Des. Automation *1*, 265–272 (1990).

[23] Brown, J. M., Bloor, M. I. G., Bloor, M. S., Wilson, M. J., Nowacki, H.: Fairness of B-spline surface approximations to PDE surfaces using the finite-element method. The Mathematics of Surfaces IV, IMA (to appear).

[24] Bloor, M. I. G., Wilson, M. J.: Representing PDE surfaces in terms of B-splines. Comput. Aided Des. *22*, 324–331 (1990).

[25] Bloor, M. I. G., Wilson, M. J.: Partial differential equations for shape generation in geometric modelling. In: Geometry and topology of submanifolds III (Verstraelen, L., West, A., eds.), pp. 32–48. Singapore: World Scientific 1991.

[26] Bloor, M. I. G., Wilson, M. J.: Generating N-sided patches with partial differential equations. In: Computer Graphics Internation 89 (Earnshaw, R. A., Wyvill, B., eds.) pp. 129–145. Berlin Heidelberg New York Tokyo: Springer 1989.

[27] Lowe, T. W., Bloor, M. I. G., Wilson, M. J.: Constraints in surface design. The Mathematics of Surfaces IV, IMA (to appear).

[28] Dekanski, C., Bloor, M. I. G., Wilson, M. J.: The design of an aerodynamical propeller using partial differential equations. In: Curves and surfaces (Laurent, P.-J., Le Mhaut, eds.), pp. 139–142. Alain & Schumaker: Academic Press 1991.

[29] Dekanski, C., Bloor, M. I. G., Nowacki, H., Wilson, M. J.: The representation of marine propeller blades using the PDE method. To appear in the Fifth International Symposium on the Practical Design of Ships and Mobile Units, 1992.

[30] Bloor, M. I. G., Wilson, M. J.: Functionality in solids obtained from partial differential equations. Computing [Suppl.] *8*, 21–42 (1992).

[31] Bloor, M. I. G., Wilson, M. J.: Interactive design using partial differential equations. In: Designing fair curves and surfaces (Sapidis, N., ed.), pp. 231–251. Philadelphia: SIAM 1994.

[32] Bloor, M. I. G., Wilson, M. J.: Local control of surfaces generated as the solutions to partial differential equations. Comput. Graphics *18*, 161–169 (1994).

Professor M. I. G. Bloor
Dr. M. J. Wilson
Department of Applied
Mathematical Studies
The University of Leeds
Leeds, LS2 9JI
U.K.

Computing Suppl. 10, 79–86 (1995)

Weight Estimation of Rational Bézier Curves and Surfaces

G.-P. Bonneau, LIMSI-CNRS, France

Abstract. Computer Aided Geometric Design has emerged from the needs of free form curves and surfaces in CAD/CAM technologies. Rational schemes are now replacing polynomial schemes in CAGD. This paper presents several methods to generate weights of rational Bézier curves and surfaces. The main idea is to find weights which minimize functionals measuring a technical smoothness of the curves and surfaces. These functionals are related to the energy of beams and plates in the sense of elasticity theory. A new result on the reparameterization of rational Bézier curves is also presented. This allows to find efficient algorithms minimizing the smoothing functionals.

Key words: Variational design, rational curves and surfaces, weight estimation.

1. Introduction

Rational curves and surfaces are now often preferred to polynomial ones in CAD technology [1]. The flexibility of these curves and surfaces is achieved through the assignment of a scalar (called weight) to each control point.

Many algorithms based on the minimization of a functional were found to model polynomial curves and surfaces. The process of minimization is often constrained by interpolation conditions ([4, 10, 13, 14]). But a least square condition is sometimes more suitable than the interpolation condition. Hagen and Santarelli ([8]) use the minimization of the integral $\int \alpha \, \| X''(t) \|^2 + \beta \, \| X'''(t) \|^2 dt$, together with a least square constraint, to obtain Bézier and B-spline polynomial curves. They extend this result to the surface case.

In the present paper, we apply the main idea of these algorithms to the rational case. New functionals are introduced, which can be used as minimization criteria to produce rational Bézier curves (Section 3), rational triangular (Section 4) or tensor-product (Section 5) Bézier patches. Efficient algorithms are presented to find weights which minimize these functionals.

To perform the minimization, we need to calculate the derivatives of the rational curves and surfaces as functions of the weights. An important idea here is to find an appropriate reparameterization for which the curve (or the surface) and all its derivatives become a polynomial function in the weights, at a particular parameter point. This result is presented in the case of rational Bézier curves in Section 2. It is quite independent of our other results, and could be used in other problems involving derivatives of rational curves and surfaces.

2. Reparameterization of Rational Bézier Curves

We assume that the reader is familiar with the definition of rational Bézier curves (see [2, 11]). Some properties of the reparameterization of these curves by rational linear functions $\left(\text{functions of the type } \dfrac{au + b}{cu + d}\right)$ are given in [3, 12, 15]. For example, one can find a rational linear function such that the reparameterized curve has its first and last weights equal to one. The next theorem generalizes this result.

Theorem 1. *For any rational Bézier curve, and for any two parameter values a, b in its parameter interval $[u_0, u_1]$, there exists a rational linear reparameterization of the curve such that,*

i) the control points of the curve are unchanged

ii) if $\bar{\omega}_0, \ldots, \bar{\omega}_n$ are the new weights, then

$$\sum_{i=0}^{n} \bar{\omega}_i B_i^n\left(\frac{a - u_0}{u_1 - u_0}\right) = 1$$

$$\sum_{i=0}^{n} \bar{\omega}_i B_i^n\left(\frac{b - u_0}{u_1 - u_0}\right) = 1$$

(1)

Proof: A rational linear reparameterization $\varphi(u)$ which preserve the parameter interval $[u_0, u_1]$ of the curve is of the form

$$\varphi(u) = \frac{\rho u_1(u - u_0) + \hat{\rho} u_0(u_1 - u)}{\rho(u - u_0) + \hat{\rho}(u_1 - u)},$$

where ρ and $\hat{\rho}$ are two non zero scalar values with the same sign, so that the denominator does not vanish. If $b_0, \ldots, b_n$ and $\omega_0, \ldots, \omega_n$ are respectively the control-points and control-weights of the curve X, then the reparameterized curve $\bar{X}$ has the following parameteric equation:

$$\bar{X}(u) = X(\varphi(u)) = \frac{\sum_{i=0}^{n} \bar{\omega}_i b_i B_i^n\left(\dfrac{u - u_0}{u_1 - u_0}\right)}{\sum_{i=0}^{n} \bar{\omega}_i B_i^n\left(\dfrac{u - u_0}{u_1 - u_0}\right)},$$

with $\bar{\omega}_i = \rho^i \hat{\rho}^{n-i} \omega_i$, $i = 0, \ldots, n$. Dividing the two Eqs. (1) by $\hat{\rho}^n$ and writing $\alpha = \rho/\hat{\rho}$, we find the following equivalent conditions to (1):

$$\exists \alpha > 0 \left/ \sum_{i=0}^{n} \alpha^i \omega_i \left[B_i^n\left(\frac{a - u_0}{u_1 - u_0}\right) - B_i^n\left(\frac{b - u_0}{u_1 - u_0}\right)\right] = 0 \right.$$

(2)

$$\frac{1}{\hat{\rho}^n} = \sum_{i=0}^{n} \alpha^i \omega_i B_i^n\left(\frac{a - u_0}{u_1 - u_0}\right)$$

$$\rho = \hat{\rho}\alpha.$$

If $a = b$ then $\rho = 0$ and $\hat{\rho} = \left(\omega_0 B_0^n\left(\dfrac{a - u_0}{u_1 - u_0}\right)\right)^{-1/n}$ is a solution.

We assume now $a \neq b$, and denote the left member of the Eq. (2) by f. $f(\alpha)$ is a polynomial of degree n, with first and last coefficients of opposite sign (because ω_0

Figure 1. Reparameterization of a rational Bézier curve

and ω_n must have the same sign, B_0^n is a strictly decreasing function, B_n^n a strictly increasing function, and $a \neq b$). So f must have at least one positive root. ■

Proposition 2. *If $a \neq b$, then for any two indices $i_0 \neq i_1$, the system of linear Eqs. (1) has a unique solution in the weights $\bar{\omega}_{i_0}, \bar{\omega}_{i_1}$.*

The Eqs. (1) assure that the denominator of the parametric equation of the curve "disappears" for the two parameter values a and b. Proposition 2 permits the solution of the system (1) for the weights $\bar{\omega}_{i_0}, \bar{\omega}_{i_1}$. At the parameter values a and b, the curve then becomes a polynomial function in the other weights. In other words, Theorem 1 states that, *for any two parameter values, there exists a reparameterization such that, in these two parameter values, the reparameterized curve and all its derivatives are polynomial functions of the weights.*

We illustrate Theorem 1 with a rational Bézier curve of degree 3, and control weights $(1, 1, 1, 20)$. We apply Theorem 1 with the parameter values $a = 0.8$ and $b = 1.0$. After the reparameteriztation, the new weights are $(16.20, 2.36, 0.34, 1.00)$. Figure 1 shows the curve before and after the reparameterization. The points plotted on the curves are images of regularly spaced parameter values in the parameter interval.

3. Weight Estimation of Rational Bézier Curves

Free-form objects are an essential part of powerful CAD systems. A major issue is the generation of smooth curves and surfaces which can be immediately supplied to the NC-process. The fundamental idea of our method is to minimize a certain functional that can be interpreted in the sense of physics/geometry.

In the case of curves, a thin elastic beam can serve as a model for a fair shape. Such a beam tends to take a position of least strain energy and the energy stored in the beam is proportional to the integral

$$\int \kappa^2(t) \| X'(t) \| \, dt. \tag{3}$$

As an approximation of the integral criterion (3), we can use

$$\int \| X''(t) \|^2 \, dt. \tag{4}$$

Following the physical analogy, we can add a jerk term to (4):

$$\int \alpha \| X''(t) \|^2 + \beta \| X'''(t) \|^2 \, dt. \tag{5}$$

Minimizing the integral (5) performs a blended optimization of energy and jerk. We adjust α and β in an interactive way. More information on the use of this integral for non-rational curves can be found in [9].

In the case of rational Bézier curves, the integral (5) is a transcendental function of the weights. Therefore we use in [5] a discretization of the integral (5) at the two endpoints of the curve:

$$\alpha(\| X''(u_0) \|^2 + \| X''(u_1) \|^2) + \beta(\| X'''(u_0) \|^2 + \| X'''(u_1) \|^2). \tag{6}$$

The results of Section 2 allow us to use a more general discretization: given any two parameter value a, b in the parameter interval $[u_0, u_1]$, we minimize the functional

$$\alpha(\| x''(a) \|^2 + \| x''(b) \|^2) + \beta(\| x'''(a) \|^2 + \| x'''(b) \|^2). \tag{7}$$

While the function (6) depends only on the first three and last three weights of the curve, function (7) involves all of the weights. Furthermore, a proper choice of the parameter values a and b in the function (7) can give better results.

We illustrate the minimization of the functions (6) and (7) on a cubic Bézier curve parameterized on $[u_0 = 0, u_1 = 1]$, with $\alpha = 0.8$ and $\beta = 0.2$. For this example, we use

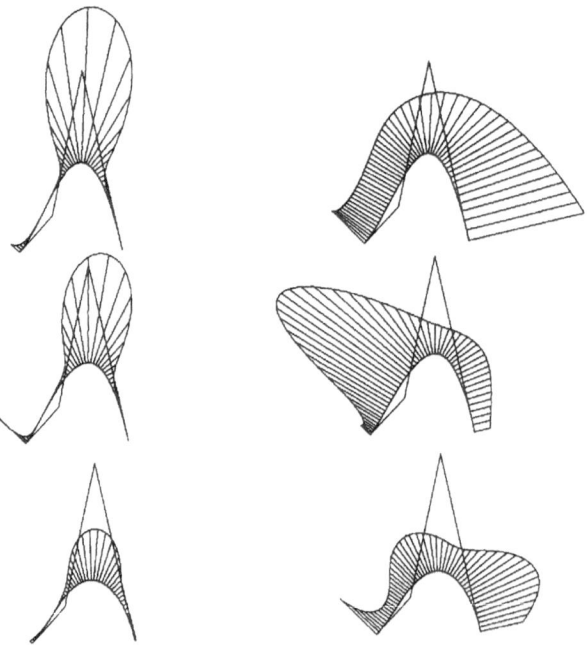

Figure 2. Weight estimation of a rational cubic Bézier curve

$a = 0.2$ and $b = 0.64$ in the function (7) (0.64 is the parameter value in which the non-rational curve reaches its highest curvature value). Figure 2 shows the non-rational curve (top), the curve with the weights minimizing (6) (middle), and the curve with the weights minimizing (7) (bottom). Each curve $(u \to X(u))$ is represented twice, to the left together with the curve $(u \to X(u) + f(u)N(u))$ where $f(u) = \kappa^2 \, \| X'(u) \|$, N is the normal vector, and to the right taking $f(u) = \alpha \| X''(u) \|^2 + \beta \| X'''(u) \|^2$. This gives a good idea of the minimization of both integrals (3) and (5).

4. Weight Estimation of Rational Triangular Bézier Surfaces

Many algorithms available for Bézier curves can be generalized to triangular Bézier surfaces. Indeed, these surfaces are the direct affine analogous of the Bézier curves. And it turns out that the smoothing functional (6) can also be efficiently generalized for triangular Bézier surfaces in the following sense. In the curve case, (6) represents the local bending energy in the unique parameter direction in the endpoints of the curve. For triangular Bézier surfaces, we minimized the local bending energy in the three parameter directions, and in the three corners (see Fig. 3).

This corresponds to the following minimization:

$$\sum_{i+j+k=1, i,j,k> \, = 0} \left(\alpha \| X_{uu}(u_{ijk}) \|^2 + \beta \| X_{uuu}(u_{ijk}) \|^2 + \alpha \| X_{vv}(u_{ijk}) \|^2 \right.$$

$$\left. + \beta \| X_{vvv}(u_{ijk}) \|^2 + \alpha \| X_{ww}(u_{ijk}) \|^2 + \beta \| X_{www}(u_{ijk}) \|^2 \right). \qquad (8)$$

To visualize the minimization of functionals defined on a surface, we use the generalized focal surfaces tool (see [7]): *Each surface X in Figs. 4–6 is represented together with the surface $X + fN$, where f is an offset function, and N is the normal vector.* We illustrate the minimization of the criterion (8) on a triangular cubic Bézier patch. For this example we choose $\alpha = 0.8$ and $\beta = 0.2$ in the function (8). The offset function f for Fig. 4 is

$$f = \alpha \| X_{uu} \|^2 + \beta \| X_{uuu} \|^2 + \alpha \| X_{vv} \|^2 + \beta \| X_{vvv} \|^2 + \alpha \| X_{ww} \|^2 + \beta \| X_{www} \|^2.$$

Therefore, *Fig. 4 shows the minimization of our criterion function* (8). In Section 5 we introduce the energy of flexure and torsion (9) of a surface. To visualize the minimization of this energy, we use in Fig. 5 the following offset function:

$$f = (\kappa_1^2 + \kappa_2^2)\sqrt{g},$$

where g is the determinant of the first fundamental form.

Figure 3. The smoothing functional for triangular Bézier surfaces

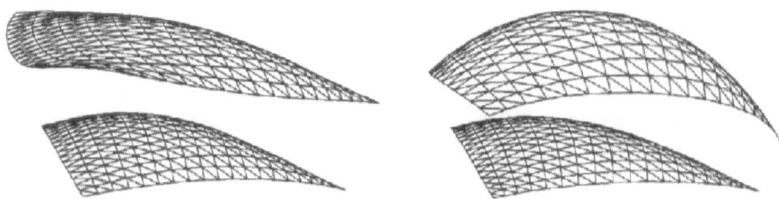

Figure 4. Weight estimation for a cubic rational triangular Bézier patch

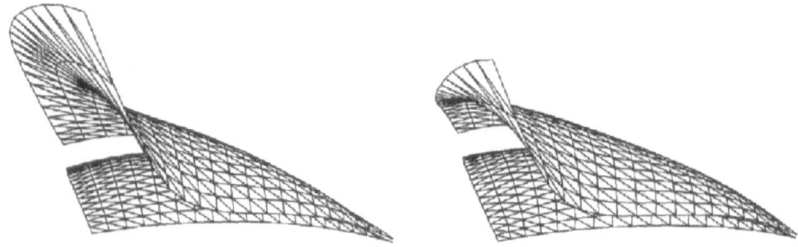

Figure 5. Weight estimation for a cubic rational triangular Bézier patch

Therefore, *Fig. 5 shows the minimization of the exact energy integral* (9).

On the left of Figs. 4 and 5 is the surface with all weights equal to one, and on the right is the surface with the weights minimizing our criteria function (8).

5. Weight Estimation of Rational Tensor-Product Bézier Surfaces

In Section 4 we minimize a generalization of the functional (6) for triangular patches: we minimize the local bending energy in the three parameter directions and in the three corners of the patch. But minimizing the local bending energy in the two parameter directions and in the four corners of a tensor-product Bézier patch leads to a functional which does not take into account geometric data from the interior of the patch. To solve this problem, we return to the physical analogy with the theory of elasticity.

In the case of surfaces, a thin elastic plate of small deflection can serve as a model for a fair shape. Such a plate tends to take a position of least strain energy of flexure and torsion. The energy stored in this plate is proportional to the integral

$$\int_S (\kappa_1^2 + \kappa_2^2)ds. \tag{9}$$

In [6], we use a quadrature of the integral (8) to find weights for rational tensor-product Bézier patches. In this paper we present another solution based on the same energy integral.

In the case of *tensor-product Bézier patches*, following [14], we can use

$$\int \int \|X_{uu}\|^2 + 2\|X_{uv}\|^2 + \|X_{vv}\|^2 \, du dv \tag{10}$$

as an approximation of the integral criterion (9).

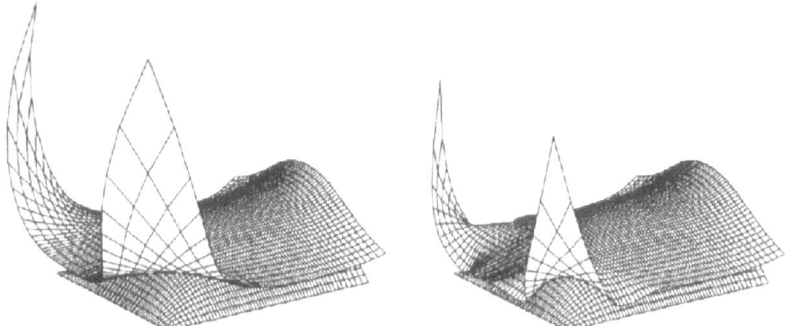

Figure 6. Weight estimation for a biquintic rational Bézier surface

For general rational tensor-product Bézier patches, the integral (10) is a transcendental function of the weights. To get a polynomial function in the weights, we use a discretization of the integral (10) at the corners of the parameter domain $[u_0, u_1] \times [v_0, v_1]$. We assume the four corners weights to be equal to one, and using the other weights, we minimize the function

$$\sum_{i=0,1; j=0,1} \| X_{uu}(u_i, v_j) \|^2 + 2\| X_{uv}(u_i, v_j) \|^2 + \| X_{vv}(u_i, v_j) \|^2. \tag{11}$$

We illustrate the minimization of the criterion (11) on a biquintic surface with four patches. The offset function for Fig. 6 is: $f(u, v) = \| X_{uu} \|^2 + 2\| X_{uv} \|^2 + \| X_{vv} \|^2$. This means that the distances between the two surfaces X and $X + fN$ along the normal vector is equal to the integrand of (10). Thus *Fig. 6 shows both the minimization of the approximated energy integral (10), and of our criterion function (11)*.

The surface with all weights equal to 1 is shown on the left of Fig. 6, the surface with the new weights which minimize the function (11) on the right.

References

[1] Farin, G.: Algorithms for rational Bézier curves. Comput. Aided Des. *15*, 73–77 (1983).
[2] Farin, G.: Curves and surfaces for computer-aided geometric design, 2nd edn. New York: Academic Press 1988.
[3] Farin, G., Worsey, A.: Reparameterization and degree elevation for rational Bézier curves. In: NURBS for curve and surface design (Farin, G., ed.), pp. 47–57. Philadelphia: SIAM 1991.
[4] Hagen, H.: Geometric spline curves. Comput. Aided Geom. Des. *2*, 223–227 (1985).
[5] Hagen, H., Bonneau, G. P.: Variational design of smooth rational Bézier curves. Comput. Aided Geom. Des. *8*, 393–399 (1991).
[6] Hagen, H., Bonneau, G. P.: Variational design of smooth rational Bézier surfaces. Computing [Suppl.] *8*, 133–138 (1993).
[7] Hagen, H., Hahmann, S.: Generalized focal surfaces: a new method for surface interrogation. In: Proc. IEEE Visualization 92 (Kaufman, A. E., Nielson, G. M., eds.), pp. 70–76 (1992).
[8] Hagen, H., Santarelli, P.: Variational design of smooth B-splines surfaces. In: Topics in surface modeling (Hagen, H., ed.) pp. 85–92. Philadelphia: SIAM 1992.
[9] Hagen, H., Schulze, G.: Variational principles in curve and surface design, In: Geometric modeling—methods and applications, pp. 161–184. Berlin Heidelberg New York Tokyo: Springer 1991.
[10] Holladay, J. C.: Smoothest curve approximation. Math. Table Aids Comput. *11*, 233–243 (1957).

[11] Hoschek, J., Lasser, D.: Grundlagen der geometrischen Datenverarbeitung, 2nd edn. Stuttgart: Teubner 1992.
[12] Lucian, M. L.: Linear fractional transformations of rational Bézier curves. In: NURBS for curve and surface design (Farin, G., ed.), pp. 131–139. Philadelphia: SIAM 1991.
[13] Nielson, G.: Some piecewise polynomial alternatives to splines under tension. In: Computer aided geometric design (Barnhill, R. E., Böhm, W., eds.), pp. 209–235 (1974).
[14] Nowacki H., Reese, D.: Design and fairing of ship surfaces. In: Surfaces in CAGD (Barnhill, R. E., Böhm, W., eds.), pp. 121–134 (1983).
[15] Patterson, R. R.: Projective transformations of the parameter of a Bernstein-Bézier curve. ACM Trans. Graphics 4, 276–290 (1985).

Dr. G.-P. Bonneau
Chargé de Recherche
LIMSI-CNRS
B.P. 133
F-91403 Orsay Cedex
France

Computing Suppl. 10, 87–99 (1995)

The Use of Multiple Knots for B-spline Finite Element Approximations to PDE Surfaces

J. M. Brown, M. I. G. Bloor, M. S. Bloor, and **M. J. Wilson,** Leeds

Abstract. Previous papers have examined the method and accuracy of generating B-spline approximations to PDE surfaces; such approximations being obtained using the finite element method with appropriate B-spline basis functions as the basis. This paper looks at the use of multiple knots in the B-spline surface approximation to a PDE surface with a view to allowing discontinuities in the surface normal at certain locations. The method is applied to the surface of a yacht hull where an analytic solution is known, and hence comparison using approximations to the Euclidean and maximum norms can be made.

Key words: B-spline surface, finite element method, multiple knots, PDE surface.

1. Introduction

The PDE method of surface design ([1, 2]) generates surfaces by solving a boundary value problem. A PDE surface, $\underline{S} = (x(u, v), y(u, v), z(u, v))$, is obtained by solving a partial differential equation in u, v-space, subject to conditions on $\underline{S}$ and its derivatives with respect to u and v. In this paper the following partial differential equation is considered:

$$\left(\frac{\partial^2}{\partial u^2} + a^2 \frac{\partial^2}{\partial n^2} \right)^2 \underline{S}(u, v) = 0 \tag{1}$$

where a is constant. Boundary conditions on $\underline{S}$ give the shape of the curve bounding the surface and those on $\underline{S}_u$ and $\underline{S}_v$ give the 'speed' and direction of the isoprarametric lines leaving the boundary. The 'physics' of a surface (e.g. for a yacht the wave drag) may also be incorporated in the design to improve the performance of the object ([3]).

Once the required PDE surfaces have been generated it may be desirable, especially for the exchange of data between different CAD systems ([4, 5]), to obtain an approximation $\underline{\tilde{S}}(u, v)$ in the form of a B-spline surface ([6–8]):

$$\underline{\tilde{S}}(u, v) = \sum_{i=0}^{m} \sum_{j=0}^{n} \underline{p}_{i,j} B_{i,k}(u) B_{j,l}(v) \tag{2}$$

where $\underline{p}_{i,j}$ are $(m + 1) \times (n + 1)$ control vertices for the B-spline approximation to the PDE surface, and $B_{i,k}(u)$ and $B_{j,l}(v)$ are polynomials of degree $k - 1$ and $l - 1$ respectively (for a given surface k and l being fixed).

Such an approximation can been obtained using the finite element method ([9–11]) or by using the collocation method [12]. This paper uses the former method, with the product of B-spline basis functions, $B_{i,k}(u)B_{j,l}(v)$, as the basis, and using Lagrange multipliers to apply the boundary conditions. The solution is in the form of the $(m+1) \times (n+1)$ mesh of control vertices, $p_{i,j}$ which gives a B-spline surface consisting of $(m-k+2) \times (n-l+2)$ patches. The method is briefly outlined in Section 2, and given in detail in [13, 14].

The accuracy of these B-spline surface approximations to a blend and the fairness of such surfaces has been assessed in [16] and [15] respectively. In practice, the required accuracy will depend on the context of the application.

This paper considers the example of a yacht that is only G^0 continuous at the bow, and for which the analytic solution is known. The potential use of multiple knots in the approximating B-spline surface is examined by incorporating them in the basis functions for the finite element method. The control vertices for the yacht are the result of solving the boundary value problem (1) using the finite element method with such basis functions.

Of crucial importance is the error made in approximating the PDE surface with a particular B-spline surface. In Section 3 two methods of quantifying this error are considered. In particular, the Euclidean and maximum norms are used as the basis for a measure of the error, both having a clear physical significance. In the next section, however, we give a brief outline of the use of finite element method in generating a B-spline approximation to a PDE surface.

2. Generating B-spline Approximations to PDE Surfaces Using the Finite-Element Method: a Brief Outline

2.1 The Finite Element Method

The finite element method finds piecewise approximations to continuous functions, in this case the solution to a PDE.

For the sake of example consider a finite region Ω of a two dimensional (u, v) parameter space. Suppose we have a function $\phi(u, v)$ that satisfies the following PDE,

$$L(\phi) = f(u, v) \qquad (3)$$

where $L(\ )$ is a partial differential operator and f is a function of u and v. The aim of the method is to approximate ϕ by an expression of the form

$$\tilde{\phi} = \sum_{i=1}^{n} c_i \psi_i \qquad (4)$$

where $c_i (i = 1, \ldots, n)$ are unknown constants and ψ_i are functions, the basis functions, (usually polynomials) of u & v. In general the approximation (4) does not satisfy (3) exactly, and thus we have a 'residual' $r(\tilde{\phi})$ defined by the equation

$$r(\tilde{\phi}) = L(\tilde{\phi}) - f \qquad (5)$$

(Note that if ϕ_∞ is the exact solution $r(\phi_\infty) = 0$). The aim of the method is to find, for a given set of basis functions, the c_i's that minimise the residual (5). There are various ways of doing this. Here the Galerkin method is used, where the residual is weighted by the basis functions ψ_i themselves, and the procedure is to set the following integral to zero for each of the basis functions in turn.

$$\int\!\!\int_\Omega r(\phi)\psi_i \, du \, dv = 0 \tag{6}$$

This results in n equations which can be solved for the n parameters c_i.

The integral in (6) is evaluated by dividing Ω into elements, with only a small number of basis functions being non-zero over each element. Thus for each element there is a corresponding set of linear equations which can be combined to given equations for the whole system of the form

$$\underline{K}\,\underline{C} = \underline{F} \tag{7}$$

where the $n \times n$ matrix $\underline{K}$ is known as the overall stiffness matrix, $\underline{C}$ is the vector of n unknowns c_i, and $\underline{F}$ is a vector of known quantities.

Various choices are possible for the set of basis functions ψ_i. What distinguishes this work from usual implementations of the finite element method is that B-splines are used for the basis functions. Additionally we are now examining a mechanism for allowing discontinuities in the surface normal along lines in the surface to be incorporated directly in the finite-element formulation of the problem.

2.2 Outline of the Finite Element Method as Implemented in the Present Work

In this work we obtain an approximate solution to Eq. (1) of the form (2) for the x, y and z components for the surface independently in turn. For the sake of example, we will consider only the x component. Consider one element, where x will be approximated by

$$\tilde{x} = \underline{N}\,\underline{p}^e \tag{8}$$

where

$$\underline{N} = (N_1, N_2, \ldots, N_{k \times l})$$
$$= (B_{i,k}(u)B_{j,l}(v),\ B_{i,k}(u)B_{j+1,l}(v), \ldots, B_{i+k-1,k}(u)B_{j+l-1,l}(v)) \tag{9}$$

is a vector formed from the B-spline basis functions that are non-zero over the element, and

$$\underline{p}^e = \begin{pmatrix} p^x_{i,j} \\ p^x_{i,j+1} \\ \vdots \\ p^x_{i+k-1,j+l-1} \\ p^x_{i+k-1,j+l-2} \end{pmatrix} \tag{10}$$

is a corresponding vector of the associated x components of the unknown control vertices.

Formulating the problem using Galerkin's method gives, for each basis function N_i $(i = 1, \ldots, k \times l)$

$$\int\int N_i \left(\frac{\partial^4 \tilde{x}}{\partial u^4} + 2a^2 \frac{\partial^4 \tilde{x}}{\partial u^2 \partial v^2} + a^4 \frac{\partial^4 \tilde{x}}{\partial v^4} \right) du dv = 0 \tag{11}$$

Then, substituting for $\tilde{x}$ and applying Green's theorem (ignoring the boundary integrals as the boundary conditions are of the essential type, see [11], for $i = 1, \ldots, k \times l$, (11) becomes

$$\int\int \left(\frac{\partial^2 N_i}{\partial u^2} \frac{\partial^2 N}{\partial u^2} + 2a^2 \frac{\partial^2 N_i}{\partial u \partial v} \frac{\partial^2 N}{\partial u \partial v} + a^4 \frac{\partial^2 N_i}{\partial v^2} \frac{\partial^2 N}{\partial v^2} \right) du dv \, \underline{p}^e = 0 \tag{12}$$

The integrations are carried out numerically, giving equations for the element of the form

$$\underline{k}^e \underline{p}^e = 0 \tag{13}$$

Each element's contribution $\underline{k}^e$ is added to the overall stiffness matrix $\underline{K}$. The boundary conditions, which are of the form

$$\underline{B} \underline{P}^x = \underline{h} \tag{14}$$

are applied using Lagrange multipliers. Then, the complete system of equations may be written

$$\begin{pmatrix} \underline{K} & \vdots & \underline{B}^T \\ \cdots & \cdots & \cdots \\ \underline{B} & \vdots & \underline{0} \end{pmatrix} \begin{pmatrix} \underline{P}^x \\ \cdots \\ \underline{\lambda} \end{pmatrix} = \begin{pmatrix} \underline{0} \\ \cdots \\ \underline{h} \end{pmatrix} \tag{15}$$

which is solved using LU decomposition, with row interchanges due to the zeros on the diagonal. The solution $\underline{P}^x$ is the required vector of x components of the control vertices. The Lagrange multipliers λ_i are necessary to impose the boundary conditions.

The process is repeated for the y and z components of the surfaces.

3. Measures of Accuracy

3.1 Euclidean Norm

For the surface, $\underline{S}$, the Euclidean norm can be expressed as follows

$$\| \underline{S}(u, v) - \underline{\tilde{S}}(u, v) \|_2 = \left(\frac{1}{R} \int_{u_{min}}^{u_{max}} \int_{v_{min}}^{v_{max}} ((\underline{S}(u, v) - \underline{\tilde{S}}(u, v))^2 \, du dv \right)^{1/2}, \tag{16}$$

where $\underline{S}$ is the analytic solution, $\underline{\tilde{S}}$ is the B-spline approximation and $R = (u_{max} - u_{min})(v_{max} - v_{min})$. In this case, root mean square distance between points of the same value of u and v is used as a measure of the difference between the two surfaces. For the case when $u_{min} = v_{min} = 0$, an approximation to this norm is obtained using the formula

$$RMS = \left(\frac{1}{(N_u \times N_v) - 1} \sum_{i=0}^{N_u-1} \sum_{j=0}^{N_v-1} (\underline{S}(ih_u, jh_v) - \tilde{\underline{S}}(ih_u, jh_v))^2 \right)^{1/2}, \qquad (17)$$

where N_u and N_v are the number of sample points, and h_u and h_v are the distances between consecutive sample points in the u and v directions respectively. The cases when $u_{min} \neq 0$ and $v_{min} \neq 0$ are a straightforward extension; u_{min} is added to ih_u and v_{min} is added to jh_v.

3.2 Maximum Norm

For $\underline{S}$, the maximum norm is

$$\| \underline{S}(u, v) - \tilde{\underline{S}}(u, v) \|_\infty = \max_{\substack{u \in [u_{min}, u_{max}] \\ v \in [v_{min}, v_{max}]}} |\underline{S}(u, v) - \tilde{\underline{S}}(u, v)|, \qquad (18)$$

where, in this case, the maximum distance between points of the same parameter is used as a measure of accuracy. For $u_{min} = 0$ and $v_{min} = 0$ this is approximated by

$$MAX = \max_{\substack{i \in [0, \ldots, N_u - 1] \\ j \in [0, \ldots, N_v - 1]}} |\underline{S}(ih_u), jh_v) - \tilde{\underline{S}}(ih_u, jh_v)|, \qquad (19)$$

Again, the cases where $u_{min} \neq 0$ and $v_{min} \neq 0$ are straightforward and dealt with as indicated above.

3.3 General Comments on the Measures

For both measures, with an increase in the number of points sampled on the two surface there is a corresponding increase in the closeness with which the two dicrcte sums, RMS and MAX, approximate the continuous norms, i.e. in the limit they tend to their respective continuous norms. With any discrete mesure it is possible for the surfaces to be 'near' to each other at the points sampled, yet much further apart away from these points. Hence, it is necessary to ensure that the sample is large enough for RMS and MAX to be a 'reasonable' approximation to the norms. Due to the polynomial form of the B-spline surface and the fact that only $(k \times l)$ control vertices have influence over one patch, there is a limit to the variation that can occur within the patch, so the knot values can be used as a rough 'scale' by which to determine the sample mesh. To obtain reasonable results it should be ensured that some points are sampled between each pair of consecutive non-equal knots. In this work a 240×240 sample mesh was used (i.e. $Nu = Nv = 240$), with the number of knot intervals being 3×11.

4. A PDE Yacht Hull

A yacht hull is obviously primarily a functional surface. In the computer-aided design of ships, generally, the design process of a vessel is initiated by a set of curves that lie on the hull of the vessel, which are used in some way, usually by interpolation or approximation, to generate a CAD surface description. This surface may be modified to take account of properties such as fairness (see [17, 18]).

Some PDE yacht hull designs are described by Bloor and Wilson [1], their form being determined by conditions at the deck line and the base of the keel, and by the PDE shape parameters. Optimisation of the design has been considered by Lowe et al. [3]. The place of this work in the design process is to obtain a good representation of the optimised PDE yacht hull, with previously determined parameters, in terms of a B-spline surface, thus allowing data transfer.

4.1 Equations for the Yacht Hull

Bearing in mind that the parameters $u\&v$ vary in the ranges $0 \le u \le 1$, $-\pi/2 \le v \le \pi/2$, the boundary conditions on $\underline{S}(u, v)$ are as follows:

On $u = 0$

$$x(0, v) = 6.0 \cos(v) \tag{20}$$
$$x_u(0, v) = 2.0 - 3.0 \cos(v)$$
$$y(0, v) = \sin(2v) - 0.1 \sin(4v) - 0.06 \sin(6v)$$
$$y_u(0, v) = 0.0$$
$$z(0, v) = 0.0$$
$$z_u(0, v) = -3.0$$

On $u = 1$

$$x(1, v) = 3.7 + 0.5 \cos(2v) \tag{21}$$
$$x_u(1, v) = 2.0$$
$$y(1, v) = 0.005 \sin(2v)$$
$$y_u(1, v) = 0.0$$
$$z(1, v) = -1.5$$
$$z_u(1, v) = -4.5$$

The analytic solution for the boundary value problem with the PDE given in (1), with $a = 0.1$, is of the form,

$$x(u, v) = X_0(u) + X_1(u) \cos(v) + X_2 \cos(2v) \tag{22}$$
$$y(u, v) = Y_2(u) + Y_4(u) \sin(4v) + Y_6 \sin(6v)$$
$$z(u, v) = Z_0(u)$$

where

$$X_0(u) = sX_0 u + (3x_f - 2sX_0 - sX_f)u^2 + (-2x_f + sX_0 + sX_f)u^3 \tag{23}$$
$$Z_0(u) = -sZ_0 + (-3d + 2sZ_0 + sZ_f)u^2 - (sZ_f) + sZ_0 - 2d)u^3$$
$$X_i(u) = \alpha_{i1}e^{\sigma u} + \alpha_{i2}ue^{\sigma u} + \alpha_{i3}e^{-\sigma u} + \alpha_{i4}ue^{-\sigma u}$$
$$Y_i(u) = \beta_{i1}e^{\sigma u} + \beta_{i2}ue^{\sigma u} + \beta_{i3}e^{-\sigma u} + \beta_{i4}ue^{-\sigma u}$$
$$Z_i(u) = \gamma_{i1}e^{\sigma u} + \gamma_{i2}ue^{\sigma u} + \gamma_{i3}e^{-\sigma u} + \gamma_{i4}ue^{-\sigma u}$$

$\sigma = ia$, and $\alpha_{i1}, \alpha_{i2}, \alpha_{i3}, \alpha_{i4}, \beta_{i1}, \beta_{i2}, \beta_{i3}, \beta_{i4}, \gamma_{i1}, \gamma_{i2}, \gamma_{i3}$, and γ_{i4} are calculated using the boundary conditions.

Note that the surface is infinitely differentiable every where except at the bow ($v = \pm \pi/2$) where it is only G^0 continuous in the v-direction.

5. B-spline Finite Element Approximation of the Yacht Hull

Initially, a B-spline surface approximation with uniform knots was obtained and then, in an effort to obtain an improved B-spline representation of the yacht hull, a multiple knot was used at the sharp bow. The reason this was done is because at the bow of the yacht the surface is only G^0 continuous in the v-direction (i.e. the unit tangent vectors are not continuous in the v-direction). For a B-spline surface, a knot of multiplicity l-1 at the corresponding value of v introduces such a discontinuity, so such a multiple knot was introduced at the bow. As the B-spline surface approximation is periodic and the bow lies at $v = \pi/2$ and $v = -\pi/2$, there were knots of multiplicity l-1 at both these values. In order to have a fair comparison between the uniform and multiple knot cases the total number of knots and the number and position of the boundary conditions were the same for both cases, the only difference being that the neighbouring knot on each side of $-\pi/2$ was moved to $-\pi/2$ (and likewise, due to periodicity, the knot on each side of $\pi/2$ was moved to $\pi/2$), compare Fig. 3a with 6a. The improvement in the representation can be seen by comparing Fig. 3b with 6b.

6. Results

This section shows how the use of multiple knots can improve the accuracy with which a B-spline surface can represent a PDE surface having a surface normal discontinuity. As mentioned earlier, the example considered has a known analytic solution with which the B-spline approximation is compared. The finite element mesh size is 3×11 and the B-spline parameters are $k = l = 4$.

Table 1 shows the value of the error measures for the case were uniform knot spacing is used (Figs. 1–3) and for the case where a knot of multiplicity three is used at the bow (Figs. 4–6). These two representations are referred to as Yacht 1 and Yacht 2 respectively. The knot of multiplicity 3 ($= l - 1$) was used at the bow to introduce a line of G^0 continuity [6]. It can be seen from the table that the use of multiple knots reduces MAX by 97% and RMS by 96%, which is a very significant reduction.

Table 1. Values of MAX and RMS for the uniform
and multiple knot B-spline representations

V-knots	MAX	RMS
Uniform	3.20×10^{-1}	5.89×10^{-2}
Multiple	8.59×10^{-3}	2.21×10^{-3}

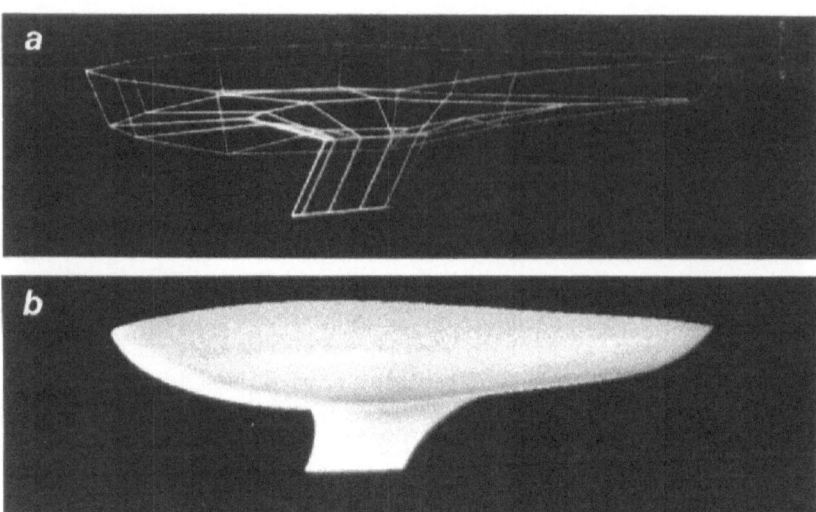

Figure 1. a Control vertices for the yacht hull, with uniform knot spacing. **b** B-spline yacht generated by the control vertices in **a**

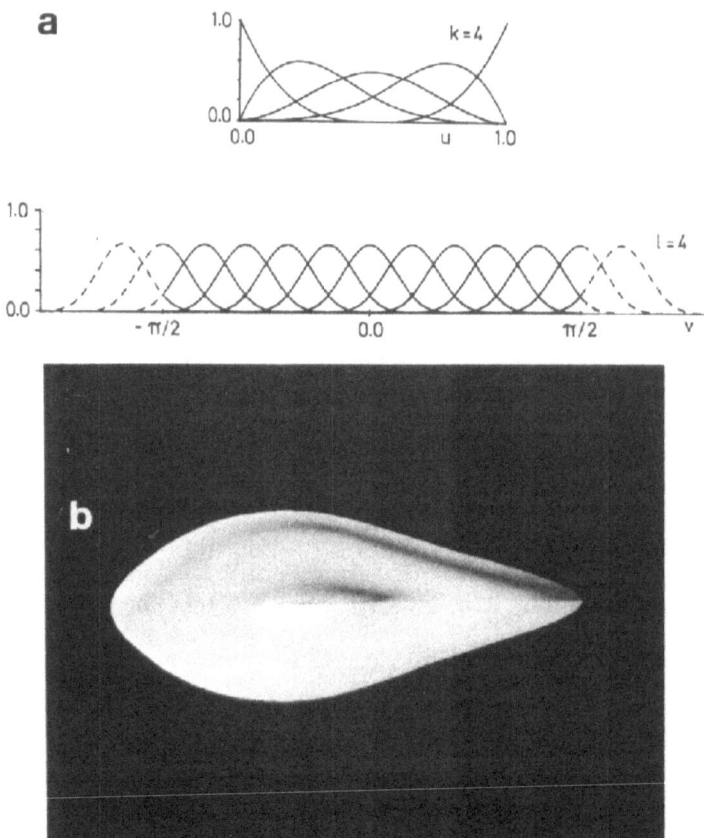

Figure 2. a B-spline basis functions used for the yacht hull in Figs. 1 and 2b. **b** View from above of the B-spline surface generated by the control vertices in Fig. 1a

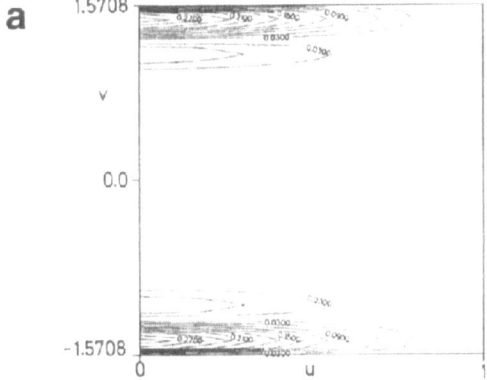

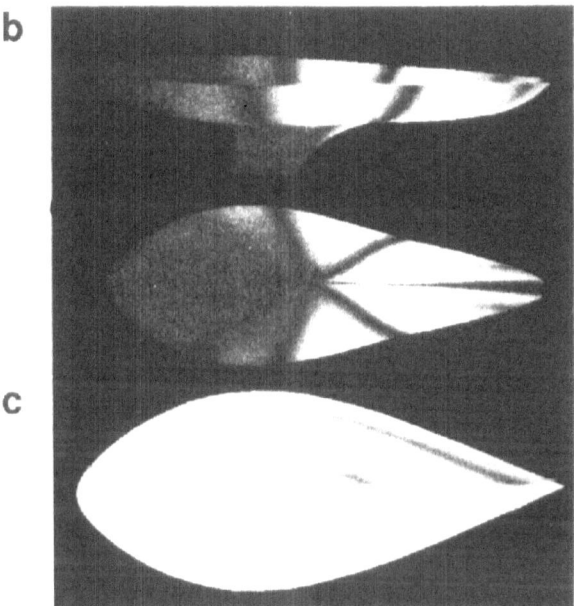

Figure 3. The error between the analytic and uniform B-spline yacht shown as **a** a contour plot over the u, v-plane, and **b** a colour coded plot over the surface (viewed from the side and from above) which can be compared with **c** the analytic solution

Observing the contour plots and colour coded plots of the error distribution for Yacht 1 in Fig. 3 it can be seen that the greatest difference between Yacht 1 and the analytic yacht is in the bow region. The lower numerical values in Fig. 6a as compared with Fig. 3a shows the improvement in the approximation of the bow region that results from the use of multiple knots. It can also be seen in Fig. 6 that the errors are more evenly distributed over the surface; they are a minimum at the v-knot values (where the discrete boundary conditions are applied) and rise then fall between consecutive v-knots.

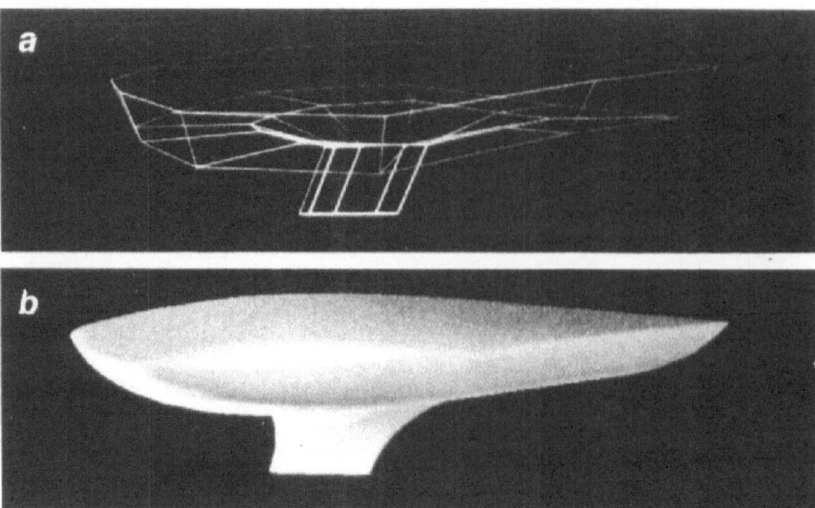

Figure 4. a Control vertices for the yacht hull with multiple knots. **b** B-spline yacht generated by the control vertices in **a**

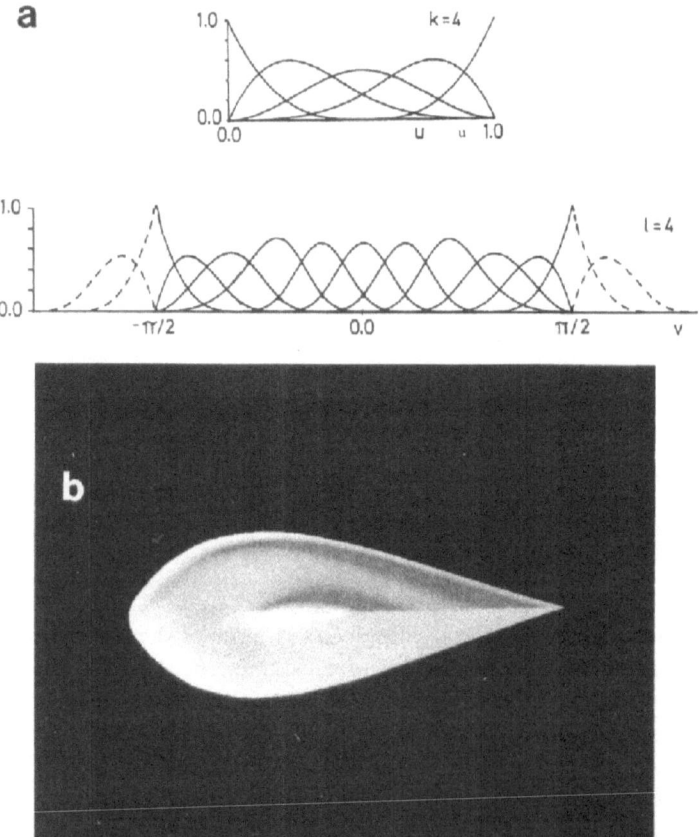

Figure 5. a B-spline basis functions used for the yacht hull in Figs. 4 and 5b. **b** View from above of the B-spline surface generated by the control vertices in Fig. 4a

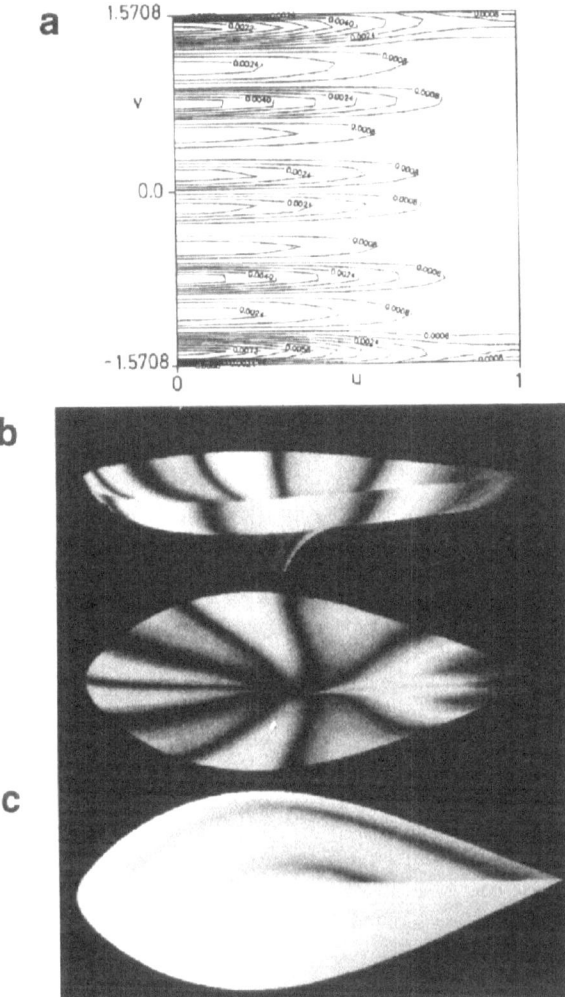

Figure 6. The error between the analytic and B-spline yacht with multiple knots shown as **a** a contour plot over the u, v-plane, and **b** a colour coded plot over the surface (viewed from the side and from above) which can be compared with **c**, the analytic solution

This section briefly illustrates the improvement in the accuracy obtained using multiple knots for a surface that possesses a line across which it is only G^0 continuous. The use of multiple knots is not just a way of improving the accuracy of a representation, more importantly, using the PDE method in this way represents a new approach to generating B-spline surfaces that contain lines at which there is only G^0 continuity.

The properties that multiple control vertices possess can also be exploited to introduce points at which there is G^0 continuity. This is an area for future research.

7. Conclusions

It has been demonstrated that incorporating multiple knots into the B-spline basis functions for a finite-element solution of the PDE surface boundary-value problem, shows a great improvement in the accuracy of representation of objects having surface gradient discontinuities. This corresponded in the PDE boundary-value problem to the inclusion of singularities at such tangent-plane discontinuities.

For the case of the yacht hull, a reduction in errors of 97% for MAX and 96% for RMS demonstrated the enormous improvement in accuracy. There is clearly a great potential for making use of the properties of B-spline representations such as multiple knots and multiple control vertices in approximation of PDE Surfaces using the finite element method, where there are changes in geometric continuity over the surface. The use of different basis functions for the finite element method allows discontinuities to be produced in a solution that would otherwise be smooth.

Acknowledgement

Joanna M. Brown was supported by a studentship from the Science and Engineering Research Council.

References

[1] Bloor, M. I. G., Wilson, M. J.: Generating blend surfaces using partial differential equations. Comput. Aided Des. *21*, 165–171 (1989).
[2] Bloor, M. I. G., Wilson, M. J.: Using partial differential equations to generate free form surfaces. Comput. Aided Des. *22*, 202–212 (1990).
[3] Lowe, T. W., Bloor, M. I. G., Wilson, M. J.: Functionality in surface design. In: Advances in design automation 1990, vol. 1. Computer aided and computational design (Ravani, B., ed.), pp. 43–50. New York: ASME 1990.
[4] ANSI Y14.26M: Digital representation for communication of product definition Data, Version 4.0, Section 3.17, 126–127 (1988).
[5] ISO 10303 Standard for the Exchange of Product Model Data, Part 42: Integrated resources: geometric and topological representation, and part 205: application protocol for mechanical design using surface representation (1992).
[6] Bartels, R. H., Beatty, J. C., Barsky, B. A.: An introduction to splines for use in computer graphics and geometric modelling. Los Altos: Morgan Kauffmann 1987.
[7] Faux, I. D., Pratt, M. J.: Computational geometry for design and manufacture. Chichester: Ellis Horwood 1979.
[8] Mortenson, M. E.: Geometric modelling. New York: J. Wiley 1985.
[9] Davies, A. J.: The finite element method: a first approach. Oxford: Oxford University Press 1980.
[10] Strang, G., Fix, G. J.: An analysis of the finite element method. Englewood Cliffs: Prentice Hall 1973.
[11] Zienkiewicz, O. C.: The finite element method, 3rd edn. New York: McGraw-Hill 1977.
[12] Bloor, M. I. G., Wilson, M. J.: Representing PDE surfaces in terms of B-splines. Comput. Aided Des. *22*, 324–330 (1990).
[13] Brown, J. M., Bloor, M. I. G., Bloor, M. S., Wilson, M. J.: Generation and modification of non-uniform B-spline surface approximations to PDE surfaces using the finite element method. In: Advances in design automation, 1. Computer aided and computational design (Ravani, B., ed.), pp. 265–272. New York: ASME 1990.
[14] Brown, J. M., Bloor, M. I. G., Bloor, M. S., Wilson, M. J.: Generating B-spline approximations of PDE surfaces. Math. Engineering Ind. (in press).
[15] Brown, J. M., Bloor, M. I. G., Bloor, M. S., Nowacki, H., Wilson, M. J.: Fairness of B-spline surface approximations to PDE surfaces using the finite element method, In: The Mathematics of Surfaces IV, (Bowyer, A., ed.), pp. 335–348. Oxford: IMA 1994.
[16] Brown, J. M., Bloor, M. I. G., Bloor, M. S., Wilson, M. J.: The accuracy of B-spline finite element approximations to PDE surfaces (in preparation).

[17] Dill, J. C., Rogers, D. F.: Color graphics and ship hull curvature. In: Computer Applications in the Automation of Shipyard Operation and Ship Design 1V (Rogers, D. F., Nehrling, B. C., Kuo, C., eds.), pp. 197–205. Amsterdam: North-Holland 1982.
[18] Nowacki, H., Reese, D.: Design and fairing of ship surfaces. In: Surfaces in CAGD (Barnhill, R. E., Boehm, W., eds.), pp. 121–134. Amsterdam: North-Holland 1983.

J. M. Brown[1,2]   M. I. G. Bloor[1]
M. S. Bloor[2]   M. J. Wilson[1]
[1] Department of Applied Mathematical Studies
[2] Department of Mechanical Engineering
The University of Leeds, Leeds LS2 9JT, U.K.

Computing Suppl. 10, 101–115 (1995)

Geometric Design with Trimmed Surfaces

G. Brunnett, Kaiserslautern

Abstract. Problems of geometric design with trimmed tensor product surfaces include Boolean sum operations of solids modeled with trimmed patches as well as smooth blending and rendering of trimmed patches. This article reviews existing strategies that provide solutions to these problems and presents a new method that uses Coons' patches for geometric redefinition of trimmed tensor product surfaces.

Key words: CAD, geometric design, trimmed surfaces, Coons' patch.

1. Introduction

For this article, a surface patch will be a one-to-one mapping x of a closed set A of $\mathbf{R}^2$ into $\mathbf{R}^3$. The most popular tool to model sculptured surfaces in geometric design is the tensor product surface scheme where the set A is chosen to be a rectangular subset of $\mathbf{R}^2$ and where the map x is of the form:

$$x(u, v) = \sum_{i=0}^{n} \sum_{j=0}^{m} b_{i,j} F_i(u) G_j(v).$$

In this representation $b_{i,j}$ denote control points which form a rough approximation of the surface. The functions F_i and G_j are called blending functions and are used to build up the surface description in terms of the control points. Examples of tensor product surfaces are Bezier patches where the blending functions are Bernstein polynomials and B-spline surfaces where B-splines are employed as blending functions (see [5]).

A trimmed tensor product surface is a regular tensor product surface, but with certain areas marked as invalid or invisible. These areas are defined by trim curves in the parameter domain of the patch (see Fig. 1). The trim curves themselves may be given in various ways, e.g. via control points of a (piecewise) polynomial curve or simply by a dense sequence of points approximating the shape of the trim curve.

There are two main reasons to consider trimmed surfaces in geometric modeling:

(a) tensor product patches can only be applied to data that are inherently rectilinear while trimmed surfaces are not restricted in their topology.
(b) the introduction of trimmed patches has significantly extended the geometric coverage (i.e. the ability to represent shape) of solid modelers (see [3]).

Figure 2 shows an example for the use of trimmed surfaces in car body design, further examples can be found in [5]. A sculptured solid bounded by trimmed tensor product surfaces is shown in Fig. 3.

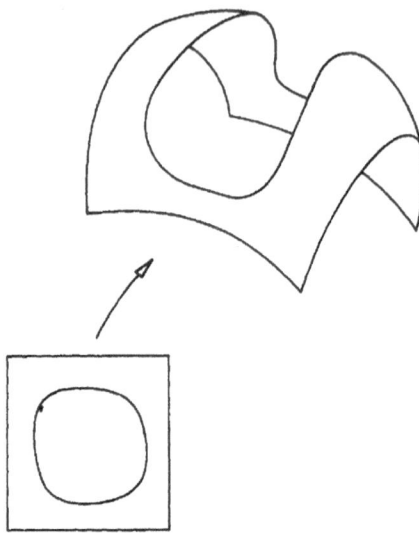

Figure 1. Trimmed surface

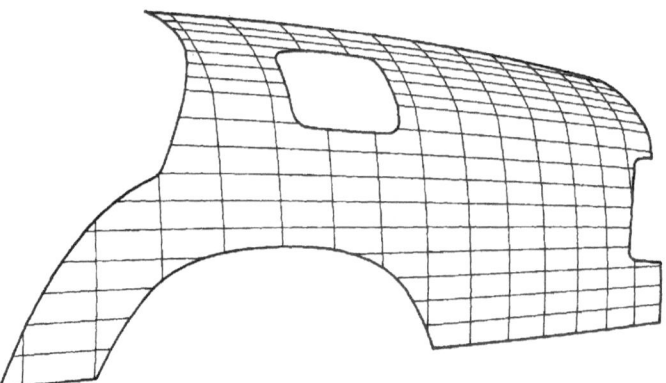

Figure 2. Trimmed car body part

The processing of trimmed surfaces involves many problems due to the fact that standard algorithms of geometric modeling do not apply to surfaces with arbitrarily modified domains. Therefore algorithms like intersection, smooth blending and rendering of tensor product patches must be extended to cover the more general situation of trimmed patches. The research done on trimmed surfaces focuses on three major issues:

- solid modeling with solids bounded by trimmed surfaces
- efficient rendering of trimmed surfaces
- geometric redefinition of trimmed tensor product patches into wider surface schemes on standardized domains.

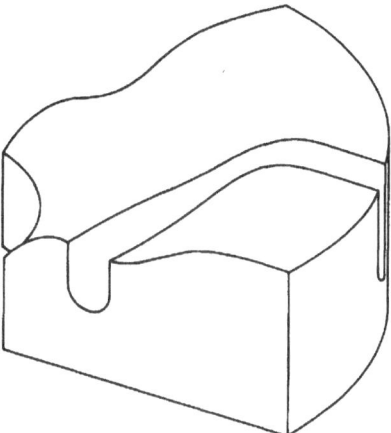

Figure 3. Solid bounded by trimmed surfaces

A major tool in working with trimmed surfaces are algorithms for surface–surface intersection. A review of such methods can be found in [8]; for a detailed description of an advanced intersection algorithm see e.g. [1]. Another related subject that is not covered in this paper is the triangulation of trimmed surfaces (see [16]).

We will begin this survey on trimmed surfaces with a review of algorithms for sculptured solids modeled with trimmed tensor product patches (see [2–6, 10]). We will then summarize two different approaches for the efficient rendering of trimmed surfaces. While Rockwood's algorithm is based on a view driven tesselation of the patch into a polygonal surface (see [12]), the method of Shantz et al. pursues a scan line oriented approach that involves an adapative forward difference scheme to render lines on the surface (see [9, 14, 15]). Finally, research of the author on geometric redefinition of trimmed patches is summarized. In this work transfinite Coons' patches are used to represent trimmed tensor product surfaces exactly.

2. Solid Modeling with Trimmed Surfaces

Free form solids modeled with trimmed patches have been introduced to solid modeling in order to increase the geometric coverage (i.e. the range of realizable geometric shape) of modern solid modelers. In this section, we review the approach of [2] and [3], which is implemented in the system PATRAN and is based on a parametric polynomial representation of the trimmed patches. The boundary of the solid is represented by a continuous collection of trimmed patches, each of which is represented by a pair of data structures: the coefficients of the patch mapping and the 2D CSG tree defining the domain of the mapping. For an alternative approach where a dual representation of trimmed patches (i.e. both parametric and implicit) is used see [6].

The trimmed patch representation of solids supports both Boolean sum operations and free form geometry and uses concepts from three popular forms of solid representation: boundary representation (B-rep), constructive solid geometry (CSG) and analytical solid modeling (ASM) [18]. The outer faces of the solid are stored as in boundary representation; Boolean sum operation symbols and half spaces are stored as in CSG-representation and the faces have the shape of bicubic patches. In contrast to a conventional CSG-tree, however, the tree does not define the 3D solid itself but the trimmed 2D domain of the patch mapping as a collection of half spaces and Boolean operations. Vertices and edges of the trimmed domain can be computed from the 2D CSG-tree using an algorithm similar to the one that evaluates a CSG representation, and the patch mapping can be applied to obtain the vertices and edges of the boundary representation.

The result of a Boolean operation between 3D solids in this representation is a collection of 2D sets which are added to the 2D CSG-tree by other 2D Boolean operations. The set operation algorithm consists of three main steps:

- Proximity test and intersection of the surface patches
- Half-space construction
- Construction of the 2D CSG tree.

The proximity test and the intersection algorithm are performed on the untrimmed surface patches. In an additional step the computed intersection curves are trimmed to the domains of the intersecting surfaces. Segments of the intersection curve in the domain of surface x are trimmed to the domain of surface y by determining the set membership classification (SMC) of corresponding points in the trimmed domain of y. SMC is a function that determines if a critical point lies in the interior, on the boundary or outside of a given closed set. The position of the point in relation to the set is found by counting the number of intersections of the boundary of the set with a ray extended from the critical point. An even number of intersections means that the point is exterior to the set and an odd number implies that the point is interior.

In theory the intersection of a patch with a solid results in curves whose preimages are closed or intersect the boundary of the domain A of x. In practice the result of an intersection algorithm of one surface patch x with all trimmed surfaces bounding the opposing solid is just a collection of linear segments. The linear segments are linked together to closed curves by comparison of the endpoints for proximity. If any of the segment endpoints coincides with the boundary of A, additional points must be introduced to form a closed curve. After creating a set of closed curves in the parameter domain A of x, it has to be determined whether the inside or the outside of these curves has to be discarded. To make this decision a 3D analog of the 2D SMC algorithm must be performed for the image $Q = x(P)$ of a single point P in the domain A. The only restriction on P is that this point should not lie on or too close to any of the curves in the parameter space.

While a 2D SMC requires only straight line intersections, in 3D the intersection of a ray and trimmed bicubic patches must be computed. As a first step in this procedure the patch $(x(u, v), y(u, v), z(u, v))$ is transformed into a coordinate system whose z-axis is the ray. Then a recursive subdivision of the patch is performed in order to find the (u, v) coordinates where the projected patch $(\bar{x}(u, v), \bar{y}(u, v))$ crosses the origin. Since the patches are trimmed, an additional 2D SMC must be carried out. The intersection is not counted if the point does not lie in the domain of the trimmed patch.

After the SMC of Q relative to the opposing solid has been completed, the regions of the plane to be discarded or kept are determined according the following table that specifies the "keep-status" with respect to the Boolean operation being performed.

Set membership	Boolean operation	Keep status
inside	union	discard
inside	intersection	keep
inside	$A-B$	discard
inside	$B-A$	keep
outside	union	discard
outside	intersection	keep
outside	$A-B$	discard
outside	$B-A$	keep
boundary	N/A	generate new point

When the regions of the parameter space to be kept are found, a CSG tree can be built to represent these regions. The algorithm to construct the tree forms a union of all areas to be kept (islands) and subtracts from each island all of its unwanted areas (lakes). Mutually exclusive regions are identified by 2D SMC, and such regions are islands if they are path connected, i.e. if it is possible to travel continuously from one region to another without crossing any of the curves in the parameter space. Once a region to be kept has been identified, all lakes are found by searching for path-connected regions interior to the island.

The complete CSG tree is created by combining the new tree with existing trees from previous Boolean operations using the intersection operation. After completing this procedure for each surface, the collection of modified trimmed patches represents the new solid resulting from the Boolean operation.

3. Rendering of Trimmed Surfaces

Popular rendering methods for tensor product surfaces use evaluation algorithms (like de Boor's algorithm for B-splines and de Casteljau's algorithm for Bézier surfaces) to evaluate the surface at equal increments in the parameter space. This yields a polygonal approximation to the surface which is then rendered.

A trimmed surface patch can be rendered with the same strategy: tesselation of the trimmed domain with consecutive rendering of a piecewise planar approximation. However, since the domain of the surface is not standardized, it is much more difficult to provide a tesselation of the domain that is fine enough to create a high quality image and at the same time fast enough to allow real-time performance.

A modular approach to render trimmed surfaces in real time by a uniform view driven tesselation per patch that provides high quality images was described by

Rockwood et al. in [12]. This method converts all surfaces into individual trimmed Bézier surfaces that can be processed in any order or in parallel. A trimming region of a patch is specified by closed loops of trimming curves. A region is called uv-monotone if any line of constant u or v parameter has a convex intersection with the region. In this method the trimming region is subdivided into uv-monotone regions, each of which is then tesselated in parameter space a grid of rectangular tiles trimmed by triangular coves along curve boundaries.

Two different algorithms for the subdivision into uv-monotone regions are given. A simple strategy makes use of the fact that regions bounded on the left and right by monotone curves and on the top and bottom by horizontal lines are uv-monotone. A patch can therefore be decomposed into proper regions by casting a line of constant v through curve endpoints and extremal points. This algorithm is simple but inefficient in that it creates regions that could be combined.

A more efficient strategy uses special extremal points of the trim curve to sub-divide the region. A point on a trim curve is refered to as v-critical if it is a local maximum (minimum) and the trimming region lies above (below) it. The u-critical points are defined in a similiar way. If the region is split vertically at each v-critical point and horizontally at the u-critical points, the resulting pieces will be uv-monotone.

The final result of both methods is a list of circular lists of curves, where each circular list surrounds a uv-monotone region.

In the next step each uv-monotone region will be tesselated into four- or three sided polygons small enough to satisfy the screen space tolerance TOL specified by the user. This is done by superimposing a lattice of horizontal and vertical lines over the uv-monotone region. The distance of the vertical u-lines is denoted by d_u while d_v denotes the distance between two horizontal v-lines. Trimming curves are segmented into points by evaluating them in increments of d_t. These step sizes are computed in the following way. Let $x(u, v)$ be a Bezier patch of polynomial degree n in u and degree m in v with bezier points $b_{i,j}$. In homogeneous coordinates the bezierpoints form the quadrupel

$$(b_{i,j}, 1) = (x_{i,j}, y_{i,j}, z_{i,j}, 1)$$

which is mapped by viewing and perspective transformations to

$$(r_{i,j}, w_{i,j}) = (\bar{x}_{i,j}, \bar{y}_{i,j}, \bar{z}_{i,j}, w_{i,j}).$$

If π denotes the projection from image space to screen space, the number of steps in the respective u and v directions which guarantee sample points on the path closer in screen space than TOL are given by:

$$n_u = n\sqrt{2}\max(\| w_{i,j}\pi r_{i,j} - w_{i+1,j}\pi r_{i+1,j} \|)/(\text{TOL}\cdot\min(w_{i,j}))$$

for $1 \leq i \leq n-1$ and $1 \leq j \leq m$ and

$$n_v = m\sqrt{2}\max(\| w_{i,j}\pi r_{i,j} - w_{i,j+1}\pi r_{i,j+1} \|)/(\text{TOL}\cdot\min(w_{i,j}))$$

for $1 \leq i \leq n$ and $1 \leq j \leq m-1$.

The step sizes in parameter space are obtained from $d_u = 1/n_u$ and $d_v = 1/n_v$.

Suppose the trim curve c is a Bézier curve of polynomial degree s with control points c_i. A step size $d_t = 1/n_t$ which partions the curve c into segments smaller than d_u or d_v is then given by

$$n_t = s \max(\| c_i - c_{i+1} \|)/d,$$

for $1 \leq i \leq s - 1$, where $d = d_u$ or $d = d_v$. A trimming curve with constant v-parameter (u-parameter) should be segmented based on $d = d_u$ ($d = d_v$). For arbitrary trimming curves the maximum value of n_t based on the two cases should be chosen.

Using the calculated step sizes, each uv-monotone region is tesselated into a grid of rectangles connected by triangles to points evaluated along the curves. In a first step the uv-monotone region is divided along the v-lines of the lattice into horizontal slices of height d_v. If the slice contains two points that lie on one of the u-lines of the lattice, the algorithm chops out the rectangular center section of the slice, generating a strip of tiles. The two regions remaining on the left and right are triangulated using a special tringulation algorithm (see [12]). If the slice does not have two points that lie on the same u-line of the lattice, a general purpose algorithm can be used to triangulate the (monotone) polygon.

An example of a trimming region tesselated with this algorithm is shown in Fig. 4. The polygons created in this way are transformed into facets in object space by evaluating their vertices with the patch mapping. Each facet is then transformed to screen space, clipped, lighted, smooth shaded and z-buffered using standard 3D graphics hardware.

The performance of this algorithm implemented on a SG IRIS-4D GTX work-station was found to be ca. 15 000 triangles per second for a surface with rather complicated trim curve compared to 60 000 triangles per second for untrimmed surfaces.

A different, scan line oriented approach for rendering trimmed surfaces that is particularly suited for hardware implementation was developed by Shantz, Lien

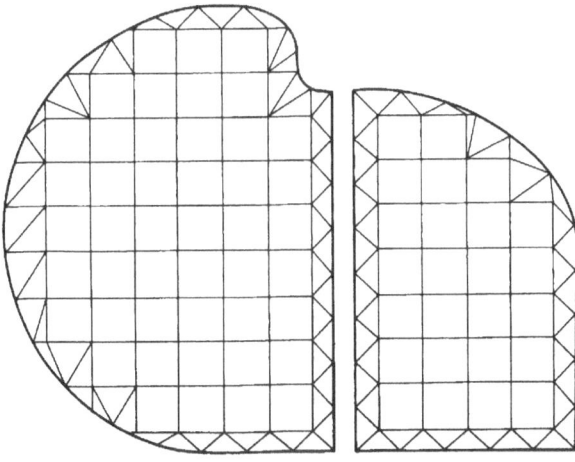

Figure 4. Tesselated trim region

et al. ([9, 14, 15]). This method is based on adaptive forward differencing (AFD) which is an extension of ordinary forward differencing and is related to adaptive subdivision methods in that it adjusts the step size to the next pixel by transforming the equation of the curve to an identical curve with different parameterization. AFD differs from recursive subdivision or standard forward differencing by generating points sequentially along the curve while adjusting the parameterization to give pixel sized steps.

To render a cubic curve with AFD the curve is represented in the form

$$C(t) = C_0 B_0(t) + C_1 B_1(t) + C_2 B_2(t) + C_3 B_3(t),$$

where B_i denotes the forward difference basis

$$B_0(t) = 1, B_1(t) = t, B_2(t) = (1/2)t(t-1), B_3(t) = (1/6)t(t-1)(t-2).$$

The choice of the basis is motivated by the computationally optimal representation of the linear reparameterizations

$$E: EC(t) = C(t+1)$$
$$L: LC(t) = C(t/2)$$
$$L^{-1}: L^{-1}C(t) = C(2t)$$

in this basis (see [9]). The process of rendering curve C with forward differencing begins with the choice of a very small initial segment D of C (e.g. $D = L^n C$) and continues with the generation of the remaining segments by ED, $E^2 D$, $E^3 D$, etc. A disadvantage of forward differencing is that it may not traverse C with uniform velocity. The method of AFD tries to avoid this problem by inserting an occasional L or L^{-1} into the stream of E's whenever the velocity is too great or too small respectively.

This algorithm was implemented in hardware called an AFD unit. An AFD unit is a third order digital differential analyzer which implements an adaptive forward difference solution to a cubic function of a parameter t that varies between 0 and 1. The step size dt for the parameter is adaptively adjusted so that the curve steps along in approximately one pixel steps in screen coordinates. A filter unit is used to control the adaptive step size. Besides other functions this unit compares the current pixel coordinates with the previous pixel coordinates generated by the AFD unit and tells the AFD unit if the increment dt must be adjusted. To generate curve points of a curve in homogeneous coordinates four AFD units are used in parallel.

Tensor product surfaces are rendered by drawing many curves spaced closely together so that no pixel gaps remain between the adjacent curves. Each curve is a univariate cubic obtained by setting one of the surface parameters to a constant, e.g. $u = u_i$. In order to determine the spacing Δu from one curve $x(u_i, v)$ to the $x(u_i + \Delta u, v)$, test trajectories in the orthogonal parameter direction (v-direction) are computed based on the AFD method. The minimum step size of these curves $x(u, v_j)$ at the parameter value $u = u_i$ is chosen as the spacing Δu for the next curve $x(u_i + \Delta u, v)$.

In [14] it is pointed out how the AFD algorithm can not only be used to evaluate a bicubic surface but also to compute the shading parameters. In the Phong lighting model

$$\text{color} = (A + D(N \cdot L)) \, \text{objectcolor} + S(N \cdot H)^m \text{lightcolor}$$

the quantities N, L and H denote the unit normal N of the surface, the unit light vector L and the unit hightlight vector H. The scalar products between these unit vectors and the power involve the most expensive computations of this method. Fast shading of bicubic patches using AFD becomes possible if either the un-normalized products $\bar{N} \cdot L$, $\bar{N} \cdot H$ and $\bar{N} \cdot \bar{N}$ or the normalized products $N \cdot L$, $N \cdot H$ themselves are approximated by bicubic polynomials. Approximation formulas for both versions can be found in [14].

Trimmed patches are rendered by scan converting the trimming region in u, v space using the Δu scanline width (see e.g. [7] for graphics terms). Note that a scanline in the u, v parameter space is different from a scanline in screen space. Rendering a parametric polynomial surface in parameter space order (rather than in scanline order of the screenspace) greatly reduces the difficulty of trimming, since the trim curves can be scan converted in parameter space order using standard techniques for scan converting polygons.

In a first step the trimming curves are converted to the forward difference basis. The trimming curves are then scan converted in u, v space using AFD to increment the curves in order to find their intersections with the first u-curve $x(u_0, v)$. $x(u_0, v)$ is then rendered between one or more intersection pairs v_{min}, v_{max} according to a trimming curve winding rule. This process of scan converting and rendering is repeated for the next u-curve $x(u_0 + \Delta u, v)$.

The method described has the advantage of producing picture quality equivalent to adaptive subdivision without the memory stack management overhead of recursive subdivision and is thus more suitable for hardware implementation. AFD also makes patch rendering performance competitive with polygon rendering. Since the method operates in parameter space order instead of screen scanline order, a transformation from screen space to image coordinates is not required for trimming or image mapping.

4. Geometric Redefinition of Trimmed Surfaces

A trimmed tensor product surface has been defined as a pair of data structures: one for the tensor product patch mapping x and the other to define those parts of the rectangular parameter space of x that are considered to be valid. The point of view expressed in the previous sections was that a trimmed surface should also be processed in this representation. The methods surveyed showed that complex operations such as Boolean sum operations of solids modeled with trimmed surfaces and rendering of trimmed patches can be successfully handeled in this way.

We shall now describe a situation where the presented approach towards trimmed surfaces seems to be insufficient. As pointed out by Farin in [5] trimmed surfaces are frequently used for the design of car bodies in the automotive industry, e.g. the window-opening. In this context as well as in many other industrial applications the trim curves used are of fairly simple geometry. However, to join two trimmed surfaces smoothly along a common trim curve is a very delicate problem even if the trim curve is a straight line in the parameter space. Unlike the case of untrimmed tensor product patches, there is no general way to ensure exact tangent plane continuity between the surfaces.

A different approach towards trimmed surfaces that is particularly suited for the above application is the method of geometric redefinition. The main idea of this method is to work with standardized simple domain topologies (rectangular and/ or triangular) and to increase the complexity of the surface description as opposed to increasing the complexity of the domain of the surface. Thus, a trimmed surface will be replaced by one or more surfaces of higher complexity over simple domains.

The redefinition of the geometry can be done in two different ways:

(a) a trimmed tensor product patch can be approximated by a collection of surfaces of the same form,
(b) a wider surface class can be used to represent a trimmed patch exactly or in better approximation.

As described in [13] the first approach has been studied at General Motors in '84. Nevertheless there are still several open problems related to this method. In highly non rectangular surface topologies it is not clear how many rectangular subpatches are needed or whether triangular or other non-retangular surface topologies will be required to achieve sufficient accuracy at reasonable cost.

Promising results on the use of a wider surface class to redefine trimmed tensor product surfaces have been published in '89 by the author (see [17]). In this work transfinite Coons' patches have been studied for their ability to represent exactly the geometry of a trimmed tensor product patch.

For simplicity we consider only bicubic tensor product surfaces defined on the unit square. If the functions H_0, H_1, H_2, H_3 denote the ordinary cubic Hermite polynomials, a bicubic tensor product patch x can be written as the product of the two operators

$$P_u x(u, v) = x(0, v)H_0(u) + x(1, v)H_1(u)$$
$$+ x_u(0, v)H_2(u) + x_u(1, v)H_3(u)$$
$$P_v x(u, v) = x(u, 0)H_0(v) + x(u, 1)H_1(v)$$
$$+ x_v(u, 0)H_2(v) + x_v(u, 1)H_3(v)$$

in the form

$$x(u, v) = P_u \cdot P_v x(u, v).$$

The transfinite analogon of a bicubic patch is a Coons' surface of type 2, which can be constructed by a so-called Boolean sum operation from the same operators:

$$x(u, v) = P_u x(u, v) + P_v x(u, v) - P_u \cdot P_v x(u, v).$$

The set of bicubic tensor product surfaces is a subset of the set of Coons' surfaces of type 2. This can be seen from the characterizations of the surface classes in terms of their differential equations. A bicubic satisfies the differential equations

$$\frac{\partial^4 x}{\partial u^4} = \frac{\partial^4 x}{\partial v^4} = 0,$$

while a Coons' patch of type 2 is a solution of the differential equation

$$\frac{\partial^8 x}{\partial u^4 \partial v^4} = 0.$$

Further relations between Coons' patches and tensor product patches have been given in [17].

The idea pursued in [17] was to understand the process of trimming as a reparametrization of the surface, i.e. as a composition of the form $x \circ \varphi$ where φ is a regular transformation of the unit square onto the trimming region. The problem of geometric redefinition of trimmed tensor product patches with Coons patches of type 2 is therefore equivalent to the question of how to choose φ such that the composition $x \circ \varphi$ is a Coons' patch of order 2.

If one considers no special properties of x other than being a bicubic patch, there are three different mappings φ that guarantee that $x \circ \varphi$ is a Coons' patch. These are of the form:

- $\varphi_1(u, v) = (h_1(v)u \mid h_2(v), h_3(v))$
- $\varphi_2(u, v) = (g_3(u), g_1(u)v + g_2(u))$
- $\varphi_3(u, v) = au + bv + c,$

where the functions h_i, g_i are four times continuously differentiable with $h_1(t)$, $g_1(t) \neq 0$ and $h_3'(t)$, $g_3'(t) \neq 0$ for all $t \in [0, 1]$ and where a, b, c denote vectors of $\mathbf{R}^2$.

To illustrate how the map φ_1 deforms the unit square into a closed set $A \subset [0, 1]^2$, we consider the images of the boundaries of $[0, 1]^2$ under this map. From

$$\varphi_1(u, 0) = (h_1(0)u + h_2(0), h_3(0))$$

$$\varphi_1(u, 1) = (h_1(1)u + h_2(1), h_3(1))$$

we see that the boundary curves $\varphi_1(u, i)$ are linear parameterizations of line segments parallel to the u-axis in parameter space. From

$$\varphi_1(0, v) = (h_2(v), h_3(v))$$

$$\varphi_1(1, v) = (h_1(v) + h_2(v), h_3(v))$$

G. Brunnett

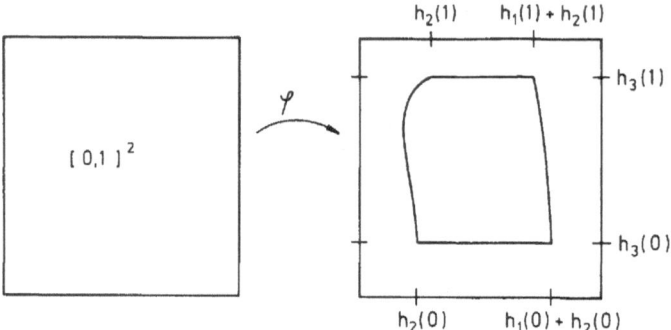

Figure 5. Domain and image of mapping φ_1

we see that the only restriction on the boundary curves $\varphi_1(j, v)$ is that they have the same second coordinate $\bar{v} = h_3(v)$ for each value of v (see Fig. 5). If, for example, h_3 is chosen to be a linear function, both curves $\varphi_1(j, v)$ are restricted in their shape to graphs of functions over the v-axis.

The second map φ_2 expresses the same possibility of trimming the unit square to a subdomain; however, the roles of images of u-lines and v-lines are switched. Here the boundaries of constant u are mapped to straight line segments, while the images of boundaries of constant v deviate from isoparameter lines in the parameter space.

The third map φ_3 reveals a different way to trim the unit square to a subset. Obviously, the image of $[0, 1]^2$ is a parallelogram which means that it is possible to deviate from isoparameter lines simultaneously with all four boundary curves.

The use of these maps is illustrated in Figs. 6–9. Figure 6 shows a bicubic patch with a trimmed curve that corresponds to a straight line segment in the parameter space of the surface. In Fig. 7 a single Coons' patch of type 2 is displayed that represents the trimmed bicubic patch exactly. A more complex trimming curve on the same patch is shown in Fig. 8 and in Fig. 9 a collection of tensor product and Coons' patches is displayed that represent the trimmed surface. Here the domain of x has been subdivided into a collection of bicubic subpatches that represent the surface exactly.

Finally, the subpatches that need to be trimmed have been replaced by Coons' patches of type 2.

The mappings $\varphi_1 - \varphi_3$ describe only a limited way to restrict the domain of a bicubic patch to a subset. However, if these possibilities are combined with a subdivision strategy, one obtains a very powerful tool for representing trimmed bicubic patches exactly with Coons' patches. Note that a tesselation of the domain along isoparameter lines using φ_3 generates a collection of bicubic patches that represents the master patch exactly. This process of subdivision can be performed until a situation has been obtained where the individual patches can be trimmed to a restricted domain via one of the mappings $\varphi_1 - \varphi_3$. Special care must be taken if the trimming region of one of these patches is of triangular topology. Although it is

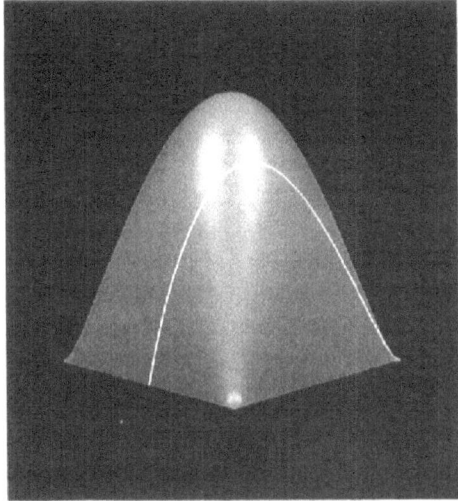

Figure 6. Bicubic patch with trim curve

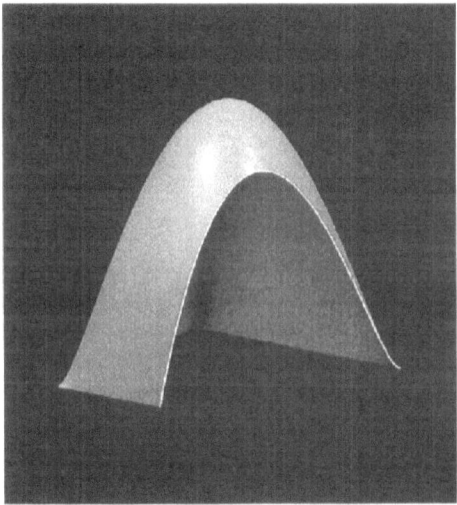

Figure 7. Coons' patch representing the trimmed surface

possible to fill into this region a degenerate quadrilateral Coons' patch where one side collapses into a point, this procedure is not always recommended for the known problems in processing such a degenerate patch. An alternative option in this situation is to approximate the triangular patch with a collection of triangular polynomial patches. Further research must be conducted on the issue of how to employ transfinite non-rectangular patches (e.g. Nielson's side-vertex method) to represent trimmed tensor product surfaces (see [11]).

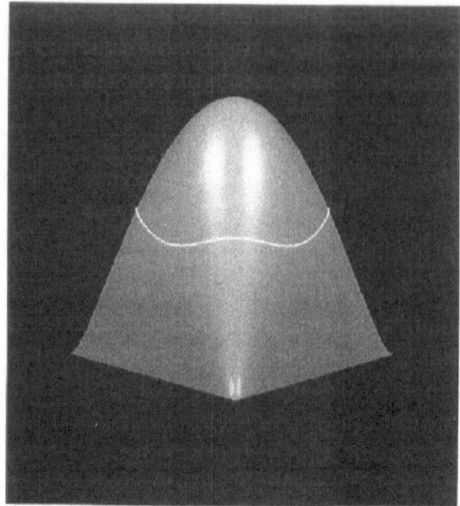

Figure 8. Bicubic patch with closed trim curve

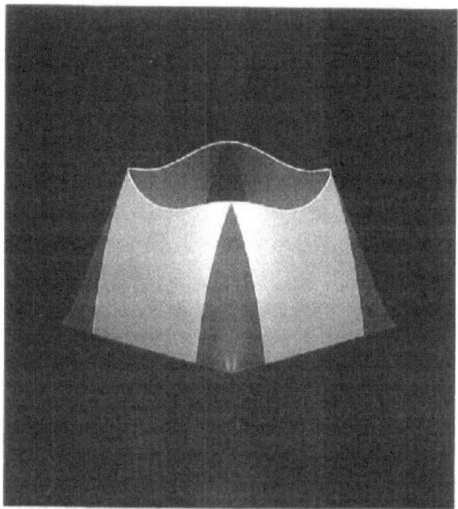

Figure 9. Collection of Coons' patches and tensor product patches

5. Conclusion

There are two different approaches towards the processing of trimmed tensor product surfaces in geometric modeling: the data structure oriented approach and the method of geometric redefinition. The first approach extends any algorithm for complete tensor product surfaces with strategies that take the restricted domain into account. Thus, these methods mainly deal with the organization of the trimming region in an appropriate data structure and the management of this data structure when the patch is processed. Using the important examples of set operations between solids bounded by trimmed patches and rendering of trimmed patches, the main principles of this approach have been presented.

The method of geometric redefinition replaces trimmed patches by one or more surfaces defined on simple standardized domains. Thus, this approach is mainly concerned with the approximation and reparametrization of segments of polynomial patches. It was pointed out that transfinite Coons' patches can be used to represent trimmed tensor product patches exactly. This procedure combined with a preprocess of subdividing the master patch into a collection of bicubic patches avoids any approximation of the trimmed surface as long no triangular trimming regions are encountered.

References

[1] Barnhill, R. E., Kersey, S. N.: A marching method for parametric surface/surface intersection. CAGD 7, 257–280 (1990).
[2] Casale, M. S.: Freeform solid modeling with trimmed patches. IEEE Comput. Graphics Appl., 33–43 (1987).
[3] Casale, M. S., Bobrow, J.: A set operation algorithm for sculptured solid models with trimmed patches. CAGD 6, 235–248 (1989).
[4] Casale, M. S., Bobrow, J. E., Underwood, R.: Trimmed-patch boundary elements: bridging the gap between solid modeling and engineering analysis. CAD 24, 193–199 (1992).
[5] Farin, G.: Curves and surfaces for computer aided geometric design, 2nd edn. New York: Academic Press 1990.
[6] Farouki, R.: Trimmed surface algorithms for the evaluation and interrogation of solid boundary representation. IBM J. Res. Dev. 31, 314–334 (1987).
[7] Foley, J., van Damme, A., Feiner, S., Hughes, J.: Computer graphics: principles and practice, 2nd edn. Reading: Addison Wesley 1992.
[8] Hoschek, J., Lasser, D.: Grundlagen der geometrischen Datenverarbeitung. Leipzig: Teubner 1992.
[9] Lien, S., Shantz, M., Pratt, V.: Adaptive forward differencing for rendering curves and surfaces. ACM Comput. Graph. 21, 111–118 (1987).
[10] Miller, J.: Sculptured surfaces in solid models: issues and alternative approaches. IEEE Comput. Graphics Appl. 6, 37–48 (1986).
[11] Nielson, G.: The side-vertex method for interpolation in triangles. J. Approx. Theory 25, 318–336 (1979).
[12] Rockwood, A.: Real-time rendering of trimmed surfaces. ACM Comput. Graph. 23, 107–116 (1989).
[13] Sarraga, R., Waters, W.: Free-form surfaces in GM solid: goals and issues. In: Solid modeling by computers from theory to applications (Pickett, M., Boyse, J., eds.), pp. 187–204. New York: Plenum Press 1984.
[14] Shantz, M., Lien, S.: Shading bicubic patches. ACM Comput. Graphics 21, 189–195 (1987).
[15] Shantz, M., Chang, S.: Rendering trimmed NURBS with adaptive forward differencing. ACM Comput. Graphics 22, 189–198 (1988).
[16] Sheng, X., Hirsch, B. E.: Triangulation of trimmed surfaces in parametric space. CAD 24, 437–444 (1992).
[17] Schulze-Brunnett, G.: Segmentation operators on Coons' patches. In: Mathematical methods in CAGD (Lyche, T., Schumaker, L., eds.), pp. 561–572. New York: Academic Press 1989.
[18] Zeid, I.: CAD-CAM theory and practice. New York: McGraw-Hill 1991.

Prof. Dr. G. Brunnett
Universität Kaiserslautern
AG CAD und Algorithmische Geometrie
Postfach 30 49
D-67653 Kaiserslautern
Federal Republic of Germany

Computing Suppl. 10, 117–128 (1995)

© Springer-Verlag 1995

The Shape of the Overhauser Spline

W. L. F. Degen, Stuttgart

Abstract. Among the well-known class of tangent-continuous parametric spline curves with cubic Bézier segments, interpolating a sequence of data points, the construction of A. Overhauser is one of the earliest examples after Ferguson [8]. Besides its extreme simplicity and robustness, this spline (in the cardinal case) deserves some interest because of its affine invariance. This allows a complete analysis of its shape-preserving properties, which is given in the present paper.

Key words: Cubic Bézier spline curve, tangent continuity, interpolation of a sequence of data points, affine invariance, shape preserving properties.

1. Introduction

When looking for a simple method to construct a C^1 spline interpolating a sequence of data points $\mathbf{p}_0, \ldots, \mathbf{p}_n$, I rediscovered a scheme given many years ago by A. W. Overhauser [18]. Most of the well-known interpolating C^1 splines composed by cubic Bézier segments can be considered as generalisations of that scheme (for more details, see [7], chap. 8, or [11], sect. 4.1.3). Though all of this seems very familiar, there are, however, some aspects which deserve a new interest in Overhauser splines:

—The unsurpassed robustness, efficiency, and simplicity of its calculation.
—The complete invariance under affine transformations.
—The shape-preserving properties.
—The idea of "blending" two curves by an affine combination, which allows many generalisations.

The present paper elaborates upon these aspects in detail. The case of surfaces is not covered here (see e.g. [10]).

2. The "Sliding Principle"

The basic idea to smooth the transition from one segment of a spline to another consists of using an affine combination to slide continuously—while the common parameter of both segments is growing—from the first segment to the second one. More generally and more precisely, the following can be stated:

Theorem 1. *Let*

$$\mathbf{y}, \mathbf{z} : [a, b] \to \mathbb{R}^d \tag{2.1}$$

be two parameter representations (of differentiability class $C^k, k \geq 2$) for two curve segments $\mathscr{C}_y, \mathscr{C}_z$, having a C^m ($1 \leq m < k$) continuity at both endpoints $\mathbf{a} := \mathbf{y}(a) = \mathbf{z}(a)$,

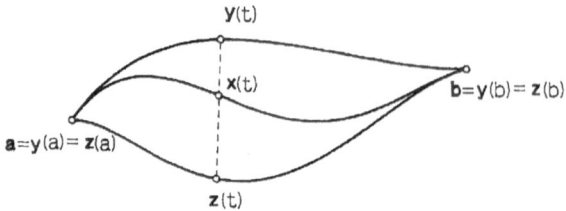

Figure 1. The sliding principle $(m = 0)$

$\mathbf{b} := \mathbf{y}(b) = \mathbf{z}(b)$, *then the segment* $\mathscr{C}_x$ *defined by the affine combination*

$$\mathbf{x} : [a, b] \to \mathbb{R}^d$$

$$\mathbf{x}(t) := \frac{b - t}{b - a} \mathbf{y}(t) + \frac{t - a}{b - a} \mathbf{z}(t), \quad t \in [a, b] \tag{2.2}$$

has a C^{m+1} *continuity with* $\mathscr{C}_y$ *at* $\mathbf{a}$ *and with* $\mathscr{C}_z$ *at* $\mathbf{b}$.

Proof: Differentiating (2.2) yields (with $h := b - a$)

$$\mathbf{x}^{j+1}(a) = \mathbf{y}^{(j+1)}(a) + \frac{1}{h}(\mathbf{z}^{(j)}(a) - \mathbf{y}^{(j)}(a)),$$

$$\mathbf{x}^{j+1}(b) = \mathbf{z}^{(j+1)}(b) + \frac{1}{h}(\mathbf{z}^{(j)}(a) - \mathbf{y}^{(j)}(a)).$$

Thus, if $\mathbf{z}^{(j)}(a) = \mathbf{y}^{(j)}(a)$ and $\mathbf{z}^{(j)}(b) = \mathbf{y}^{(j)}(b)$ for $j = 0, 1, \ldots, m$, then $\mathbf{x}^{(j+1)}(a) = \mathbf{y}^{(j+1)}(a)$ and $\mathbf{x}^{(j+1)}(b) = \mathbf{z}^{(j+1)}(b)$ for $j = 0, \ldots, m$. Furthermore $\mathbf{x}(a) = \mathbf{y}(a)$ and $\mathbf{x}(b) = \mathbf{z}(b)$ is valid by (2.2). $\square$

3. Constructing the Overhauser Spline

We now construct the Overhauser spline. We confine ourselves to the "cardinal case", where no additional information (like a knot sequence or something equivalent) besides the data points is needed (see Section 4.5).

The construction uses only the simplest case $m = 0$ from the sliding principle. It is combined with the well-known fact that there is a unique parabola interpolating three noncollinear points, say $\mathbf{a}, \mathbf{b}, \mathbf{c}$, in such a way that the tangent at $\mathbf{b}$ is parallel to $\mathbf{ac}$. The segment of this parabola from $\mathbf{a}$ to $\mathbf{c}$ can easily be represented in Bézier form as

$$\mathbf{q}(s) = \sum_{k=0}^{2} B_k^2(s) \mathbf{b}_k, \quad s \in [0, 1] \tag{3.1}$$

with the quadratic Bernstein polynomials B_k^2 and the control points

$$\mathbf{b}_0 = \mathbf{a}, \quad \mathbf{b}_1 = 2\mathbf{b} - \tfrac{1}{2}(\mathbf{a} + \mathbf{c}), \quad \mathbf{b}_2 = \mathbf{c}. \tag{3.2}$$

Now, given a sequence of points $\mathbf{p}_0, \mathbf{p}_1, \ldots, \mathbf{p}_n \in \mathbb{R}^d$ $(n \geq 2, d \geq 2)$, such a parabola segment can be constructed for every triple $\mathbf{p}_{i-1}, \mathbf{p}_i, \mathbf{p}_{i+1}$ of consecutive data points ($1 \leq i \leq n-1$). These segments will be denoted by $\mathscr{P}_i$, their Bézier representations are given by (3.1) with an additional index i on $\mathbf{q}$ and on $\mathbf{b}_k$ and, by (3.2), we have

$$\mathscr{P}_i \cdots \mathbf{q}_i(s) = \sum_{k=0}^{2} B_k^2(s)\mathbf{b}_{ik}, \quad s \in [0,1],$$

$$\mathbf{b}_{i0} = \mathbf{p}_{i-1}, \quad \mathbf{b}_{i1} = 2\mathbf{p}_i - \tfrac{1}{2}(\mathbf{p}_{i-1} + \mathbf{p}_{i+1}), \quad \mathbf{b}_{i2} = \mathbf{p}_{i+1}. \tag{3.3}$$

By this procedure we have *two segments* joining each pair $\mathbf{p}_i, \mathbf{p}_{i+1}$ ($1 \leq i \leq n-2$) of data points, namely the second half-segment of $\mathscr{P}_i(s \in [\tfrac{1}{2}, 1])$ and the first half-segment of $\mathscr{P}_{i+1}(s \in [0, \tfrac{1}{2}])$. Since the endpoints coincide, we can apply the sliding principle (with $m=0$) to them, obtaining a new sequence of segments

$$\mathscr{C}_i \cdots \mathbf{x}_i(t) := (1-t)\mathbf{q}_i\left(\frac{1+t}{2}\right) + t\mathbf{q}_{i+1}\left(\frac{t}{2}\right), \quad t \in [0,1] \tag{3.4}$$

($1 \leq i \leq n-2$). By Theorem 1, $\mathscr{C}_i$ has a C^1 continuity with $\mathscr{P}_i$ at $\mathbf{p}_i$ and with $\mathscr{P}_{i+1}$ at $\mathbf{p}_{i+1}$ (both with half-speed parameters). Since the relation of C^1 continuity of curves at a fixed point is transitive, every pair of consecutive segments $\mathscr{C}_i$ and $\mathscr{C}_{i+1}$ join at $\mathbf{p}_{i+1}$ with C^1 continuity (the half-speed factor cancels, going back to $\mathscr{C}_i$ and $\mathscr{C}_{i+1}$).

On the other hand, by (3.4), the segments $\mathscr{C}_i$ are *cubic* since $\mathbf{q}_i$ and $\mathbf{q}_{i+1}$ are quadratic. Thus we obtain a C^1 spline $\tilde{\mathscr{S}} = \mathscr{C}_1 + \mathscr{C}_2 + \cdots + \mathscr{C}_{n-2}$ (where $+$ means the joining procedure) from $\mathbf{p}_1$ to $\mathbf{p}_{n-1}$ interpolating all intermediate data points. It remains only to add a segment $\mathscr{C}_0$ from $\mathbf{p}_0$ to $\mathbf{p}_1$ and a segment $\mathscr{C}_{n-1}$ from $\mathbf{p}_{n-1}$ to $\mathbf{p}_n$. To do this, one has to distinguish whether the curve should be closed (assuming $\mathbf{p}_0 = \mathbf{p}_n$ and $n \geq 3$ in this case) or not (in this latter case $\mathbf{p}_0$ and $\mathbf{p}_n$ may be distinct points or coincident, but then the curve is allowed to have different tangents at that common point in contrary to the closed case).

In the closed case, we simply take an additional parabola $\mathscr{P}_0 = \mathscr{P}_n$ interpolating the points $\mathbf{p}_{n-1}, \mathbf{p}_n = \mathbf{p}_0, \mathbf{p}_1$, and then we construct $\mathscr{C}_0$ and $\mathscr{C}_{n-1}$ also by (3.4); thus the entire construction becomes *periodic* and all the indices can be calculated modulo n.

In the non-closed case, we simply take the first half-segment of the first parabola $\mathscr{P}_1$ as $\mathscr{C}_0$:

$$\mathscr{C}_0 \cdots \mathbf{x}_0(t) = \mathbf{q}_1\left(\frac{t}{2}\right), \quad t \in [0,1] \tag{3.5}$$

and the second half-segment of the last parabola $\mathscr{P}_{n-1}$ as $\mathscr{C}_{n-1}$:

$$\mathscr{C}_{n-1} \cdots \mathbf{x}_{n-1}(t) = \mathbf{q}_{n-1}\left(\frac{1+t}{2}\right), \quad t \in [0,1]. \tag{3.6}$$

(If a *cubic* representation is desired throughout, one can apply the degree-elevation procedure to these two quadratic segments.)

Thus the construction of the cardinal Overhauser spline is completed; it will be denoted by $\mathscr{S}$. The additional segments join also with C^1 continuity, so we can summarize:

Theorem 2. *The cardinal Overhauser spline $\mathscr{S}$ (as described in this section) is a cubic C^1 spline interpolating the data points $\mathbf{p}_0, \mathbf{p}_1, \ldots, \mathbf{p}_n$ (for both closed and non-closed cases).*

The practical calculation of the spline follows directly from these formulas; in addition, one can define a global parameter $u \in [0, n]$ such that for $u < n$ the integer part $i := [u]$ defines the index of the segment $\mathscr{C}_i$ and the fractional part $t := u - [u]$ is the parameter value $t \in [0, 1)$ for the current point $\mathbf{x}_i(t)$ in (3.4), resp. (3.5) or (3.6).

For simplicity we write down the algorithm only for the closed case; minor alterations have to be made for the non-closed case.

Algorithm

Input: $\mathbf{p}_0, \mathbf{p}_1, \ldots, \mathbf{p}_n \in \mathbb{R}^d$ {data with $\mathbf{p}_0 = \mathbf{p}_n$}
         $u \in [0, n)$ {parameter value}

Output: $\mathbf{p}(u) \in \mathbb{R}^d$ {current point of the spline}.

Begin:

       $i := [u]$;   {largest integer $\leq u$} $t := u - i$;
       if $(t = 0)$ **then**
          $\mathbf{p}(u) := \mathbf{p}_i$ {short cut for data point}
       else
          if $(i = 0)$ **then** $\mathbf{p}_{-1} := \mathbf{p}_{n-1}$ **fi**; {cyclic $\cdots$}
          if $(i = n - 1)$ **then** $\mathbf{p}_{n+1} := \mathbf{p}_1$ **fi**;   {$\cdots$ indices}
          Calculate $\mathbf{q}_i\left(\dfrac{1+t}{2}\right)$ and $\mathbf{q}_{i+1}\left(\dfrac{t}{2}\right)$ by (3.3); {de Casteljau}
          Calculate $\mathbf{x}_i(t)$ by (3.4);
          $\mathbf{p}(u) := \mathbf{x}_i(t)$
       fi
End.

4.   Properties of the Cardinal Overhauser Spline

From the detailed description of the Overhauser spline given in the previous section, one immediately deduces the following properties:

4.1. Simplicity and Fastness

Assuming that a great number of current points have to be calculated, one does the calculation of the control points $\mathbf{b}_{i1}$ only *once*, so this step can be neglected. It remains to compute five affine combinations, two for each de Casteljau step to get $\mathbf{q}_i$ and $\mathbf{q}_{i+1}$, and one further combination (3.4). This makes the algorithm extremely simple and fast.

4.2. Robustness

The assumptions on the data points (each triple of consecutive points $\mathbf{p}_i, \mathbf{p}_{i+1}, \mathbf{p}_{i+2}$ not collinear) were made only for theoretical reasons (uniqueness of the parabola (3.3)). If such a triple is collinear this segment degenerates into a part of a straight line without any numerical instability, because only the very stable arithmetic operation of affine linear combinations (with positive factors!) have to be carried out. Even if two or more data points coincide, the algorithm remains stable (of course the resulting spline curve may no longer be regular, for example cusps may occur).

4.3. Affine Invariance

The entire construction is invariant under affine transformations of $\mathbb{R}^d$, i.e., the two procedures

1. First calculate the spline $\mathscr{S}$ for data points $\mathbf{p}_0, \dots, \mathbf{p}_n$, then apply an affine mapping A to $\mathscr{S}$
2. First apply A to each data point, and then calculate the spline $\bar{\mathscr{S}}$ for $A\mathbf{p}_0, \dots, A\mathbf{p}_n$

lead to the same result: $A\mathscr{S} = \bar{\mathscr{S}}$. This is true because, for each step, it is equivalent to constructing the parabolas (3.3) and making the affine combinations (3.4) of them:

Theorem 3. *The cardinal Overhauser spline is affinely invariant; in particular, it reflects all symmetries of the data points.*

As far as the author knows, all other later generalisations of the Overhauser spline (for references see Section 4.5) that produce a C^1 or G^1 cubic spline use some metric properties (chord lengths, distances, angles between tangents etc.) to select a unique cubic segment between consecutive data points, *thus making the scheme no longer affinely invariant.*

4.4. Locality

As can be seen from the algorithm in section 3, the current point $\mathbf{p}(u)$ of the Overhauser spline is influenced only by the four data points $\mathbf{p}_{i-1}, \mathbf{p}_i, \mathbf{p}_{i+1}$ and $\mathbf{p}_{i+2}$ ($i = [u]$). This property is expressed as *locality*. This is important for practical reasons, because changes of the data points imply changes of the spline only in a neighborhood of that data.

4.5. Bézier Representation of the Segments $\mathscr{C}_i$

We have seen that the segments $\mathscr{C}_i$ are cubic and polynomial and depend only on $\mathbf{p}_{i-1}, \mathbf{p}_i, \mathbf{p}_{i+1}, \mathbf{p}_{i+2}$. Therefore they must be representable in cubic Bézier form

$$\mathscr{C}_i \cdots \mathbf{x}_i(t) = \sum_{k=0}^{3} B_k^3(t) \bar{\mathbf{b}}_{i,k}, \quad t \in [0, 1]. \tag{4.1}$$

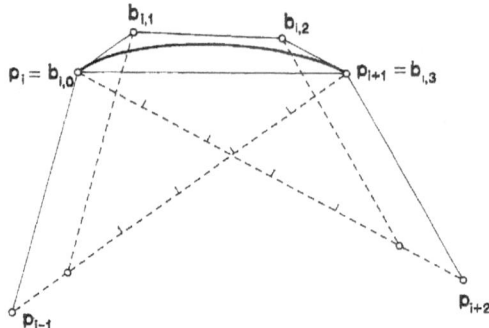

Figure 2. The Bézier control points of the cardinal Overhauser spline

Here the control points $\bar{\mathbf{b}}_{i,k}$ are determined by the given data points. Expanding (3.4) with respect to the Bernstein polynomials $B_k^3(t)$ as a basis we get

$$\bar{\mathbf{b}}_{i,0} = \mathbf{p}_i, \quad \bar{\mathbf{b}}_{i,1} = \mathbf{p}_i + \tfrac{1}{6}(\mathbf{p}_{i+1} - \mathbf{p}_{i-1}),$$
$$\bar{\mathbf{b}}_{i,2} = \mathbf{p}_{i+1} - \tfrac{1}{6}(\mathbf{p}_{i+2} - \mathbf{p}_i), \quad \bar{\mathbf{b}}_{i,3} = \mathbf{p}_{i+1}. \tag{4.2}$$

These formulas have a simple geometric meaning, which is illustrated by Fig. 2. In particular, the two tangents of the cardinal Overhauser spline at $\mathbf{p}_i$ and $\mathbf{p}_{i+1}$ are *parallel to the corresponding diagonals in the quadrangle* $\mathbf{p}_{i-1}, \mathbf{p}_i, \mathbf{p}_{i+1}, \mathbf{p}_{i+2}$ and the distances of the interior control points from the data points $\mathbf{p}_i$ resp. $\mathbf{p}_{i+1}$ are the sixth part of them.

Corollary 1. *The Bézier control points of the cardinal Overhauser spline are given by* (4.2).

Thus we have arrived at a stage which is very familiar in CAGD: to construct a tangent continuous piecewise cubic spline interpolating the data points $\mathbf{p}_0, \ldots, \mathbf{p}_n$, the segments of which are represented in Bézier form. There have been many solutions suggested (see for example [7], chapter 8 and [11], section 4.1.3). They differ only by the choice of the tangents through the data points and the placement of the interior control points $\mathbf{b}_{i-1,2}$ and $\mathbf{b}_{i,1}$ on the tangent through $\mathbf{p}_i$.

At this point we have to remark that the original construction of Overhauser is a little more general than we have discussed up to now:

Instead of taking the middle point ($s = 0.5$) on each parabola $\mathscr{P}_i$ to interpolate $\mathbf{p}_i$, one can take an *arbitrary* value s_i (restricted to $0 < s_i < 1$), obtaining the *non-cardinal* case of the Overhauser spline. But then the spline is no longer dependent only on the data points. Furthermore, it must be pointed out *that the tangents of the spline at* $\mathbf{p}_i$ *are no longer parallel to the chord* $\mathbf{p}_{i-1}, \mathbf{p}_{i+1}$, because this tangent coincides with that of the parabola $\mathscr{P}_i$ which is parallel to the chord only for $s_i = 0.5$. On the other hand, the Cutmull–Rom spline [5] has this property throughout (cardinal or not). Therefore we have:

Lemma 1. *The Overhauser spline and the Cutmull–Rom spline are identical only in the cardinal cases.*

In that case, indeed, all differences $\Delta_i = u_{i+1} - u_i$ of the knot vector (see [7] for notations and details) are equal and the formulas (8.13), (8.14) in loc. cit. coincide with (4.2). However, the remark on page 100 has to be corrected according to the lemma.

5. Affine Invariants and Local Convexity of Planar Cubic Bézier Curves

We want to investigate shape preserving properties for planar cubic splines with Bézier segments, in particular for the cardinal Overhauser spline. We focus on the question of local *convexity*. Since convexity is *affinely invariant* we can continue to work within the affine geometry (omitting all metric relations and properties).

Let $\mathscr{C}$ be a planar cubic Bézier segment and $\mathbf{b}_i$ $(i = 0, \ldots, 3)$ its control points. Obviously, $\mathscr{C}$ is determined by the three difference vectors $\Delta\mathbf{b}_i$ $(i = 0, 1, 2)$ up to a translation. An affine mapping may transform any pair of linearly independent vectors into any other pair of this kind. Assuming that $\Delta\mathbf{b}_0$ and $\Delta\mathbf{b}_2$ are lin. ind., only the two coefficients ρ, σ in the relation

$$\Delta\mathbf{b}_1 = \rho\Delta\mathbf{b}_0 + \sigma\Delta\mathbf{b}_2 \tag{5.1}$$

are essential (see also [12], [24], [6]).

Lemma 2. *A planar cubic Bézier segment $\mathscr{C}$ (with $\Delta\mathbf{b}_0, \Delta\mathbf{b}_2$ lin. ind.) is defined up to an affine mapping by the coefficients ρ, σ in (5.1) and these two real numbers can be arbitrarily prescribed.*

Thus, all information on $\mathscr{C}$ which is affinely invariant is contained in ρ and σ. In particular, we can derive the equation for an inflection point from the determinant

$$D(t) := \tfrac{1}{18}\det(\mathbf{x}'(t), \mathbf{x}''(t))$$
$$= (1-t)^2 \det(\Delta\mathbf{b}_0, \Delta\mathbf{b}_1) + (1-t)t \det(\Delta\mathbf{b}_0, \Delta\mathbf{b}_2) + t^2 \det(\Delta\mathbf{b}_1, \Delta\mathbf{b}_2). \tag{5.2}$$

Inserting (5.1) in it, we obtain

$$(1-t)^2\sigma + (1-t)t + t^2\rho = 0 \tag{5.3}$$

with the solutions

$$t_{1/2} = \frac{1}{2} + \frac{(\rho - \sigma) \pm \sqrt{\delta}}{2\omega}, \tag{5.4}$$

whereby

$$\delta = 1 - 4\rho\sigma, \quad \omega = 1 - \rho - \sigma. \tag{5.5}$$

The (excluded) case $\omega = 0$ yields a *cusp at infinity* (including the special case where $\mathscr{C}$ is degenerated into a parabola $\rho = \sigma = 1/2$, degree elevated to 3). The remaining cases lead to the well-known classification of planar cubic Bézier curves (see [25], [12], [6]; in this latter paper the sign of δ has to be corrected according to (5.5)):

Theorem 4. *The planar Bézier cubic $\mathscr{C}$ with the affine invariants ρ, σ (satisfying $\omega \neq 0$) has*

a) *a* **cusp** *iff* $\delta = 0$ *with parameter value*

$$t_0 = \frac{1}{2} + \frac{\rho - \sigma}{2\omega}, \tag{5.6}$$

b) *a* **double point** *iff* $\delta < 0$ *with parameter values*

$$t_{3/4} = \frac{1}{2} + \frac{(\rho - \sigma) \pm \sqrt{-3\delta}}{2\omega}, \tag{5.7}$$

c) *two* **real inflection points** *iff* $\delta > 0$ *with parameter values given by* (5.4).

From this theorem we conclude:

Corollary 2. *The planar cubic Bézier segment $\mathscr{C}$ with invariants ρ, σ (such that $\omega \neq 0$ and $\Delta b_0, \Delta b_2$ lin. indep.) is* **locally convex** *iff one of the following conditions is satisfied*

$$\delta < 0 \tag{5.8}$$

or

$$t_1, t_2 \text{ outside } [0, 1] \tag{5.9}$$

$(t_1, t_2$ *being the solutions of* (5.4)).

6. Shape Preserving Properties of the Cardinal Overhauser Spline

An interpolating spline should reflect as well as possible the "shape properties" of the data points $p_0, \ldots, p_n$. Many investigations are devoted to shape properties of spline *functions* (see, for example, [3, 4, 14, 15, 19, 20, 22, 23]) but much less is known for *parametric spline curves* [9, 12, 21]. In particular, for planar (data and) curves it is desired that the spline has no more inflection points than the data points enforce and that collinear data points imply the corresponding segment of the spline to be a part of a straight line. In this strong sense, no tangent-continuous spline can be shape preserving: If there is a corner between two consecutive triples of collinear data points, additional inflection points *must occur*. Figure 3 shows

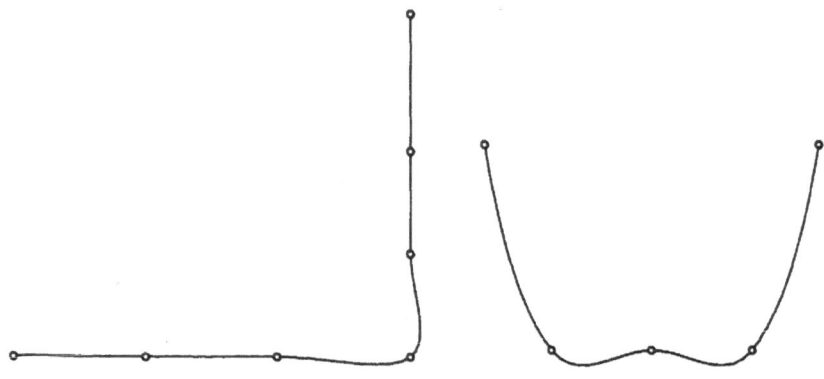

Figure 3. Impossibility of strong shape preserving

an example for this situation and the good-natured behaviour of the Overhauser spline.

On the other hand, the following weaker version of the shape preserving property can be satisfied:

Definition 1. A scheme S which assignes to each admissible sequence $\mathbf{p}_0, \ldots, \mathbf{p}_n$ of planar data points a tangent-continuous interpolating spline $\mathscr{S} = \mathbf{S}(\mathbf{p}_0, \ldots, \mathbf{p}_n)$ is called *shape preserving*, if the following conditions hold for each quadruple of consecutive data points $\mathbf{p}_{k-1}, \mathbf{p}_k, \mathbf{p}_{k+1}, \mathbf{p}_{k+2}$:

(a) If they are collinear, then (at least) the spline segment between $\mathbf{p}_k$ and $\mathbf{p}_{k+1}$ coincides with the leg from $\mathbf{p}_k$ to $\mathbf{p}_{k+1}$.
(b) If they define a (non degenerate) convex quadrangle then the spline segment between $\mathbf{p}_k$ and $\mathbf{p}_{k+1}$ is locally convex (has no inflection points).

Remarks:
—Clearly, four consecutive data points which are neither collinear nor convex enforce an inflection point of the corresponding spline segment.
—Clearly, the definition of shape preserving can be modified if the spline segment depends on more than four consecutive data points.
—It is easy to check that the known cubic C^1 and G^1 schemes satisfy condition (a) because the control points for the segment $\mathscr{S}_k$ become collinear if the data points $\mathbf{p}_{k-1}, \ldots, \mathbf{p}_{k+2}$ are so.
—Condition (b) is much more difficult. On the other hand, it is a *affinely invariant*; thus affinely invariant schemes are good candidates to check the shape preserving property.

Now it turns out that the cardinal Overhauser spline is not shape preserving (in the sense of Definition 1). Figure 4 shows a counterexample: a convex data quadrangle, the spline segment of which has an inflection point.

Nevertheless, this situation is fairly rare. To be more precise, we define

Definition 2. A (non degenerated) convex quadrangle is called *excessive*, if at least one of the ratios of the legs into which the diagonals are subdivided by their intersection point is greater than five.

We introduce the numbers α, β for the quadrangle $\mathbf{p}_{i-1}, \mathbf{p}_i, \mathbf{p}_{i+1}, \mathbf{p}_{i+2}$ and the intersection point $\mathbf{d}$ of its diagonals by

$$\overrightarrow{\mathbf{p}_i \mathbf{p}_{i+2}} = \beta \overrightarrow{\mathbf{p}_i \mathbf{d}}, \quad \overrightarrow{\mathbf{p}_{i+1} \mathbf{p}_{i-1}} = \alpha \overrightarrow{\mathbf{p}_{i+1} \mathbf{d}} \tag{6.1}$$

(see Fig. 4). Then the quadrangle is excessive, if

$$\alpha > 6 \quad \text{or} \quad \beta > 6. \tag{6.2}$$

With this notion we can state our main result as follows

Theorem 5. *If the sequence of data points does not contain any excessive quadrangle of four consecutive points, then the Overhauser cardinal spline is shape preserving (no inflection point is in the interior of the corresponding segment).*

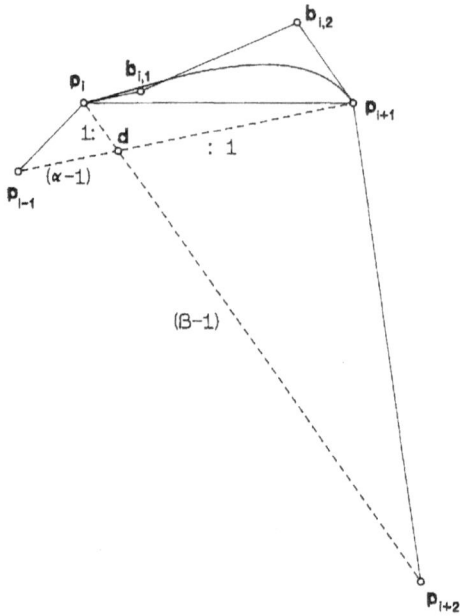

Figure 4. Counterexample of convexity preserving property for cardinal Overhauser spline

Proof: Observing (4.1) and the definition (5.1), (6.1) we obtain by elementary affine geometry

$$\rho = \frac{6 - \alpha}{\alpha}, \quad \sigma = \frac{6 - \beta}{\beta}. \tag{6.3}$$

Hence

$$\rho \geq 0, \quad \sigma \geq 0 \tag{6.4}$$

since, by assumption, the quadrangle is not excessive. On the other hand, by (5.4), the condition that at least one of the roots, say that one with sign $\varepsilon \in \{+1, -1\}$, is contained in the interior of the interval $[0, 1]$ is

$$-\frac{1}{2} < \frac{(\rho - \sigma) + \varepsilon\sqrt{\delta}}{2\omega} < \frac{1}{2}. \tag{6.5}$$

In the case

$$\omega > 0 \tag{6.6}$$

this is, by (5.5) equivalent to

$$2\sigma - 1 < \varepsilon\sqrt{\delta} < 1 - 2\rho. \tag{6.7}$$

Now we have no distinguish three further cases

(i) $\sigma \geq 1/2$
(ii) $\rho \geq 1/2$
(iii) $\rho < 1/2$ and $\sigma < 1/2$.

In the case (i) all three quantities in (6.7) must be positive, hence $\varepsilon = +1$, and the squares are in the same relation

$$(2\sigma - 1)^2 < \delta < (1 - 2\rho)^2.$$

Observing the first equation of (5.5), this simplifies to

$$\omega\rho < 0 < \omega\sigma,$$

hence $\rho < 0$ by (6.6). This contradicts (6.4), showing that this case is impossible. By the same method one can conclude $\sigma < 0$ in case (ii), thus also being impossible. In a similar way one can derive, that in the case (iii) either $\varepsilon = +1$ and

$$0 < \sqrt{\delta} < 1 - 2\rho$$

or $\varepsilon = -1$ and

$$0 < \sqrt{\delta} < 1 - 2\sigma.$$

Taking again the squares in both subcases implies

$$\rho\omega < 0 \quad \text{resp.} \quad \sigma\omega < 0,$$

likewise in contradiction to (6.4).

In the case $\omega < 0$ we obtain instead of (6.7)

$$2\sigma - 1 > \varepsilon\sqrt{\delta} > 1 - 2\rho. \tag{6.8}$$

But this inequality differs from (6.7) only by interchanging ρ and σ. Because of the symmetry with respect to these quantities and the case (iii) being excluded by $\omega < 0$, we get the same result; this completes the proof. $\qquad\Box$

References

[1] Akima, H.: A new method of interpolation and smooth curve fitting based on local procedures. J. ACM 17, 589–602 (1970).

[2] Brewer, J., Anderson, D.: Visual interaction with Overhauser curves and surfaces. Comput. Graphics 11, 132–137 (1977).

[3] Carlson, R., Fritsch, F.: Monotone piecewise bicubic interpolation. SIAM J Numer. Anal, 22, 386–400 (1985).

[4] Carlson, R., Fritsch, F.: An algorithm for monotone piecewise bicubic interpolation. SIAM J Numer. Anal, 26, 230–238 (1989).

[5] Cutmull, E., Rom, R.: A class of local interpolating splines. In: Computer aided geometric design (Barnhill, R., Riesenfeld, R., eds.), pp. 317–326. New York: Academic Press 1974.

[6] Degen W. L. F.: Some remarks on Bézier curves. Comput. Aided Geom. Des. 5, 259–268 (1988).

[7] Farin, G.: Curves and surfaces for computer aided geometric design, 3rd edn. New York: Academic Press 1992.

[8] Ferguson, J.: Multivariable curve interpolation. J. ACM 11, 221–228 (1964).

[9] Ferguson, J. C., Pruess, S.: Shape preserving interpolation by parametric piecewise cubic polynomials. CAD 23, 498–505 (1991).

[10] Hill, D. J. M., Fletcher, E. J., Moscardini, A. O., Wilkinson, T. S.: Overhauser patches for generally curved surfaces. In: Applied surface modelling (Creasy, C. F. M., Craggs, C., eds.), pp. 139–149. New York: Ellis Horwood 1990.

[11] Hoschek, J., Lasser, D.: Grundlagen der Geometrischen Datenverarbeitung, 2nd edn. Stuttgart: B. G. Teubner 1992.

[12] Liu, D. The shape control of the parametric cubic curve segment and the Bézier cubic curve. Acta Math. Appl. Sinica 4, 158–165 (1981).
[13] Maclaren, D.: Formulas for fitting a spline curve through a set of points. Appl. Math. Report 2 (1958), Boing.
[14] McAllister, D., Roulier, J.: Interpolation by convex quadratic splines. Math. Comput. 32, 1154–1162 (1978).
[15] McAllister, D., Passow, E., Roulier, J.: Algorithms for computing shape preserving spline interpolation to data. Math. Comput. 31, 717–725 (1977).
[16] McConalogue, D.: A quasi-intrinsic scheme for passing a smooth curve through a discrete set of points. Comput. J. 13, 392–396 (1970).
[17] McConalogue, D.: Algorithm 66 an automatic French-curve procedure for use with an incremental plotter. Comput. J. 14, 207–209 (1971).
[18] Overhauser, A.: Analytic definition of curves and surfaces by parabolic blending. Ford Motor Company, 1968.
[19] Passow, E., Roulier, J.: Monotone and convex spline interpolation. SIAM J. Numer. Anal. 14, 904–909 (1974).
[20] Roulier, J., Passow, E.: Monotone and convex spline interpolation. SIAM J. Numer. Anal. 14, 904–909 (1977).
[21] Roulier, J.: Bézier curves of positive curvature. CAGD 1, 59–70 (1988).
[22] Schumaker, L.: Spline functions: basic theory. New York: Wiley 1981.
[23] Schumaker, L.: On shape preserving quadratic spline interpolation. SIAM J. Numer. Anal. 20, 854–864 (1983).
[24] Su, B., Liu D.: An affine invariant theory and its applications in computational geometry. Scientia Sinica, Ser. A. 26, 259–272 (1983).
[25] Walker, R. J.: Algebraic curves. London: Dover 1950

Prof. Dr. W. Degen
Mathematisches Institut
Universität Stuttgart
Pfaffenwaldring 57
D-70569 Stuttgart
Federal Republic of Germany

Computing Suppl. 10, 129–147 (1995)

© Springer-Verlag 1995

Local Energy Fairing of B-spline Curves

M. Eck, Seattle, and **J. Hadenfeld,** Darmstadt

Abstract. An automatic algorithm for fairing B-spline curves of general order is presented. This work was motivated by a method of Farin and Sapidis about fairing planar cubic B-spline curves by subsequently removing and reinserting knots. Instead our new algorithm is based on the idea of subsequently changing one control point of a given B-spline curve so that the new curve minimizes the integral of the squared l-th derivative of the B-spline curve. How to proceed, if a tolerance is given and must be kept, is also discussed.

Key words: B-spline curves, knot removal, knot insertion, strain energy, energy integral, fairing, smoothing, distance tolerance.

1. Introduction

The usage of piecewise polynomials for the representation of curves and surfaces is very common in modern CAD systems since a lot of requirements like interpolation and approximation are simple and even numerically stable to deal with. However, the designer is not satisfied in every case with the resulting curves or surfaces obtained from such applications since the results might be *non-smooth* or even *unfair* meaning that any subjective *fairing criterion* is offered.

Unfortunately, in the literature about *fairing* a lot of different subjective fairing criteria can be found even for the much simpler case of polynomial spline curves. The two most common *definitions* for planar curves are the following:

(C1) A (spline) curve is fair if it minimizes the integral of the squared curvature $\kappa^2(s)$ with respect to the arc length (i.e. *strain energy* of a thin elastic beam); see [17].
(C2) A (spline) curve is fair if the plot of the curvature $\kappa(s)$ is continuous, has the appropriate sign, and is as close as possible to a piecewise monotone function with as few as possible monotone pieces; see [21, 23].

Now in practice these two fairing criteria are not suitable for piecewise polynomials since they lead to non-linear problems whereby explicit solutions cannot be found in general. Thus very time-consuming numerical tools are necessary to obtain approximative solutions; see [19] for criterion (C1) and some similar integral minimization problems.

Therefore, in most known fairing algorithms the criterion (C1) is linearized by the assumption that the actual parameter t of the spline curve $\mathbf{x}$ nearly represents the arc length, meaning that $|\mathbf{x}'(t)|$ is nearly constant. Then the much simpler integral

$$E_2 = \int_a^b (\mathbf{x}''(t))^2 \, dt \tag{1}$$

is minimized instead. Doing so, (1) yields a quadratic minimization problem with a unique solution if linear function spaces, like piecewise polynomials, are used.

Nevertheless, it should be kept in mind that the usage of (1) can produce strange results if the actual parameterization is far from satisfying $|\mathbf{x}'(t)|$ to be constant. Here, in [16] some exemplary effects are pointed out for spline curves with singularities (see also Section 3.5).

Let us now come back to the problem stated originally. To avoid *unfair* effects, which can originate from e.g. digitizing errors, two different principles are used in general. The first one consists of incorporating the fairness criterion (1) already into the interpolation or approximation process; see [12, 13, 20] for examples and more details. These global methods yield good results although the maximal pointwise error usually cannot be controlled without dealing with non-linear constraints.

The second way to obtain smooth spline curves is the separation of the construction and the fairing process. Therefore, the problem of fairing a given spline must be solved. Here, again two principles exist: global one-step algorithms as described in [14] or local algorithms which have to be executed iteratively; see Section 4 for details.

In the case of B-spline curves of order k, which are always notified in the present paper by

$$\mathbf{x}(t) = \sum_{i=0}^{n} \mathbf{d}_i \cdot N_{i,k}(t), \quad t \in [t_{k-1}, t_{n+1}] \tag{2}$$

with general knot sequence $T = (t_j)_{j=0}^{n+k}$ and control points $\mathbf{d}_i \in \mathbb{R}^s$, a first attempt for *local* fairing using a rather simplified version of criterion (C2) was given by Kjellander [15] in the cubic case ($k = 4$) although this method still contains a *global* part. Later on, Sapidis and Farin [21] (see also [5]) modified Kjellander's method to be really local. The main part of their algorithm is a *knot removal–knot reinsertion* step which is explained in more detail in Section 2.

Recently, Farin [7] developed another local fairing method fo cubic B-splines whereby the main part is a *degree reduction–degree reelevation* step. This method is not considered here although it yields good results, too.

In the present paper we state an alternative local fairing method which is oriented according to criterion (C1) instead of criterion (C2).

We locally minimize the linearized energy integral (1) by modifying only one control point in a local step and keeping all others fixed. This idea is also used successfully in a recent paper of Eck and Jaspert [4] where polylines are faired by minimizing discretized fairness criteria.

Furthermore, we also consider the integral over the third instead of the second derivative as described in [12, 18]. This criterion, by the way, is equivalent to the integral over $(\dot{\kappa})^2 + \kappa^2(\kappa^2 + \tau^2)$ if the underlying curve is parametrized with respect to the arc length. Here τ represents the torsion of the curve. Therefore, also the variation of the curvature is minimized in some sense as done in [19], too.

In Section 3 this iterative fairing method which is valid for B-spline curves of general order with general knot vectors in general dimension is described together with some simple modifications and extensions. It is especially outlined how to satisfy a uniform prescribed bound of the deviation which is important for applications.

Further, in Section 4 a general discussion follows where the local and global methods are compared in more detail. In addition, some considerations about local and global convergence of the iterative algorithm can be found.

Finally, some concluding remarks and future research directions are contained in Section 5.

2. Knot Removal – Knot Reinsertion Fairing

As already mentioned above the iterative local fairing method for planar cubic B-spline curves of Farin and Sapidis is based on fairing criterion (C2). In detail they observed the following: if the discontinuity

$$z_i = |\dot{\kappa}(t_i^-) - \dot{\kappa}(t_i^+)|, \quad 4 \leq i \leq n \tag{3}$$

of the slope of the curvature (now differentiated with respect to the arc length) at the inner knot t_i is small then the spline curve $x(t)$ consists of few monotone curvature pieces in the neighbourhood of the knot t_i.

In addition, they concluded that the sum

$$\xi = \sum_{i=4}^{n} z_i \tag{4}$$

of all *local criteria* z_i is a *reasonable* measure to control the number of monotone curvature pieces of the entire curve.

For the iterative fairing algorithm itself they used the property of cubic splines that, if the curve is three (instead of at most two) times differentiable at a point $x(t_j)$ then the discontinuity z_j is zero (assuming that the tangent vector does not vanish at t_j). Thus, they considered how to change the curve x locally so that the continuity order at a knot t_j is raised by one. This process, of course, can be interpreted as a knot removal step with a subsequently performed knot reinsertion of the knot t_j.

Obviously, the simplest way of doing that is to change only one control point namely d_{j-2}. Then the new location $\tilde{d}_{j-2}$ of this control point is given by

$$\tilde{d}_{j-2} = \frac{(t_{j+2} - t_j)l_j + (t_j - t_{j-2})r_j}{t_{j+2} - t_{j-2}} \tag{5}$$

with the two auxiliary points

$$l_j = \frac{(t_{j+1} - t_{j-3})d_{j-3} - (t_{j+1} - t_j)d_{j-4}}{t_j - t_{j-3}}, \tag{6}$$

$$r_j = \frac{(t_{j+3} - t_{j-1})d_{j-1} - (t_j - t_{j-1})d_j}{t_{j+3} - t_j}. \tag{7}$$

For example, in the special case of an equidistant knot vector T these formulas simplify to

$$\tilde{\mathbf{d}}_{j-2} = -\tfrac{1}{6}\mathbf{d}_{j-4} + \tfrac{2}{3}\mathbf{d}_{j-3} + \tfrac{2}{3}\mathbf{d}_{j-1} - \tfrac{1}{6}\mathbf{d}_j. \tag{8}$$

Here we note that (8) allows a surprising interpretation: consider the cubic interpolating polynomial $\mathbf{c}(t)$ satisfying $\mathbf{c}(-4) = \mathbf{d}_{j-4}$, $\mathbf{c}(-3) = \mathbf{d}_{j-3}$, $\mathbf{c}(-1) = \mathbf{d}_{j-1}$, $\mathbf{c}(0) = \mathbf{d}_j$, then we simply obtain $\tilde{\mathbf{d}}_{j-2} = \mathbf{c}(-2)$. On the other hand, a similar interpretation for general knot vectors could not be found.

We immediately notice that the (faired) curve with $\tilde{\mathbf{d}}_{j-2}$ differs from the original curve only in the interval $]t_{j-2}, t_{j+2}[$. Thus this fairing step is local.

Several other possibilities to perform this local changing are discussed in [5] by varying more than one control point in one local step whereby the changing interval is always increased.

Finally, repeating the above local fairing step subsequently at different knots (preferably each time at the knot with the largest value of z_j) Farin and Sapidis have found out in several numerical tests that the global criterion ξ from (4) always decreases if the iteration number is only large enough.

Altogether, they formulated the following *automatic* fairing algorithm:

1. Compute ξ and z_i; $4 \le i \le n$.
2. Find $z_j = \max\{z_i: 4 \le i \le n\}$.
3. Compute the new location $\tilde{\mathbf{d}}_{j-2}$ of $\mathbf{d}_{j-2}$ according to (5)–(7).
4. If a suitable criterion to stop is fulfilled then exit else goto step 1.

As possible interruptions in [21] it is suggested either to restrict the number of iterations or to stop if ξ increases resp. if the actual changing of ξ is small enough. Further, a restriction of the new location $\tilde{\mathbf{d}}_{j-2}$ can be built in if a prescribed tolerance of the deviation should be fulfilled.

Several examples in [21] illustrate that the fairing algorithm works well in most cases. Nevertheless, the following *peculiarities* and *questions* arise:

(i) the minimization of ξ is not ensured in every iteration step,
(ii) the method is restricted to cubic B-spline curves,
(iii) is the local criterion also useful for space curves?
(iv) the new location $\tilde{\mathbf{d}}_{j-2}$ only depends on the four neighbouring control points $\mathbf{d}_{j-4}$, $\mathbf{d}_{j-3}$, $\mathbf{d}_{j-1}$, $\mathbf{d}_j$ although $\mathbf{d}_{j-5}$ and $\mathbf{d}_{j+1}$ also influence the interval $]t_{j-2}, t_{j+2}[$.

All these aspects are evaded in the new iterative fairing method described in the next section.

3. Energy Fairing

In principle, the *energy* fairing works similarly to the *knot removal – knot reinsertion* method. The main difference is the relation to fairness criterion (C1) meaning the local minimization of an energy integral.

Before going into details, we introduce at first some useful notations which allow a distinction of the different stages of the curve:

1. The given curve which is to be faired:

$$\mathbf{x}(t) = \sum_{i=0}^{n} \mathbf{d}_i \cdot N_{i,k}(t), \quad t \in [t_{k-1}, t_{n+1}]. \tag{9}$$

2. The B-spline curve which has already been faired by a certain number of iterations:

$$\bar{\mathbf{x}}(t) = \sum_{i=0}^{n} \bar{\mathbf{d}}_i \cdot N_{i,k}(t), \quad t \in [t_{k-1}, t_{n+1}]. \tag{10}$$

3. The new curve in the next iteration step:

$$\tilde{\mathbf{x}}(t) = \sum_{\substack{i=0 \\ i \neq r}}^{n} \bar{\mathbf{d}}_i \cdot N_{i,k}(t) + \tilde{\mathbf{d}}_r \cdot N_{r,k}(t), \quad t \in [t_{k-1}, t_{n+1}]. \tag{11}$$

Then we consider the following local minimization problem: find a new location $\tilde{\mathbf{d}}_r \in \mathbb{R}^s$ of $\bar{\mathbf{d}}_r$ in order to minimize the energy integral depending on $\tilde{\mathbf{d}}_r$

$$E_l(\tilde{\mathbf{d}}_r) = \int_{t_{k-1}}^{t_{n+1}} \left(\frac{d^l}{dt^l} \tilde{\mathbf{x}}(t) \right)^2 dt = \int_{t_{k-1}}^{t_{n+1}} (\tilde{\mathbf{x}}^{(l)}(t))^2 dt \tag{12}$$

where $l = 2$ or $l = 3$ are appropriate choices, as already mentioned in the Introduction.

In addition, the deviation from the original curve $\mathbf{x}$ has to be controlled by a prescribed tolerance δ, so we always have to satisfy the constraint

$$\max \{ |\mathbf{x}(t) - \tilde{\mathbf{x}}(t)| : t \in [t_{k-1}, t_{n+1}] \} \leq \delta. \tag{13}$$

Now, before we can state the actual fairing algorithm it is necessary to investigate the following aspects in more detail:

a) how to compute the minimal solution $\tilde{\mathbf{d}}_r$,
b) how to fulfil the distance tolerance δ,
c) how to rank all control points $\{\bar{\mathbf{d}}_i\}$ of $\bar{\mathbf{x}}$ in order to find the location with the best fairing effect in the next step.

Finally, it should be remarked that the integrals (12) obviously enforce some kind of continuity in order to get well defined minimization problems. Doing so, the most convenient way is to allow at least $(k - l)$-fold knots in the underlying knot sequence of the spline curves since the curve is then at least C^{l-1} continuous everywhere. If the knot sequence, however, contains knots with higher multiplicity then the curve should be split at these knots by treating the resulting pieces separately.

3.1. Computation of the New Control Point

Writing down the energy integral (12) in more detail for a general point $\mathbf{d} \in \mathbb{R}^s$ we get

$$E_l(\mathbf{d}) = \int_{t_k\,_1}^{t_{n-1}} \left(\sum_{\substack{i=0 \\ i \neq r}}^{n} \bar{\mathbf{d}}_i \cdot N_{i,k}^{(l)}(t) + \mathbf{d} \cdot N_{r,k}^{(l)}(t) \right)^2 dt \tag{14}$$

representing a quadratic function in $\mathbf{d}$.

However, we note as an interesting fact that in the planar case ($s = 2$) the real contour lines of $E_l(\mathbf{d})$ are concentric circles with the unique minimum $\tilde{\mathbf{d}}_r$ as common centre. This observation (valid in a similar way in higher dimensions) will be used later on to control the deviation δ.

The unique minimum $\tilde{\mathbf{d}}_r$ itself is determined by

$$\frac{\hat{c}}{\partial \tilde{\mathbf{d}}_r} E_l(\tilde{\mathbf{d}}_r) \overset{!}{=} 0 \tag{15}$$

and inserting (14) we obtain from (15)

$$\sum_{\substack{i=0 \\ i \neq r}}^{n} \bar{\mathbf{d}}_i \cdot \int_{t_{k-1}}^{t_{n+1}} N_{i,k}^{(l)} \cdot N_{r,k}^{(l)} \, dt + \tilde{\mathbf{d}}_r \cdot \int_{t_{k-1}}^{t_{n+1}} (N_{r,k}^{(l)})^2 \, dt \overset{!}{=} 0. \tag{16}$$

This single (vector valued) *normal equation* can be solved simply. If we additionally take advantage of $N_{r,k}^{(l)} \equiv 0$ for $t \notin [t_r, t_{r+k}]$ the solution is

$$\tilde{\mathbf{d}}_r = \sum_{\substack{i=i_0 \\ i \neq r}}^{i_1} \gamma_i \cdot \bar{\mathbf{d}}_i \tag{17}$$

with weighting factors γ_i of the form

$$\gamma_i = -\frac{\int_a^b N_{i,k}^{(l)} \cdot N_{r,k}^{(l)} \, dt}{\int_a^b (N_{r,k}^{(l)})^2 \, dt}. \tag{18}$$

Here, some further abbreviations have been already incorporated which are necessary to restrict the integration to the curve interval $[t_{k-1}, t_{n+1}]$ only:

$$a = \max\{t_r, t_{k-1}\} \quad \text{and} \quad b = \min\{t_{r+k}, t_{n+1}\}, \tag{19}$$

$$i_0 = \max\{0, r-k+1\} \quad \text{and} \quad i_1 = \min\{r+k-1, n\}. \tag{20}$$

Moreover, we notice that $\tilde{\mathbf{d}}_r$ is computed in (17) by an affine combination of the neighbouring control points $\{\bar{\mathbf{d}}_i\}$ because we can deduce the following identity from (18)

$$\sum_{\substack{i=i_0 \\ i \neq r}}^{i_1} \gamma_i = 1. \tag{21}$$

However, it seems to be a disadvantage that no simple explicit formula depending only on the actual knot vector T is known for the *weights* γ_i in (18). Nevertheless, the weights γ_i can be computed exactly in a finite number of operations since we are here dealing with piecewise polynomial basis functions. For example, we can use a numerical integration method like *Gauss quadrature* on each interval which works exactly for polynomials of degree $2(k - l - 1)$ in order to calculate the appearing integrals in (18). For more details on integrating products of B-splines we refer to [24].

Again, in the very special case of an equidistant knot vector T and cubic B-spline curves we can state the formula (17) explicitly:

$$\bar{\mathbf{d}}_r = \begin{cases} -\frac{1}{16}\bar{\mathbf{d}}_{r-3} + \frac{9}{16}\bar{\mathbf{d}}_{r-1} + \frac{9}{16}\bar{\mathbf{d}}_{r+1} - \frac{1}{16}\bar{\mathbf{d}}_{r+3}, & \text{if } l = 2; \\ \frac{1}{20}\bar{\mathbf{d}}_{r-3} - \frac{6}{20}\bar{\mathbf{d}}_{r-2} + \frac{15}{20}\bar{\mathbf{d}}_{r-1} + \frac{15}{20}\bar{\mathbf{d}}_{r+1} - \frac{6}{20}\bar{\mathbf{d}}_{r+2} + \frac{1}{20}\bar{\mathbf{d}}_{r+3}, & \text{if } l = 3. \end{cases} \quad (22)$$

Here, it is especially remarkable for $l = 2$ in (22) that the two control points $\bar{\mathbf{d}}_{r-2}$ and $\bar{\mathbf{d}}_{r+2}$ do not appear what is of course not valid for general knot vectors T. Nevertheless, similar interpretations as possible for formula (8) can also be derived in the present cases:

- consider the interpolating cubic polynomial $\mathbf{c}(t)$ satisfying $\mathbf{c}(-3) = \bar{\mathbf{d}}_{r-3}$, $\mathbf{c}(-1) = \bar{\mathbf{d}}_{r-1}$, $\mathbf{c}(1) = \bar{\mathbf{d}}_{r+1}$, $\mathbf{c}(3) = \bar{\mathbf{d}}_{r+3}$, then, for $l = 2$, (22) is equivalent to $\bar{\mathbf{d}}_r = \mathbf{c}(0)$;
- consider the interpolating quintic polynomial $\mathbf{q}(t)$ satisfying $\mathbf{q}(-3) = \bar{\mathbf{d}}_{r-3}$, $\mathbf{q}(-2) = \bar{\mathbf{d}}_{r-2}$, $\mathbf{q}(-1) = \bar{\mathbf{d}}_{r-1}$, $\mathbf{q}(1) = \bar{\mathbf{d}}_{r+1}$, $\mathbf{q}(2) = \bar{\mathbf{d}}_{r+2}$, $\mathbf{q}(3) = \bar{\mathbf{d}}_{r+3}$, then, for $l = 3$, (22) is equivalent to $\bar{\mathbf{d}}_r = \mathbf{q}(0)$.

3.2. Distance Tolerance

If we replace the control point $\bar{\mathbf{d}}_r$ by $\tilde{\mathbf{d}}_r$ as computed above in (17) then the originally posed constraint (13), controlling the prescribed δ-distance between $\mathbf{x}$ and $\tilde{\mathbf{x}}$, is not yet taken into consideration.

In order to overcome this problem we will use the very simple condition

$$|\mathbf{d}_r - \tilde{\mathbf{d}}_r| < \delta \quad (23)$$

which guarantees by help of the convex-hull property of B-spline curves that (13) is fulfilled even if the local fairing step is repeated several times at different control points.

Condition (23) represents a good upper bound for the actual pointwise error in (13), at least, if the polynomial degree of the spline is small; c.f. [1, 22] where some alternative bounds are discussed.

Now, with respect to (23) we have to distinguish two cases. The first one is obvious, namely condition (23) is already fulfilled. Then $\tilde{\mathbf{d}}_r$ is lying within the δ-sphere with the centre $\mathbf{d}_r$. In the second case the new control point is situated outside this sphere and we are therefore interested in that alternative position $\tilde{\mathbf{d}}_r^*$ on the boundary of the distance sphere in which the value of the energy integral $E_l(\tilde{\mathbf{d}}_r^*)$ is as small as possible.

Here, we can use the fact already mentioned that the isolines of the energy integral are concentric spheres with centre $\tilde{\mathbf{d}}_r$ in order to compute the location $\tilde{\mathbf{d}}_r^*$ by

$$\tilde{\mathbf{d}}_r^* = \mathbf{d}_r + \delta \cdot \frac{\tilde{\mathbf{d}}_r - \mathbf{d}_r}{|\tilde{\mathbf{d}}_r - \mathbf{d}_r|} \quad (24)$$

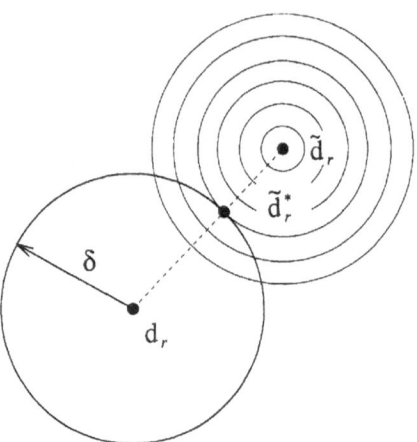

Figure 1. Satisfying the δ-distance constraint

or equivalently as the intersection point of the connecting line from $\mathbf{d}_r$ to $\tilde{\mathbf{d}}_r$ and the δ-sphere around $\mathbf{d}_r$. In Fig. 1 this geometrical situation is illustrated in the planar case.

3.3. Ranking Number

In the two previous subsections we have learnt how to change the curve locally by considering a δ-distance. Now, the question is: at which point should we fair a given B-spline curve?

Here, a *natural* answer is to fair the curve at that control point $\overline{\mathbf{d}}_r$ where the largest improvement of the energy integral is to be expected.

Thus, we introduce as a *local fairness criterion* the following non-negative number

$$z_r = E_l(\overline{\mathbf{d}}_r) - E_l(\tilde{\mathbf{d}}_r) \geq 0 \tag{25}$$

or in more detail with help of the notations (10) and (11)

$$z_r = \int_a^b (\overline{\mathbf{x}}^{(l)}(t))^2 \, dt - \int_a^b (\tilde{\mathbf{x}}^{(l)}(t))^2 \, dt. \tag{26}$$

After some calculations, one verifies directly that (26) can simply be expressed as

$$z_r = (\overline{\mathbf{d}}_r - \tilde{\mathbf{d}}_r)^2 \cdot \int_a^b (N_{r,k}^{(l)}(t))^2 \, dt \tag{27}$$

and we recognize that the local criterion (ranking number) is just the squared change of the control point $\overline{\mathbf{d}}_r$ weighted by the integral of the squared and l-times differentiated basis function $N_{r,k}(t)$.

Finally, following the performances of the previous subsection we can also modify (27) in order to take care of the distance tolerance δ already in the ranking number by using

$$z_r^* = (\bar{\mathbf{d}}_r - \tilde{\mathbf{d}}_r^*)^2 \cdot \int_a^b (N_{r,k}^{(l)}(t))^2 \, dt \tag{28}$$

if $|\mathbf{d}_r - \tilde{\mathbf{d}}_r| > \delta$, although (28) is not equivalent to $E_l(\bar{\mathbf{d}}_r) - E_l(\tilde{\mathbf{d}}_r^*)$.

3.4. The Algorithm

We are now ready to summarize the previous developments to a complete *automatic* fairing algorithm which, of course, has the same structure as the algorithm of Farin and Sapidis already described:

1. Compute the ranking numbers z_i; $0 \le i \le n$.
2. Find $z_r = \max\{z_i : 0 \le i \le n\}$.
3. Compute the new location $\bar{\mathbf{d}}_r$ (resp. $\tilde{\mathbf{d}}_r^*$ satisfying the δ-distance constraint).
4. If a suitable criterion to stop is fulfilled then exit else goto step 1.

There are only two interruptions possible in the present algorithm: either the number of iterations is restricted or the algorithm stops if the actual improvement z_r is smaller than a certain amount. In the examples which we shall present afterwards, we have always limited the number of iterations. Please note that we cannot consider the increase of the *global* fairness criterion E_l as a possible interruption since its decrease is ensured in every iteration step.

It is remarkable that the algorithm, as stated above, allows the modification of all control points. This might not be convenient if we are dealing with B-splines which itself are part of a composed spline curve. Thus we always exchange the expression $0 \le i \le n$ by $2 \le i \le n - 2$ to achieve C^l-continuity between the original and the faired B-spline curve at the two end points. Similar arguments are valid for continuities of different order.

If the number of control points is large, a faster strategy to get a *smooth* curve is to determine a so-called *ranking list* by ordering all points according to their ranking number. Then this ranking list can be used as the computational order for the fairing algorithm. In most cases, it will be sufficient to use only the first quarter of the ranking list. Afterwards the ranking list should be rebuilt. This process is repeated *rank_repeat*-times whereby *rank_repeat* is a variable which will be set individually for every example in the next subsection. Moreover we realized that it is further advantageous to limit the number each point can be changed during the algorithm. Doing so, a only local fairing effect due to the restrictions (23) is avoided.

Finally it should be remarked that the ranking list has to be calculated only in the first step for all points which can be changed. The reason is that the control point $\tilde{\mathbf{d}}_r$ influences only the ranking numbers z_i with the indices $r - k + 1 \le i \le r + k - 1$ and only these ranking numbers have to be recalculated in the next step.

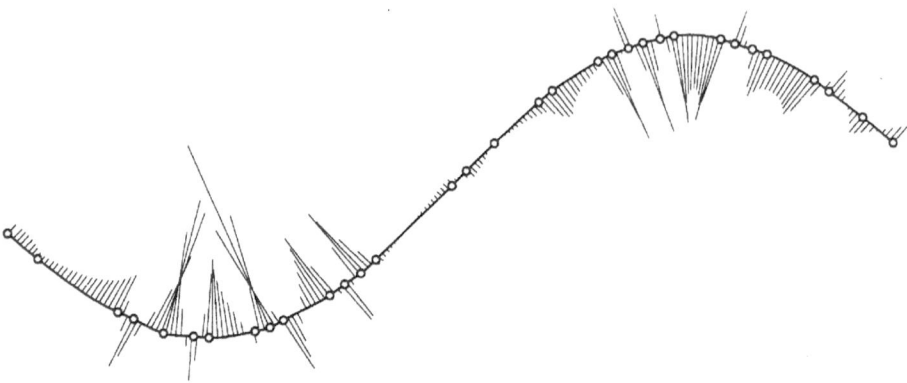

Figure 2. Given B-spline curve of example 1

3.5. Examples

We will now demonstrate the behaviour and some special effects of our fairing algorithm in three examples. Here, the first two curves are constructed in an somewhat academical manner whereas the third example is a real-life curve coming out of a CAD system. All current computations have been done on a HP 9000/735 machine.

In order to construct a non-uniform B-spline curve containing some maybe *unwanted* wiggles we proceeded as follows: We started with a planar cubic B-spline curve ($k = 4$) having the 61 control points

$$\mathbf{d}_i = \begin{pmatrix} x_i \\ \sin x_i \end{pmatrix}, \quad x_i = -3 + \frac{i}{10}, \quad 0 \le i \le 60,$$

and being defined over an equidistant knot vector with k-fold knots at the two ends. Then 57 inner control points were disturbed with help of random numbers whereby the maximal allowed disturbance of each control point was 0.02 absolute. Next, we removed 26 interior knots using an automatic algorithm recently developed by the two authors [2] in order to end up with a non-uniform disturbed B-spline. Here the bound of the maximal allowed knot removal error was preselected to 0.01 whereby C^1 continuity was preserved at both end points. The final curve is shown in Fig. 2, and only for completeness we also list the underlying knot vector:

$$T = (3, 3, 3, 3, 5, 10, 11, 13, 15, 16, 19, 20, 21, 24, 25, 26, 27, 32, 33, 35,$$
$$38, 39, 42, 43, 44, 45, 46, 47, 50, 51, 52, 53, 56, 57, 59, 61, 61, 61, 61).$$

Following criterion (C2), the curvature of the curve is drawn as *porcupines* to make all the wiggles and unfair regions visible. We have chosen these porcupines, because we only want to illustrate the shape of the curvature here. The porcupines are scaled large enough and equal in every picture. Further in the curves shown in this section all knots are marked.

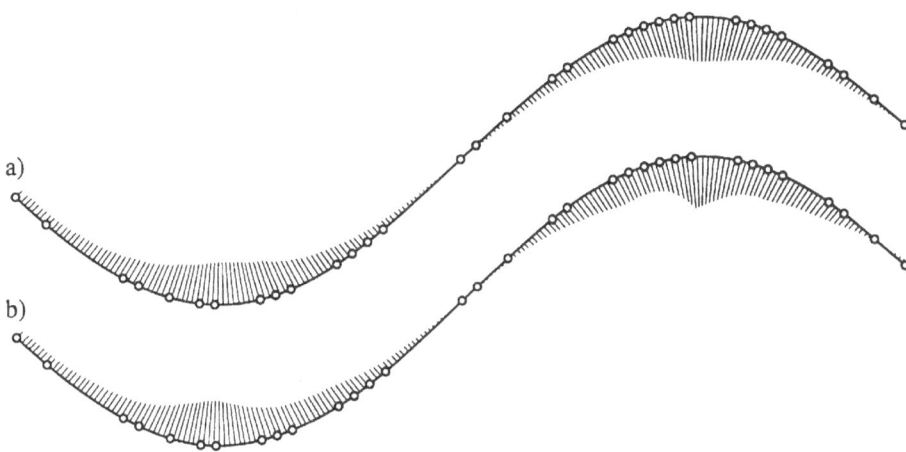

a)

b)

Figure 3. B-spline curve faired with criterion E_2 (a) and E_3 (b)

For the actual fairing of that curve we selected the two open parameters to be *rank_repeat* = 34 and $\delta = 0.03$ what exactly is identical to the summed amount of the disturbance and the knot removal error. The fairing results (each obtained in 0.09 CPU-seconds after 272 control point changes) are as follows: In Fig. 3a the fairness criterion E_2 is applied whereas in Fig. 3b criterion E_3 is used.

It is obvious that both curves are much *smoother* than the disturbed curve in Fig. 2 and at least the porcupines in Fig. 3a are in a rather good shape now. The curve in Fig. 3b is not totally convincing since the porcupines are more uneven or *wavy*. Moreover an unexpected inflection point is appearing at the right end point.

However, such end point inflections can always happen here because we let the two control points at each boundary unchanged during the fairing process. So in Fig. 3b this inflection point would vanish if we only would fix one control point at each boundary.

Next we look at the fairing results if the number of ranking list loops is increased considerably. So for *rank_repeat* = 1000 and again $\delta = 0.03$ we obtain the results illustrated in Fig. 4 (each obtained in 0.37 CPU-seconds after 15500 total control point changes). Now we realize that both curves are in really good shape according to the porcupines. Moreover, the result from E_3 is even better than the one from E_2. Also the end point inflection has vanished.

This observation however seems to be true in general. In order to get satisfying results using criterion E_3 one has to execute the algorithm a large number of times. But then the results may (not in every case, as we will see later) even be better than the one obtained with E_2. That means that the velocity of convergence is much smaller in the case of E_3. This indolent behaviour is not advantageous for the CPU-time behaviour of our algorithm. However, to be able to compare the two criteria we will choose large numbers for *rank_repeat* also in the remaining examples.

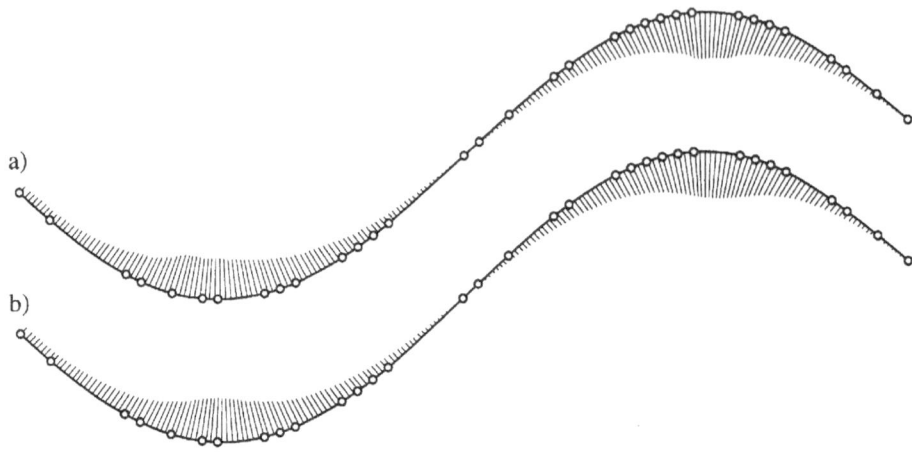

Figure 4. B-spline curve faired with criterion E_2 (**a**) and E_3 (**b**)

In (17) and (18), it is obvious that not only the control points but also the knot vector with respect to the quality of the parameterization is important for the fairing results. To illustrate this effect we took the control points of example 1 and changed manually 9 entries of the knot vector to

$$T = (3, 3, 3, 3, 5, 10, 11, 13, 15.7, 16, 20.99, 20.999, 21, 24, 25, 26, 27, 32.9, 33, 35, 38, 39,$$
$$43.000001, 43.000002, 43.000003, 45, 46, 47, 51, 51.000001, 52, 53, 56.8, 57, 59,$$
$$61, 61, 61, 61)$$

in order to create the second example. The resulting curve is shown in Fig. 5 where the bad knot distribution on the curve now causes additional wiggles. The changed knots are displayed as filled dots. Please note that we could not introduce double knots here because of the assumption of at least $(k - l)$-fold knots, made before.

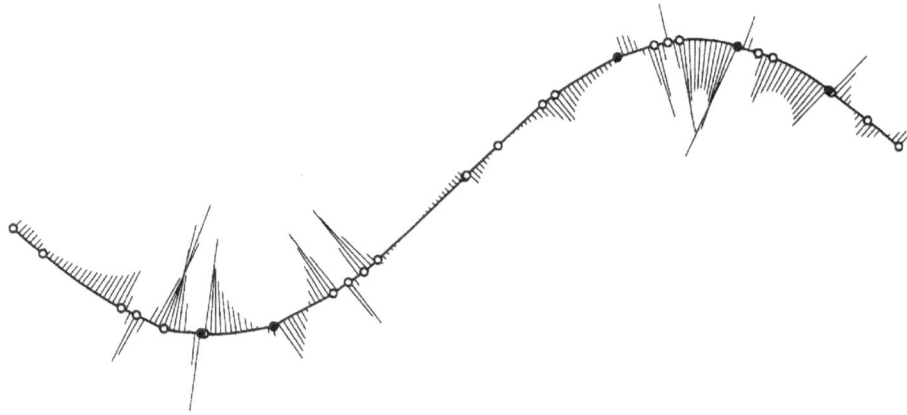

Figure 5. Given B-spline of example 2

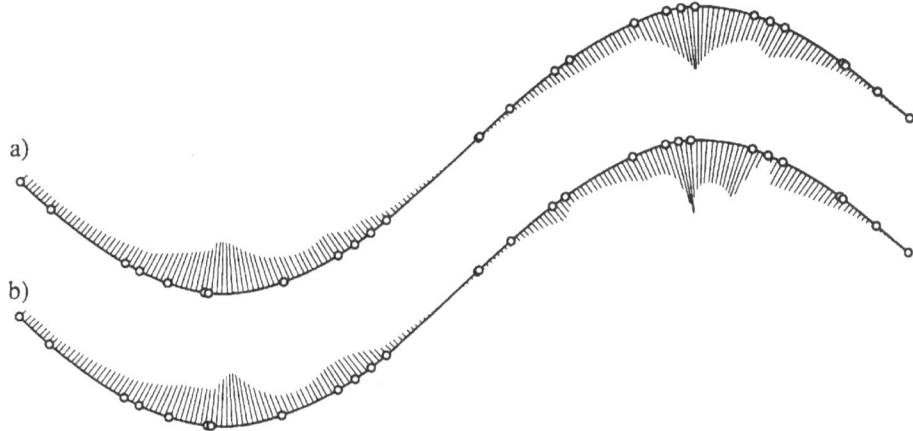

Figure 6. B-spline curve faired with criterion E_2 (a) and E_3 (b)

Then this new B-spline curve is faired with *rank_repeat* = 1000 and $\delta = 0.03$. The result is again obtained in 0.37 CPU-seconds after 15500 iterations (see Fig. 6). Here, it is obvious that the faired curves are not in such a good shape as before what clearly is an effect of the unfavourable parameterization of the underlying curve.

Now, the first curve (derived from E_2) is obviously better. Here all porcupines point at least into the expected direction. That is even not the case for the faired curve with E_3 which contains two (nearly invisible) inflection points in the interior.

So, our somewhat subjective conclusion, obtained from many practical tests, is that the energy integral E_2 has in general less problems with these kind of parameterizations. Altogether, the resulting curves appear more *stiff* or *stable* in such situations at least for cubic splines which we mainly considered.

The third example presented here is a boundary curve of a Bézier spline surface named *Surf B* and originates from the car body industry; see [3] for details. This Bézier spline curve of degree $k = 5$ has at first been converted to a B-spline curve with simple interior knots (see Fig. 7a). Then 14 inner control points are additionally disturbed with help of random numbers (see Fig. 7b) whereby the maximal disturbance at each control point is at least 3.0 absolute. The disturbance in this case is larger than in the first two examples because the distance from the first to the last control point is nearly 367 units.

Using the fairness criteria E_2 and E_3 with $\delta = 3.0$ and *rank_repeat* = 1000, the fairing results (obtained in 0.44 CPU-seconds after 7000 iterations) can be seen in Fig. 8. Here, both results are nearly perfect meaning that the curvature distribution is obviously even better than the one of the original curve in Fig. 7a.

Altogether we can recommend the usage of the fairing criterion E_2 because it is not so susceptible for bad parameterizations and moreover a suitable result is obtained

a)

b)

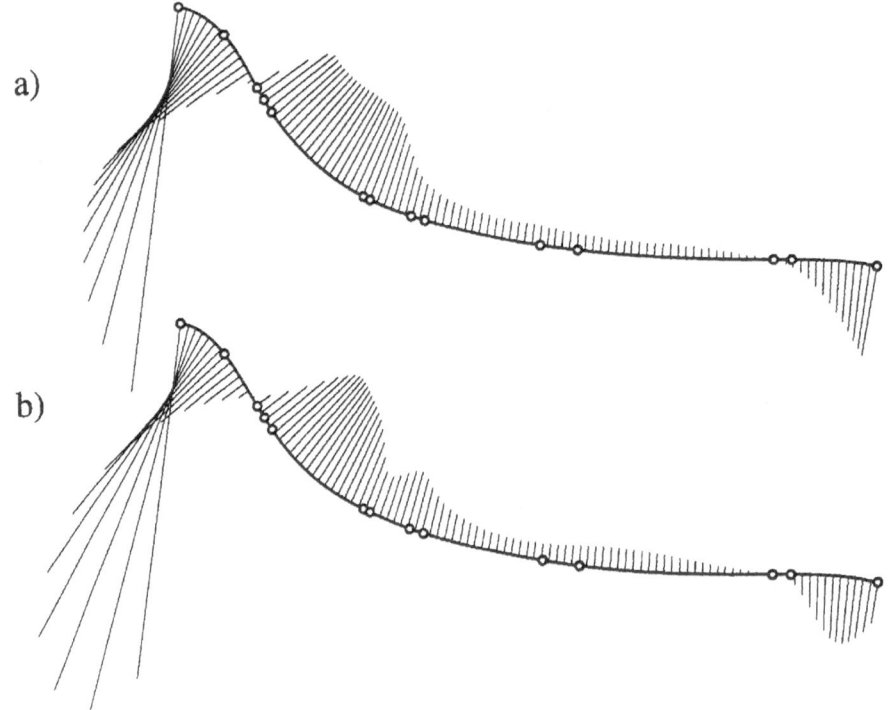

Figure 7. Given B-spline curve (**a**) and with a maximal disturbance of 3.0 (**b**)

after fewer iteration steps (meaning less CPU-time). However, it is always worth to try also the criterion E_3 if (nearly) real-time behaviour is not required.

3.6. A Simple Extension of the Algorithm

We have seen in the last part that the two cases of minimization with respect to E_2 and to E_3 can be interpreted as two *extreme* cases.

This behaviour is also documented in [18] where the variation of the curvature is considered. Therefore, for practical applications Meier proposed the following mixture of both energy integrals

$$M = \alpha \cdot E_2 + \beta \cdot E_3, \quad \alpha, \beta \geq 0, \quad \alpha + \beta = 1 \tag{29}$$

which is claimed to be technically optimal. Here, the values α and β must be preselected by the user.

Now, in our algorithms we can also minimize M with respect to $\tilde{\mathbf{d}}_r$ in every iteration step. Then it follows by simple calculations that the new location of $\tilde{\mathbf{d}}_r$ is computed by

$$\tilde{\mathbf{d}}_r = \bar{\alpha} \cdot \tilde{\mathbf{d}}_r'' + \bar{\beta} \cdot \tilde{\mathbf{d}}_r''', \quad \bar{\alpha} + \bar{\beta} = 1 \tag{30}$$

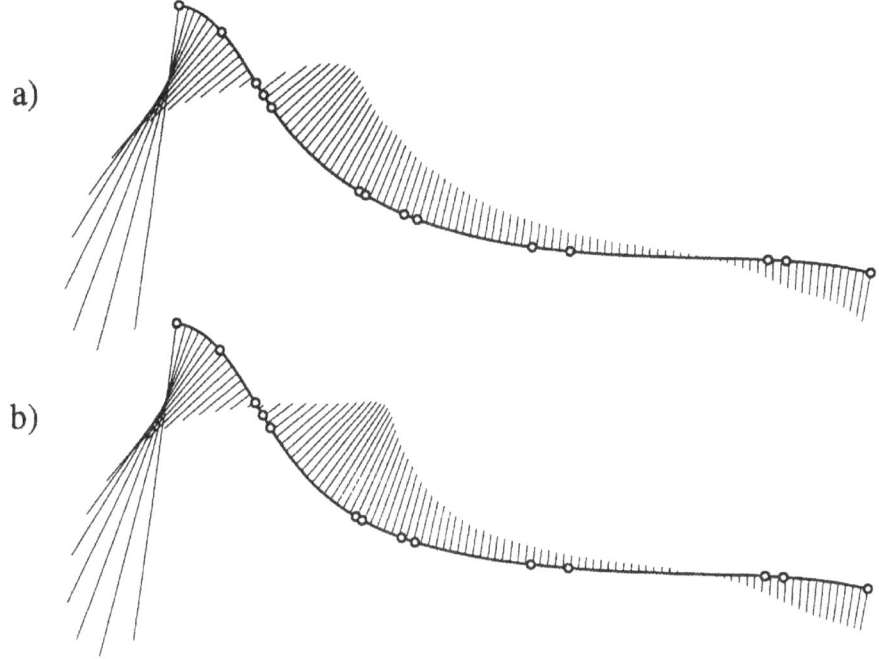

Figure 8. B-spline curve faired with criterion E_2 (a) and E_3 (b)

whereby $\tilde{\mathbf{d}}_r''$ resp. $\tilde{\mathbf{d}}_r'''$ represents the solution with respect of E_2 resp. E_3. Further the weight factor $\bar{\alpha}$, for instance, is given by

$$\bar{\alpha} = \frac{\alpha \cdot \int_a^b (N_{r,k}^{(2)})^2 \, dt}{\alpha \cdot \int_a^b (N_{r,k}^{(2)})^2 \, dt + \beta \cdot \int_a^b (N_{r,k}^{(3)})^2 \, dt}. \tag{31}$$

Now, from (30) we further notice that every point of the connecting line of $\tilde{\mathbf{d}}_r''$ and $\tilde{\mathbf{d}}_r'''$ can be interpreted uniquely as a solution of (29) for certain fixed values of α and β.

This observation can be used to keep the changing of the curve as small as possible in every iteration step of our fairing algorithms. Doing so, we define $\tilde{\mathbf{d}}_r$ to be that point on the connecting line which is as close as possible to $\mathbf{d}_r$ whereby $\tilde{\mathbf{d}}_r$ always has to lie between $\tilde{\mathbf{d}}_r''$ and $\tilde{\mathbf{d}}_r'''$. This idea is illustrated in Fig. 9, for the planar case.

So, obviously a different value of α (resp. β) is chosen in every step. Therefore, we have found an automatic determination of these values which are usually preselected. But because the value α is changed in every iteration step, the convergence of the algorithm is not ensured. Nevertheless, we obtained good results with this procedure.

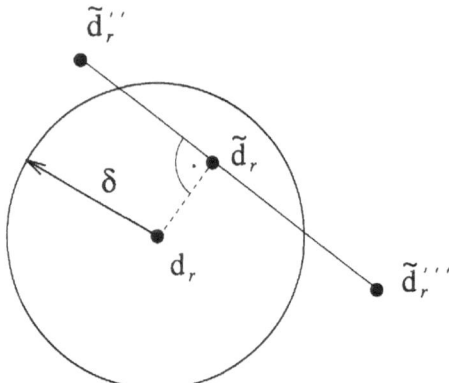

Figure 9. The idea of the mixed energy method

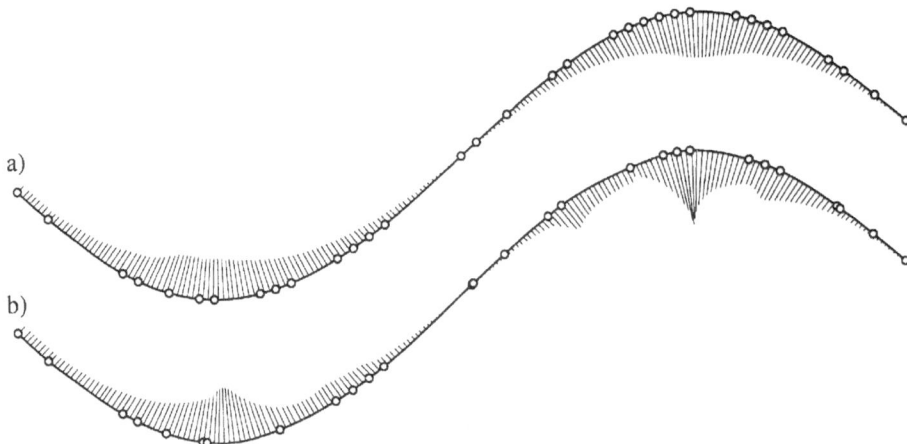

a)

b)

Figure 10. Automatic fairing with the mixed energy method

In Fig. 10 the fairing results of the mixed energy method for the curves as given in Fig. 2 and Fig. 5 are demonstrated. We used the same B-spline curves to be able to compare the results.

4. General Discussion

The algorithm described thoroughly in the previous section automatically fairs a B-spline curve $\mathbf{x}$ of general order k as given in (2). In more detail, our method computes an approximative local solution to the following problem for preset values $\delta \geq 0$ and $\alpha, \beta > 0$:

Find a curve $\tilde{\mathbf{x}}$, contained in the set

$$\chi_{\delta}^{\alpha,\beta} = \left\{ \sum_{i=0}^{\alpha-1} \mathbf{d}_i \cdot N_{i,k}(t) + \sum_{i=\alpha}^{n-\beta} \tilde{\mathbf{d}}_i \cdot N_{i,k}(t) + \sum_{i=n-\beta+1}^{n} \mathbf{d}_i \cdot N_{i,k}(t) \,\middle|\, \|\tilde{\mathbf{d}}_i - \mathbf{d}_i\| \leq \delta \right\} \quad (32)$$

of B-splines in the δ-neighborhood of $\mathbf{x}$ with agreeing boundary values, which satisfies

$$\int_{t_{k-1}}^{t_{n+1}} (\tilde{\mathbf{x}}^{(l)}(t))^2 \, dt \le \int_{t_{k-1}}^{t_{n+1}} (\mathbf{y}^{(l)}(t))^2 \, dt \quad \text{for all } \mathbf{y} \in \chi_\delta^{\alpha,\beta}. \tag{33}$$

Now, trying to find an (approximative) solution $\tilde{\mathbf{x}}$ with help of well-known optimization methods for nonlinear inequality-constrained problems, as described e.g. in [8], is certainly very time consuming for large n since such algorithms require iteratively the solution of a large linear system. Hence the need of any kind of simplification is obvious.

One possibility can be found in the already mentioned paper [14] where the quadratic functional

$$(1 - \mu) \cdot \int_{t_{k-1}}^{t_{n+1}} (\mathbf{y}^{(l)}(t))^2 \, dt + \mu \cdot \sum_{i=\alpha}^{n-\beta} (\tilde{\mathbf{d}}_i - \mathbf{d}_i)^2 \tag{34}$$

with any preset value $0 \le \mu \le 1$ is minimized for $\mathbf{y} \in \chi_\infty^{\alpha,\beta}$ (meaning the set of all B-splines with same boundary values as $\mathbf{x}$). Here, the solution is found by inverting a linear system of magnitude $(n - \alpha - \beta + 1) \times (n - \alpha - \beta + 1)$.

Nevertheless, two disadvantages occur in this method. At first, for $\mu \ne 0$ a different functional is minimized as originally stated so that the resulting smoothing effect is not optimal. Secondly, the restrictions (23) are not necessarily fulfilled for the preset value μ so that one or more iterations with raised values of μ might be necessary.

So, in contrast to Kallay's method the idea used in the present paper seems to be more practical because the right functional is minimized in every step and the restrictions (23) are always hold. Further, the implementation is much simpler because no linear equation solver is used.

However, our method only produces a local approximative solution to the above addressed problem. Nevertheless the (weak) convergence of the iterative process is at least guaranteed since the always non-negative value of the energy is lowered in every step, producing a bounded decreasing sequence.

Moreover, any kind of statement about convergence of the algorithm to a global solution can be given only in the unrestricted case, meaning $\delta = \infty$, with $\alpha, \beta \ge l$ and $k \ge 2l$.

In that case the problem of minimizing the positive definite quadratic functional

$$E_l(\mathbf{y}) = \int_{t_{k-1}}^{t_{n+1}} (\mathbf{y}^{(l)}(t))^2 \, dt \tag{35}$$

for $\mathbf{y} \in \chi_\infty^{\alpha,\beta}$ is reduced to the computation of the unique solution of a linear system of magnitude $(n - \alpha - \beta + 1) \times (n - \alpha - \beta + 1)$ as mentioned above. Note that the resulting solution only depends on the boundary values of the given curve $\mathbf{x}$.

Now, it is easy to see that our local fairing algorithm tends in the limit to the same global solution. The reason for this behaviour is the possible comparison of our local fairing step to the usual Gauss-Seidel iteration for linear systems [9]. There in each single step also only one of the unknowns is altered in order to satisfy one equation of the entire linear system exactly. Then in the next step the most recently available information is used to alter the next variable.

However, it follows from (25) that the limit of our algorithm is characterized by $z_r = 0$ for $r = \alpha, \ldots, n - \beta$ since otherwise the value of the energy could be lowered by changing some further points. Hence we can deduce from (27) that $\bar{\mathbf{d}}_r = \tilde{\mathbf{d}}_r$ for $r = \alpha, \ldots, n - \beta$, meaning that every equation of the underlying linear system is satisfied. Thus the obtained limit curve is identical to the global unique solution.

For instance in the cubic case $(k = 4)$ the minimization of $E_2(\mathbf{y})$ with $\mathbf{y} \in \chi_\alpha^{2,2}$ gives for general knot vectors T that cubic Hermite polynomial as limiting curve which interpolates the boundary values $\mathbf{x}(t_{k-1})$, $\mathbf{x}'(t_{k-1})$, $\mathbf{x}(t_{n+1})$ and $\mathbf{x}'(t_{n+1})$.

Finally, we mention that our algorithm converges for the same reasons also to a global minimum if the restrictions $\alpha, \beta \geq l$ and $k \geq 2l$ are not satisfied. Here, it is much more difficult to predict the limit curve in cases where the global problem has no unique solution since then the limit curve mainly depends on the order the control points are changed.

5. Conclusion

We presented in this paper an algorithm which automatically fairs B-spline curves of general order. The key idea was to minimize the integral of the squared l-th derivative of a given curve iteratively by changing only one control point in every step. The control points at which the B-spline curve has to be faired, are selected automatically by a fairness criterion. Furthermore, the algorithm is very fast because we need to calculate the appearing integrals only once.

We tested this algorithm in many examples with given disturbances and we drew our conclusion that if the knots are shared *well* the fairing results are very good. This means, the influence of the parametrization to the fairing process is an area for more investigation.

Finally, we want to mention that the method obviously can be extended to surfaces. In [11] it is especially outlined that in case of tensor product B-spline surfaces very good results can be obtained which are again comparable to those in [10, 14].

Acknowledgement

This work was finished during a postdoctoral fellowship supported by the German Research Foundation (DFG).

References

[1] de Boor, C.: A practical guide to splines. Berlin Heidelberg New York: Springer 1978.
[2] Eck, M., Hadenfeld, J.: Knot removal for B-spline curves. Comp. Aided Geom. Des. (1995), to appear.

[3] Eck, M., Hadenfeld, J.: A stepwise algorithm for converting B-splines. In: Curves and surfaces in geometric design (Laurant, P. J., Le Méhauté, A., Schumaker, L. L., eds.), pp. 131–138. AK Peters (1994).

[4] Eck, M., Jaspert, R.: Automatic fairing of point sets. In: Designing fair curves and surfaces (Sapidis, N., ed.), pp. 45–60. Philadelphia: SIAM 1994.

[5] Farin, G., Rein, G., Sapidis, N., Worsey, A. J.: Fairing cubic B-spline curves. Comp. Aided Geom. Des. *4*, 91–103 (1987).

[6] Farin, G.: Curves and surfaces for computer aided geometric design. A practical guide, 2nd edn. New York: Academic Press 1991.

[7] Farin, G., Degree reduction fairing of cubic B-spline curves. Geometry Processing for Design and Manufacturing, SIAM 87–99 (1992).

[8] Gill, P. E., Murray, W., Wright, M. H.: Practical optimization. New York: Academic Press 1981.

[9] Golub, G. H., van Loan, C. F.: Matrix computations, 2nd edn. The John Hopkins University Press 1989.

[10] Greiner, G.: Variational design and fairing of spline surfaces. Comput. Graphics Forum *13*, 143–154 (1994).

[11] Hadenfeld, J., Local energy fairing of B-spline surfaces. Submitted to M. Dæhlen, T. Lyche, L. L. Schumaker (eds.) Mathematical Methods in CAGD III (1995).

[12] Hagen, H., Schulze, G.: Extremalprinzipien im Kurven- und Flächendesign. In: Geometrische Verfahren der Graphischen Datenverarbeitung (Encarnação, J. L., Hoschek, J., Rix, J., eds.), pp. 46–60. Berlin Heidelberg New York Tokyo: Springer 1990.

[13] Hosaka, M.: Theory of curves and surfaces synthesis and their smooth fitting. Inf. Proc. Jpn. *9*, 60–68 (1969).

[14] Kallay, M.: Constrained optimization in surface design. Manuscript, 1993.

[15] Kjellander, J.: Smoothing of cubic parametric splines. Comp. Aided Des. *15*, 175–179 (1983).

[16] Lee, E. T. Y.: Energy, fairness, and a counterexample. Comp. Aided Des. *22*, 37–40 (1990).

[17] Mehlum, E.: Nonlinear splines. In: Computer aided geometric design (Barnhill, R. E., Riesenfeld, R. F., eds.), pp. 173–208. New York: Academic Press 1974.

[18] Meier, H.: Der differentialgeometrische Entwurf und die analytische Darstellung krümmungsstetiger Schiffsoberflächen. Fortschrittsberichte VDI-Reihe 20, No. 5, VDI-Verlag, 1987.

[19] Moreton, H. P.: Minimum curvature variation curves, networks, and surfaces for fair free-form shape design. Phd Thesis, Univ. of California, 1993.

[20] Nowacki, H.: Mathematische Verfahren zum Glätten von Kurven und Flächen. In: Geometrische Verfahren der Graphischen Datenverarbeitung (Encarnadão, J. L., Hoschek, J., Rix, J., eds.), pp. 23–45. Berlin Heidelberg New York Tokyo: Springer 1990.

[21] Sapidis, N., Farin, G.: Automatic fairing algorithm for B-spline curves. Comp. Aided Des. *22*, 121–129 (1990)

[22] Schaback, R.: Error estimates for approximations from control nets. Comp. Aided Geom. Des. *10*, 57–66 (1993).

[23] Su, B., Dingyuan, L.: Computation geometry. New York: Academic Press 1989.

[24] Vermeulen, A. H., Bartels, R. H., Heppler, G. R.: Integrating products of B-splines. SIAM J. Sci. Stat. Comput. *13*, 1025–1038 (1992).

M. Eck
Department of Computer Science
and Engineering, FR-35
University of Washington
Seattle, WA 98195
U.S.A.

J. Hadenfeld
Fachbereich Mathematik
Technische Hochschule Darmstadt
D-64289 Darmstadt
Federal Republic of Germany

Computing Suppl. 10, 149–161 (1995)

Integrating Analysis Tools into the Design Process through Constrained Parametric Structures

R. Fischer, Wolfsburg, and **A. S. Vieira,** Darmstadt

Abstract. The efficient integration of analysis tools in early stages of the design process has been regarded as a strategic methodology to ensure quality and to reduce time and costs during the product development cycle. In this paper, an investigation into the nature of this integration problem is performed, followed by the specification of a series of requirements on CAD systems to support it. Guided by these pragmatic requirements, we present an approach based on constrained parametric structures and on an analysis-flow graph. Finally, we discuss our experience with a prototype implementation on the top of the *GeneSys* CAD system.

Key words: Computer-aided geometric design, computer-aided design, feature-based design, dimensional-driven design.

1. Introduction

In the product development cycle tasks like drafting, modeling, virtual analysis[1], simulation, NC programming and many others, generate series of problems related with the conversion of data, the connection of different hardware and software platforms and user interfaces. Moreover, the knowledge of handling and managing each of these tasks and their logical relationships is left to experts, arising other semantically communication problems.

It is a common agreement that the integration of virtual analysis tools in early phases of the product development cycle provide great benefits [14,23]. The development time and costs decrease, and the quality increases. Additionally, technological improvements made in different engineering areas can be more effectively used in industrial sites. Figure 1 presents a typical pipeline for the development of a general product. Looking in more detail, steps 1 through 4 perform the following activities:

1. The specification of general product requirements (functional constraints) and engineering requirements (engineering constraints);
2. Sketching of the first guidelines for the design adding general geometric requirements;
3. Producing design versions using geometric primitives and constraints (form constraints), or using already constrained functional primitives (design features);

---

[1] The term "virtual analysis" is used to define analysis methods applied to symbolic representations of a part in a computer-aided environment and not to the physical prototype.

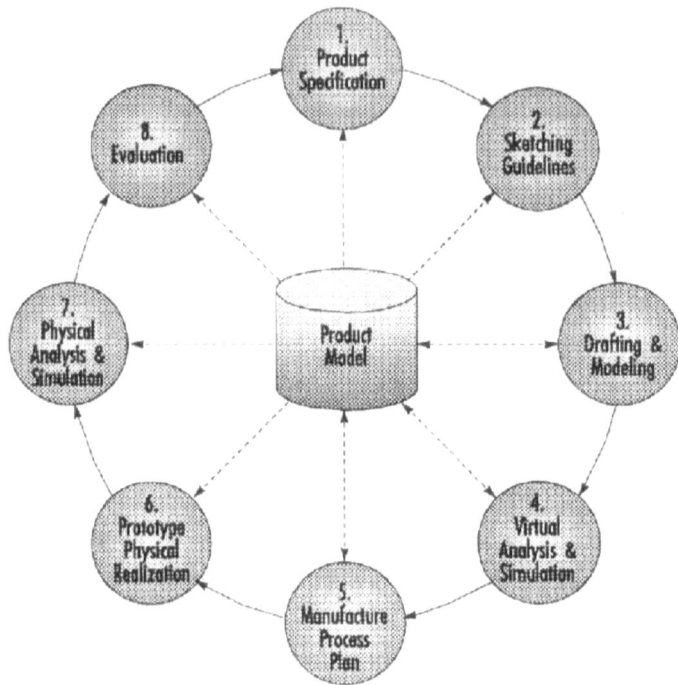

Figure 1. A typical pipeline flow for the development of a product

4. Running virtual analysis methods and simulations to validate and optimize the given design versions based on the degrees of freedom (DOF's) provided by designers[2] and engineers.

In practice, these tasks are not executed in the given order. Usually, engineering constraints[3] are defined much later by the same persons who are responsible for running the analysis and interpreting its results. At this point we observe one of the most classical problems. The group of persons who define the functional requirements, who make the product design and who run and interpret the results of the virtual analysis, are not the same. More than this, normally they do not have the same knowledge background, they do not speak the same language, they do not know the project as a whole and they tend to believe that their part is the most important one. As a result of this multidisciplinary problem, time and costs during the product development cycle increase and the quality, in general, decreases.

Actually, the group of persons who are really responsible for given initial solutions to the related product are the designers. They must produce their work based on the

---

[2] In this context, we understand a "designer" as being a CAD user without a deep background or interest, in the engineering theory behind analysis and simulation tools. In the development cycle he is principally responsible for aspects like functionality, ergonomy and form.
[3] A precise definition for the term "constraint" will be given in the following section.

functional constraints and at the same moment, being able to understand requirements coming from very complex engineering analysis and simulations. At the end of the day, most of the changes in the project affect the current geometric form. Sometimes, the engineering constraints are so strong that the final version has nothing to do with the original intended design [12]. This frustration is very serious since it reduces the creative potential of the designer.

2. Integrating Analysis Tools into the Design Process

Many CAD systems offer the functionality to run a restricted set of analysis tools during the design phase. Some of these tools require no additional pre- or post-processing steps. Mass-Properties Analysis (MPA), Tolerance Analysis (TOA) and Clash Detection (CLD) are some examples. All they need is basically the geometry and topology of the parts being considered. Therefore, they can be very easily integrated in the design environment. In contrast, other analysis tools like Finite Element Analysis (FEA) need not just the geometry and topology description of the part, but also a considerable pre- and post-processing effort to be executed [19]. Normally, only expert engineers are able to drive the system in these pre- and post-processing stages. Only a few commercial systems have some of these analysis tools fully integrated.

Even if through a series of simplification steps a designer could drive the pre- and post-processing tools, he will probably be unable to decide which changes are necessary. Actually, even when the user is dealing with analysis tools that do not need the pre- and post-processing overhead, he can have great difficulties to decide which changes should be done. That is more or less the struggling point—how to insert engineering knowledge into a CAD system so that it can help the designer to produce analysis conformed products. As an example, take a simple mechanical part and try to make its volume be conformed to a specific value. Even if the user has the facility to inquire the current volume, he will probably not know what to change on the geometry of the part to reach the target value. In the same sense, if we leave the system the whole responsibility to propose solutions to the problem, it can also spend an infinite number of CPU-clock-cycles suggesting an unlimited number of solutions.

For a really full-integration, and consequently, for a full-use of any coupled analysis tools, the output must have a direct influence in the shape representation. This influence by itself should be presented to the designer in a form that he can understand the changes and also, be able to decide which alternative, from his point of view, is the best one. With this motivation in mind, we can list some more detailed requirements on a CAD system to support this integration, which are:

1. A large number of analysis tools must be provided for solving and optimizing models;
2. The analysis tools must suggest changes that can reflect directly on the model;
3. The solution changes must be represented and expressed in a way which is syntactically and semantically compatible with the tools available on the system;

4. The environment must provide the flexibility for a "move", at any stage, from one solution instance to another.
5. Solutions for one problem must be re-usable in similar situations;
6. In the case that an expert for one application field is using the system, the system should be able to "learn" (at least by example) the applied strategy for solving the specific problem [3, 4, 11].

The integration methodology promoted by these rules is basically to stimulate the communication process between the different persons involved in the product development process. The idea is to link the functional constraints on the desired project as soon as possible with the geometric properties [12]. The introduction of these constraints has the following effects on the design process:

- They remind the designer about the basic requirements of the product;
- They serve as a common communication language between the ones who state these constraints and the designers;
- As it will be clear in the next section, constrained parametric structures can be a very efficient way of running optimization tasks that perform automatic changes.

The constraints actually handled by the designer are normally different than the ones passed to him. Therefore, a semantic conversion has to be done. Software tools can help the designer to define and to remember these relations. The most important aspect is that a relation must be establish as soon as possible. Following, the designer will add his own constraints on the geometric form and the resulting set of requirements can now be linked with the analysis methods. At this stage, the product information model must be updated with the analysis specific data, the analysis/simulation must be executed, the results must be interpreted and finally, the original geometric and topological model must be changed according to the application-specific engineering know-how. Again, the approach is to use the same set of constraints to achieve a common communication language. If the related optimization task works on the given set of constraints, a link can be expressed in a common semantical way. Additionally, the solutions used in this optimization steps can be stored and used later [17, 15].

3. The Analysis-Flow Graph

Based on the target product, the development team has to define a series of analysis tools that have to be executed in order to optimize and validate the product during the design process. The formal description of this analysis flow can be obtained through a directed graph structure (the Analysis-Flow Graph). The nodes of this graph represent the analysis tools and contain a complete description how this tool should be used and what are the desired target requirements. For instance, the analysis tool description for the first node in Fig. 2 could have the description table shown in Table 1.

Figure 2 shows an example of this graph structure. The connection arrows indicate the sequence in which the tools have to be called and give them the necessary

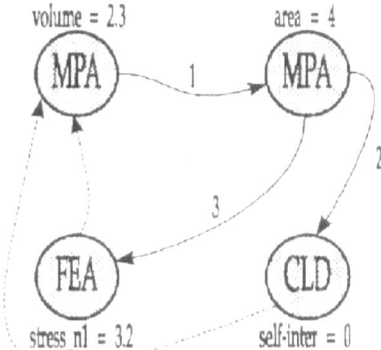

Figure 2. An Analysis-Flow Graph representing one possible optimization flow

Table 1. Analysis node description

• Analysis base class	*Mass Properties Analysis*
• Analysis target class:	*Volume Optimization*
• Target value:	*2.3*
• Maximum error allowed:	*0.1%*
• Desired base classes of solution:	*(i) Change of Parametric Structures* *(ii) Uniform Scaling*
• Target classes of solution and data:	*(i) Single Parameter Change: {p1, p4, p7},* *Multiple Parameter Change: {p1&P4}* *(ii) Scaling Directions: {xy, xyz}*

adjacency information for a conformance with neighbor tools (local conformance). Dashed connection lines mean implicit sequences that have to be defined in order to force the resulting object to be in conformance with all analysis requirements (global conformance).

For every well executed analysis, the connection arrow is marked and the next node is executed. This procedure is continued until all requirements have been fulfilled. The whole optimization process can be stopped if an infinite loop or an error is detected. Note that the analysis node description table requires the definition of a finite number of "base-classes" and "target-classes" of solution. This avoids an infinite search for solutions and also accelerates the whole optimization/validation process.

4. Constrained Parametric Structures

In the previous sections we spoke about different kinds of constraints that would provide us with a common communication base for integrating different activities during the product development cycle. In this section we will define a form constraint and will show how it can be associated to the traditional Boundary Representation scheme (BRep).

A constraint alone has no semantical meaning and can map different kinds of requirements. In the context of form constraints, that is, constraints placed by

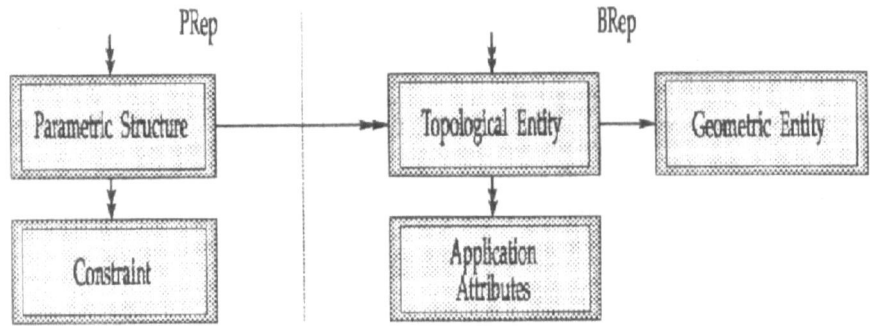

Figure 3. The parametric structure between the BRep scheme and the form constraints

a designer (or by the system when using design features) to reflect semantics on the geometry and also on the topology, we can establish a very clean link with the traditional BRep scheme.

This association can be done through the use of a parametric structure. In this context, a parametric structure is an object with different semantic meanings depending on which topological/geometric entities it acts, and on the set of constraints it encloses. The proposed model is shown in Fig. 3. A common error is to perform a direct link between the parametric structures and the geometric entities [7]. There are many reasons why one should not follow this approach. A topological entity can have different geometric representations and therefore, the direct mapping can be ambiguous or at least not well defined. On the other hand, topological entities are well defined and contain explicitly and implictly adjacency information. Furthermore, they have been recognized as being the key entities in providing a full associative kernel representation. This means that other application-dependent attributes are also linked to them. Besides, a simple change of a parametric value can not only affect the geometric attributes of a part, but also its topological representation. In this case, the topological information is required for updating the model. Figure 4 presents a simple example showing a part under the influence of a change in the related parametric structures.

Each parametric structure has a *functional* and a *semantical* meaning. The *functional* meaning is to associate constraints to the main shape representation, to enable quick and clean changes to the model and also, to enable the realization of Assembly constraints and Hierarchically organized constraints [12].

The *semantical* meaning of a parametric structure is partially defined by the set of topological entities it is related to. In the above example, the parametric structure *p1* acts on two distinct faces, *p2* on two edges and *p3* on two vertices. The constraints placed on these parametric structures will define how they influence the model. Here we have to distinguish between the constraints that define how the entities referenced by the parameter should react to changes (*dynamic* constraints), and the ones

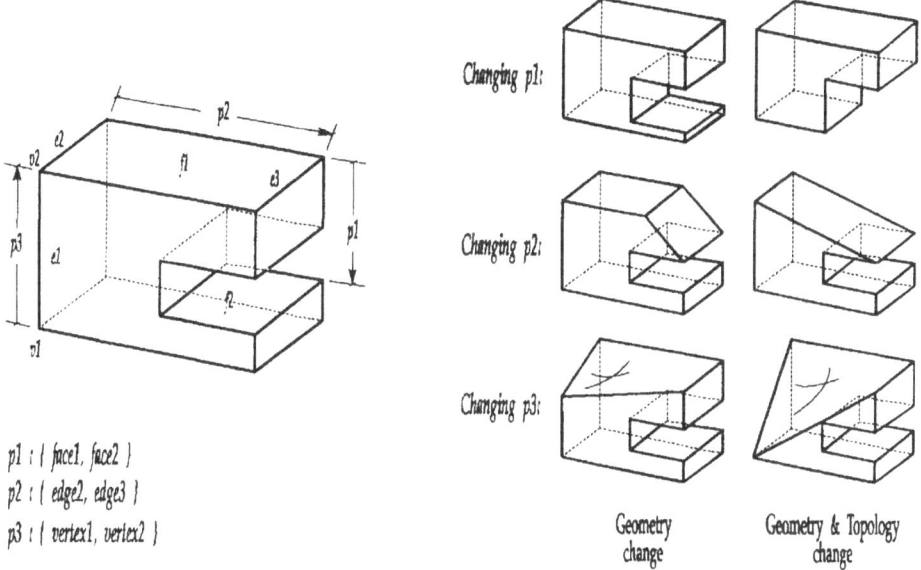

$p1 : \{ face1, face2 \}$
$p2 : \{ edge2, edge3 \}$
$p3 : \{ vertex1, vertex2 \}$

Figure 4. A parametric constrained part showing how parameter changes can affect the geometry and the topology configuration

Table 2. Constraint attributes

Static	Dynamic
• bounds and precision (*min, max, double, tolerance*)	• orientation (*directed, bi-directed, centered,...*)
• type of values allowed (*linear, step, discrete_set,...*)	• origin placement (*free, xyz, topological_entity,...*)
• explicit math formulation with a symbolic representation (*equation*)	• implicit geometry constraints (*parallel, perpendicular, fixed_angle, reflected...*)

that say what are the rules the parameter value must satisfy (*static* constraints)—see Table 2.

The *static* constraints on the parametric structure *p1* define, for a single-value parameter, its bounds (*min* and *max* values), the type of values it can have and the tolerance. For example, *p1* could accept any real value (*linear*), values in defined increments (*step*: 0.1, 0.2, 0.3,...) or also, values extracted from a pre-defined input set (*discrete_set*: 0.3, 3.9, 5,...). The *static* constraints give the related parametric object a consistency that reflects some of the real world requirements. Normally, they are defined by engineers and are associated to a mathematical formulation through engineering equations.

The *dynamic* constraints express the geometric flexibility and intention given to the part by the designer. They include, for example, statements that one face should be held always parallel or perpendicular to another one. They define also the orientation of the parameter in respect to the related topological entities. This information will indicate for instance, that face $f1$ should always be held fixed while face $f2$ should change when the parameter $p1$ changes.

Concluding, *static* constraints are more related to the engineering requirements, while the *dynamic* constraints express the designer's requirements on the geometric form. Together, they build the second *semantical* meaning for the parametric structure.

Another high-level *semantical* interpretation for a constrained parametric structure can be extracted from the application context it is being used. A parametric structure created and referenced by the representation of a design-by-features application has a different meaning than one placed by an analysis program. In the first case, a constrained parametric structure can be used to complete and to check the consistency of a specific design feature class [10, 18, 21], while in the second case, it is normally used to study an optimization strategy. Nevertheless, when fully-defined, the parametric object can express different high-level requirements in a common way.

Different constraints applied to the same parametric structure of a form feature, define different design features[4]. As an example, a *bore* feature and a *cylindrical pocket* can be distinguished through the maximum bound of the parameter radius. The *bore* will later be drilled while the *pocket* must be milled. Besides, the specification of what is valid or invalid for a certain feature is many times based on the constrained parametric structures it associates. During the feature creation phase if the designer enters a value for the radius of the *bore* bigger than the specified bound limit, the application can verify if the value satisfies the parametric constraints of a similar (in form) feature and suggest him the creation of e.g. a *cylindrical pocket*. It is also interesting to mention the known possibility to use the constrained parametric structures to detect and solve feature interactions [20].

With the complete constrained parametric representation two known strategies can be adopt to solve/evaluate the BRep model. The first approach is to build a system of constraint equations to be subsequently solved by interactive numerical methods [9, 17]. This is known as the Algebraic Approach [12]. Another idea is based on Artificial Intelligence methods and is known as the Deductive Approach. Rather than converting the constraints to a system of equations, the Deductive Approach uses rule-based reasoning to deduce analytic coordinate representation and consequently, the geometric configuration [12, 22]. It should be notice, that both methods act only on some of the constraint attributes presented in the above table, more precisely, on the implicit geometry constraints, on the explicit math formulation and on the value itself. A more general approach that works on the complete constrained parametric model is not known.

---

[4] Design features are not pure shape concepts, but rather have non-shape connotations from their application domain. When not specified, the term "feataure" is used to refer design features. On the other side, form features have no application connotation. They refer generic shape properties of a product [9].

Perhaps the designer will never be able to understand the engineering equations that are referenced by a *static* constraint, but its influence on the form will be clear for him (from the *dynamic* constraints). In the same sense, the engineer can associate these constrained parametric structures in the definition of the analysis-flow graph. Consequently, assuming that all attributes of the related parameters have been well defined, an optimization process can suggest changes that conform to the designer's and engineer's intentions. The parametric structures define all degrees of freedom on the model and enable quick and clean changes. At the same time, the encapsulated constraints related to the parametric structures will guarantee the overall conformance to the designer's and engineer's requirements. In addition to this, if one picks an object that includes this extra constrained parametric representation, he can capture the design and engineering decisions and re-use them in similar situations.

5. The Realization in *GeneSys*

In the last 30 years software packages have been developed to assist engineers and designers in each of the tasks depicted in Fig. 1. Most of them are based in the concept of a black-box system with built-in solutions and methods for solving specific problems. Thus, in these systems the user is almost driven by the system according to the limitations of the implemented methods. Some of these limitations can be classified under the following topics:

1. Pure limitation of the static and dynamic representation schemes for the design;
2. No associativity or a restricted information exchange mechanism between different tools;
3. Distinct and machine-driven user interfaces;
4. Poor set of modeling tools and analysis programs;
5. No concept for influencing the design process based on the virtual analysis results;
6. No concept for representing and extracting the engineering knowledge.

5.1. *The* GeneSys *System*

GeneSys is a third generation Computer-Aided Design system [2, 3, 5, 6]. The know-how required to develop *GeneSys* involves research efforts done mainly during the last 20 years in the field of: Computer Aided Geometric Design, Computer Graphics, Computer Engineering, Artificial Intelligence, and many others. It has been originally developed to solve problems related with the topics 1 and 2. Later, the development was oriented towards a good user interface and naturally, for a set of modeling and analysis tools that could profit from a rich representation kernel (topics 3 and 4). Finally, following the needs to integrate analysis tools, *GeneSys* was redesigned focusing problems presented by topics 5 and 6.

As a result of this development process a very nice and innovative full 3D interactive system came up. The application fields of *GeneSys* are very broad, ranging from

Graphic Design, Industrial Design, Architecture, to Mechanical and Civil Engineering. Nevertheless, to reach such a high degree of flexibility, new technological improvements have been introduced and many state of the art research results have been used. Some of them are:

- The Solid Modeling approach based on a hybrid representation kernel integrating an extended Boundary Representation Scheme (BRep), a modified Constructive Solid Geometry (CSG) modeling tree, Sweep and a Constrained Parametric Representation (PRep) based on the concept explained in the last section. The domain of all representations is restricted to physical realizable manifold lamina bodies [5, 6, 16];
- The concept of a Modeling Process with invertible high-level modeling operations based on a set of low-level basic operations [1, 2, 3, 6]. This technique ensures, during all phases of the modeling process, a consistent topological and geometrical representation kernel. Additionally, it offers the required support for a Design-by-Features approach;
- A Design Representation Tree that offers a semantical framework for the integration of fields such as Cooperative Work, Concurrent Engineering, Design Knowledge Representation, Design-by-Features and Analysis-Aided Geometric Design [2, 3, 6, 20];
- Parametric Design bound with traditional interactive modeling techniques [17];
- A configurable 3D Interactive User Interface Environment based on the perception and cognitive needs of the user and on the requirements of the process being performed [6].

5.2. The Integration of Analysis Tools

Like in other systems, in *GeneSys* the user can perform standalone analysis during the modeling process. Mass Properties Inference, Clash Detection, Topology and Geometric Consistency Checks are some of the available analysis modules totally integrated in the system. Relying on the rich information model of *GeneSys*, these analysis can be performed very efficiently leading to interactive simulations and decisions.

Guided by the more general integration requirements listed in section 2, *GeneSys* stores for each analysis tool, a decision tree where the possible build-in classes (and sub-classes) of solution are described. This tree contains the basic information necessary to define the Analysis Node Description Table for the Analysis-Flow Graph. All classes and sub-classes of solution are based on the constrained parametric structures. In the moment, this Analysis-Flow Graph is being implicitly defined through menu options. Following, the user is asked to define other analysis requirements like target value, maximum allowed error, etc. Based on the resulting information model, the system starts optimization methods provided by each analysis tool, and for each consistent solution it proposes a series of modifications in the current design.

During the modification phase the system has the capability to represent explicitly the solutions through the Design Representation Tree [6]. All changes are expressed by a sequence of user-known high-level operations. At the end of the optimization process the user gets under the current design node a sequence of new branches, each of them representing one consistent solution. All solutions are carefully tested in regard to the constraints associated with the modified parameter and also, in regard to implicit automatic analysis calls, like: Clash Detection and Topology Analysis. This means that for a solution to be regarded as being really consistent, it can not admit self-intersecting objects or objects with an inconsistent topological state.

An illustrative example is the Volume Analysis integrated in *GeneSys*. With this module the user is capable of optimizing the volume of a 3D body. In the startup phase the user in prompted with the current volume and is asked for the new desired value. Following, the system asks him to specify the classes of solution he is interested, for instance, parametric changes, scaling, etc. The system then starts an interactive analysis process selecting possible solutions and putting them under the current volume optimization node. When finished, the user gets under this node, branches with different consistent solutions. Later, he can browse through these solutions and select the most appropriated one (regarding other project constrains).

Figure 5 shows a practical example of a cheese package that has been modeled and optimized for the industry using the *GeneSys* system. The functional requirements include the number of cheeses per package, the cheese's density and the total liquid

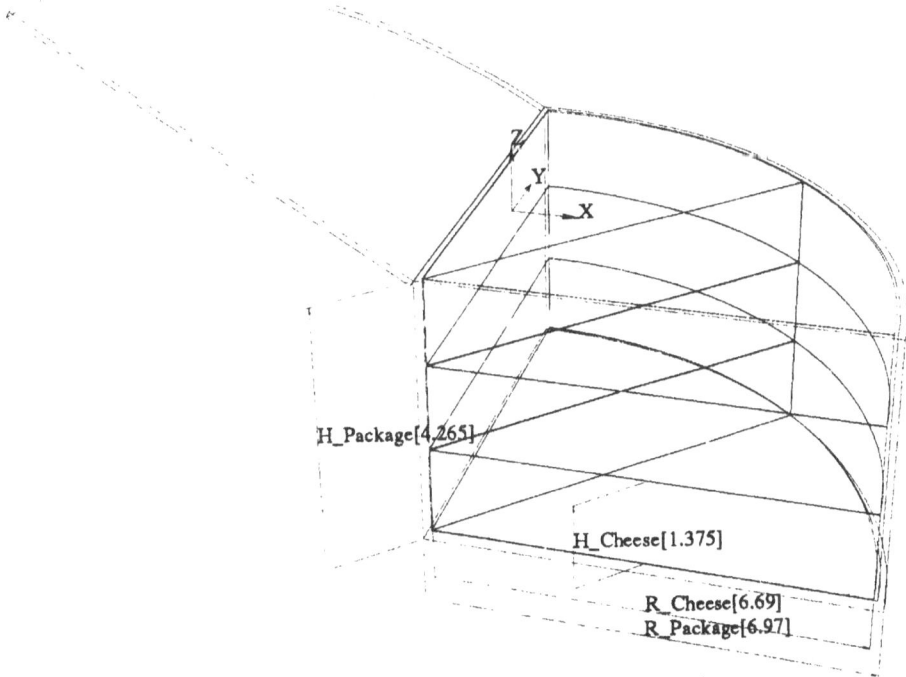

Figure 5. The cheese package designed and optimized through the constrained parametric structures

weight of the package. The requirements on the form specified that it should have the $\frac{1}{4}$ pie form, two levels of cheese and a minimum distance between each cheese (for manufacturing reasons). The system has optimized the individual cheese dimensions and the sizes of the package, based on the following analysis tools: Volume Analysis, Mass Analysis, Clash Detection and Topology Analysis.

6. Conclusions

Recalling the requirements listed in Section 2, for a CAD systems to support the full integration of Analysis tools in the design process, we claim that points 2, 3, 4 and 5 have been fulfilled in the current realization. Far from being a complete implementation of the whole concept, the current implementation serves as an insight into the breakthrough that this new technology can represent if more analysis modules and classes of solution are provided (point 1).

For the future, based on the rich semantical information mechanism of the Design Representation Tree, different solutions to similar problems could effectively be compared, arising ideas beyond the traditional build-in solution strategy (point 6). If an expert user decides to try his own solution, and if the system is now capable of comparing this new solution with those it knows already, it could probably store this new approach, increasing its knowledge [4, 11]. Consequently, designers with no knowledge about FEA methods could produce realizable products.

Concluding, given a flexible constrained parametric hybrid representation kernel, a modeling process based on high-level operations and a Design Representation Tree, the integration of Analysis Tools into the Design Process can be realized and proves to be a straightforward strategy to integrate the engineering knowledge into Design systems. Running an analysis process that produces concrete suggestion changes, instead of just an output log, makes current users really believe in the new ideas behind Intelligent Computer-Aided Design systems. Nevertheless, such an associative engineering framework requires also an outstanding user interface based mainly on the perception and cognitive needs of the user [3, 6]. Otherwise, most of the efforts spent integrating different disciplines and users, are lost in a bunch of communication problems.

Acknowledgements

This research is being supported by *CNPq—Brazilian National Research Council* under grant 200496/91-3 at the *Institut für Konstruktionslehre Maschinen- und Feiwerkelemente* (Braunschweig, Germany).

References

[1] Chiyokura, H.: Solid modeling with DESIGNBASE—theory and implementation. Reading: Wesley 1988.
[2] Feijó, B., Fischer, R., Dreux, M.: Better criteria for the development of solid modeling software. Comput. Mech. *4*, 32–43 (1991).

[3] Feijó, B.: Perception and cognition in intelligent CAD systems. Lectures in the Advanced School, CISM, Italy, July 1992.

[4] Findler, N. V.: Associative networks: representation and use of knowledge by computer. New York: Academic Press 1979.

[5] Fischer, R.: Representation schemes for 3D objects. Final course thesis in computer engineering, Intelligent CAD Laboratory, Computer Science Department, PUC-Rio/Rio de Janeiro, November 1989 (in Portuguese).

[6] Fischer, R.: Genesys—a hybrid solid modeling system. MSc. thesis, Intelligent CAD Laboratory, Computer Science Department, PUC-Rio/Rio de Janeiro, August 1991 (in Portuguese).

[7] Hsu, C., Bruderlin, B.: Constraint objects—integrating constraint definition and graphical interaction; 2nd ACM Solid Modeling, Montreal/Canada, May 1993, pp. 467–468.

[8] ISO 1992: Industrial automation systems, product data representation and exchange—Part 48. Integrated generic resources: form features, working draft, ISO TC/84/SC4/WG3 N102(P5), January 1992.

[9] Light, R. A., Gossard, D.: Modification of geometric models through variational geometry. Comput. Aided Des., 14, 209–14 (1982).

[10] Lin, W., Myklebust, A.: A constraint-driven solid modeling open environment. 2nd ACM solid modeling, Montreal/Canada, May 1993, pp. 233–242.

[11] Maher, M. L., Li, H.: Automatic learning preliminary design knowledge from design examples. Microcompat. Civil Eng. 7, 73–80 (1992).

[12] Martini, K., Powell, G. H.: Geometric modeling approaches for civil engineering and building design. Microcomput. Civil Eng. 7, 3–14 (1992).

[13] Nakai, S., Katukura, H., Ebihara, M., Niimi, K., Hirose, K.: A knowledge-based structural analysis based on a object-oriented approach. Microcomput. Civil Eng. 7, 15–28 (1992).

[14] Pratt, M. J.: Solid modeling and the interface between design and manufacture. IEEE Comput. Graph. Appl. 6, 52–59 (1984).

[15] Pérez, A., Serrano, D.: Constraint based analysis tools for design. 2nd ACM solid modeling, Montreal/Canada, May 1993, pp. 281–290.

[16] Requicha, A. G.: Representations of rigid solids—theory, methods, and systems. ACM Comput. Surv. 12, 437–464 (1980).

[17] Roller, D.: Advanced methods for parametric design. Geometric modeling: methods and applications (Hagen, H., Roller, D., eds), pp. 251–266. Berlin Heidelberg New York Tokyo: Springer 1990.

[18] Roller, D.: Parametrische Formelemente als Basis für intelligentes CAD. Proceedings of GI-Fachgesprächgraphik und KI, Königswinter bei Bonn, April 1990, Springer-Verlag (in German).

[19] Rudd, B. W.: Impacting the design process using solid modeling and automated finite element mesh generation. Comput. Aided Des. 20 (1988).

[20] Shah, J. J.: Conceptual development of form features and feature modeler. Res. Eng. Des. 2, 93–108 (1990).

[21] Sheu, L. C., Lin J. T.: Representation scheme for defining and operating form features. Comput. Aided Des. 25, 44–58 (1993).

[22] Sunde, G.: Specification of shape by dimensions and other geometric constraints. IFIP WG 5.2 on geometric modeling; Rensselaerville, NY, May 1986.

[23] Autorenkollektiv: Verbundprojekt CAD-Referenzmodell, Gestaltung zukünftiger computergestützter Konstruktionsarbeit: aktueller Stand der CAD-Technik und der rechnergestützten Konstruktionsarbeit; FZI, Karslruhe, Germany, 1992 (in German).

R. Fischer
Volkswagen AG
Forschung und Entwicklung, EZTI
D-38436 Wolfsburg
Federal Republic of Germany

A. S. Vieira
Fraunhofer Institut für Graphische
Datenverarbeitung
D-64283 Darmstadt
Federal Republic of Germany

Computing Suppl. 10, 163–176 (1995)

Localized Radial Basis Methods Using Rational Triangle Patches

T. A. Foley, S. Dayanand, Tempe, and **D. Zeckzer,** Kaiserslautern

Abstract. Two of the most effective methods for interpolating scattered data are the multiquadric method and the thin plate spline method. A negative aspect of these radial basis methods is that they are not local and they are computationally expensive and unstable if there are a large number of data points. We present a localized interpolation method that involves partitioning the data into arbitrary overlapping triangular regions based on arbitrary points, forming local radial basis interpolants to the data in each region, and then blending the local interpolants with rational triangle patches so that the resulting function is a locally defined C^1 interpolant. When the triangular regions form a triangulation of the scattered data locations, the resulting interpolant can be considered as a new local triangle patch method that generally has effective derivatives at the vertices and cross-boundary derivatives along the edges of the triangulation.

Key words: Scattered data interpolation, geometric modeling, multivariate data fitting.

1. Introduction

The problem of fitting a function to data sampled at arbitrarily located positions in a planar domain arises often in scientific and engineering applications. Data of this type commonly occur in applications where the data site locations are restricted. Some data sets come from experimental tests, the measurement of physical quantities, and from computational values such as those output from a finite element solution of a partial differential equation. Examples occur in meteorology, mining, optics, medicine, oceanography, fluid flow analysis, and numerous other places in the broader fields of physics, chemistry and engineering.

For a precise statement of the problem, suppose that we are given N distinct points (x_i, y_i) and N real values f_i. The interpolation problem that we address is to construct a smooth function $F(x, y)$ that satisfies $F(x_i, y_i) = f_i$ for $i = 1, \ldots, N$. Since the scattered data interpolation problem has so many important applications, a great deal of research has been done in this area and the survey papers [1, 2, 18, 19, 34] contain descriptions of many methods that solve this problem. The text books [25, 26] also discuss several methods and they give some applications and examples.

Two of the most effective and simplest methods to implement are Hardy's multiquadric (MQ) method [22, 23] and the thin plate spline (TPS) method [7]. In the critical testing of methods in [15, 16], the multiquadric and thin-plate spline methods consistently produced visually pleasing results with very small observed errors on known test functions. A negative aspect of these radial basis methods is that if there are a large number of data points, the linear system of equations can be ill conditioned and costly to solve. One remedy is to localize the interpolant using techniques similar to those used in [17] for thin plate splines. This technique involves partitioning the data into overlapping rectangular regions, forming local interpolants to the data in each region, and then blending the local interpolants so that the resulting function is a C^1 interpolant. Since the blending functions are locally defined, data points in one region have no effect on the interpolant on non-neighboring regions. One drawback of this approach is the limitation of using a tensor product grid structure to define the local regions.

We present a generalization of this approach which defines local regions that depend on an arbitrary triangulation of an arbitrary collection of points. Weight functions are defined as C^1 composite rational triangular patches and the interpolant is a local rational blending of local radial basis interpolants. Section 2 describes the basic localized technique and Section 3 gives a brief background on the multiquadric and thin plate spline interpolation methods. The rational triangle weight functions are discussed in Section 4, while Section 5 contains strategies for selecting local regions and it gives several examples. One interesting example is where the arbitrary points of the triangular regions are equal to the scattered data points. In this case, the resulting interpolant can be considered as a new local triangle patch method that generally has effective derivatives at the vertices and cross-boundary derivatives along the edges of the triangulation. This local triangle approach uses cross-boundary derivatives that are computed together with the derivatives at the vertices, whereas most other local triangle methods separate these two steps.

2. Localized Techniques

The basic technique used in Franke [17] for localizing a global interpolation method uses the following process. Define *local domain regions* S_j, $j = 1, \ldots, K$, such that the union of these regions contains the domain of interest for evaluating the final interpolant. In general, the S_j should be bounded regions in the plane and they should not be mutually disjoint. Define *weight functions* $W_j(x, y)$ such that

$$\sum_{j=1}^{K} W_j(x, y) = 1$$

for all (x, y) in the domain and $W_j(x, y) = 0$ for (x, y) outside of S_j. Franke[17] also assumes that $W_j(x, y) \geq 0$ and that $S_j = \{(x, y) : W_j(x, y) > 0\}$, but neither of these two assumptions are required. Define *local interpolants* $Q_j(x, y)$ that interpolate to only the data points in S_j. More precisely, let $I_j = \{i : (x_i, y_i) \in S_j\}$, and let $Q_j(x, y)$ be an interpolant that satisfies $Q_j(x_i, y_i) = f_i$ for all $i \in I_j$. Finally, define the *localized interpolant* to be

$$F(x, y) = \sum_{j=1}^{K} W_j(x, y) Q_j(x, y).$$

Since the weight functions sum to one, we have that $F(x_i, y_i) = f_i$ for $i = 1, \ldots, N$. If the $W_j \in C^a$ and $Q_j \in C^b$, then $F \in C^{\min(a,b)}$. The interpolant is local in the sense that $F(x, y)$ only depends data points (x_i, y_i, f_i) for $i \in J_{(x,y)}$, where $J_{(x,y)} = \{j : (x, y) \in S_j\}$.

Since the weight functions have compact support, the localized interpolant can be evaluated as

$$F(x, y) = \sum_{j \in J_{(x,y)}} W_j(x, y) Q_j(x, y).$$

In Franke [17], the local regions are chosen to be rectangles formed by overlaying the domain with a rectangular grid and the weight functions are standard C^1 piecewise bicubic Hermite basis functions. The local interpolants are thin-plate splines and the resulting localized interpolant is very effective on data that is somewhat uniformly distributed in the domain. However, if the data is dense in some regions and sparse in other regions, it is possible to have no data points in some regions S_j, while other local regions may have a large number of data points. To remedy this situation, which is caused by the constrained tensor product grid structure, we use regions defined as the union of triangles.

3. Global Interpolation Methods to be Localized

The local interpolants $Q_j(x, y)$ used in this paper are the multiquadric (MQ) and thin plate spline (TPS) scattered data interpolation methods, which are radial basis functions of the form

$$F(x, y) = \sum_{i=1}^{N} a_i C_i(x, y) + \sum_{k=1}^{M} c_k D_k(x, y), \tag{1}$$

where $D_1(x, y), D_2(x, y), \ldots, D_M(x, y)$ from a basis for the space of polynomials of degree $< m$. The multiquadric method is a C^∞ function with basis functions $C_i(x, y) = \sqrt{d_i^2(x, y) + R^2}$, where $R^2 > 0$, and $d_i^2(x, y) = (x - x_i)^2 + (y - y_i)^2$. For the results in this paper, a constant precision MQ is used with $M = 1$ and $D_1(x, y) = 1$. A discussion of the effects of the parameter R^2 are given in [5] and for the localized interpolant, a different value for the parameter R^2 can be used for each region S_j. The TPS method is a C^1 interpolant where $C_i(x, y) = d_i^2(x, y) \log(d_i(x, y))$ and it has linear precision with $M = 3$ in (1) where the three polynomial basis functions D_k are $1, x$ and y. The thin-plate spline is so named because it is the function in an appropriate Sobelev space that minimizes the strain energy in a clamped elastic plate given by

$$\int \int_{R^2} \left(\frac{\partial^2 F}{\partial x^2} \right)^2 + 2 \left(\frac{\partial^2 F}{\partial x \partial y} \right)^2 + \left(\frac{\partial^2 F}{\partial y^2} \right)^2 dx\, dy,$$

subject to the interpolation conditions.

The coefficients $a_1, \ldots, a_N$ and $c_1, \ldots, c_M$ are computed from the system of $N + M$ linear equations

$$F(x_j, y_j) = \sum_{i=1}^{N} a_i C_i(x_j, y_j) + \sum_{k=1}^{M} c_k D_k(x_j, y_j) = f_j, \quad \text{for } j = 1, 2, \ldots, N, \quad \text{and}$$

$$\sum_{i=1}^{N} a_i D_k(x_i, y_i) = 0, \quad \text{for } k = 1, 2, \ldots, M.$$

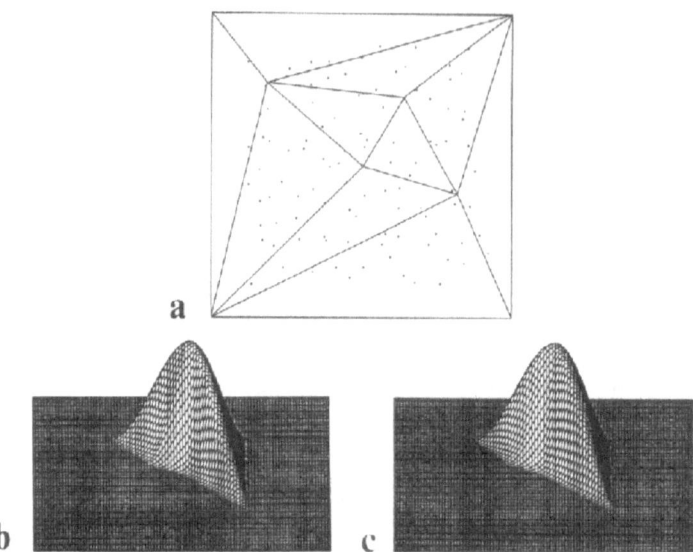

Figure 1. a Local regions determined by triangles, **b** a weight function using linear varying derivatives and **c** a weight function that minimizes (7)

Further discussion of these effective global methods can be found in [5, 13, 16, 18, 23, 31, 32].

4. Local Weight Functions

The local domain regions used here depend on an arbitrary triangulation of an arbitrary set of 2D points V_j, for $j = 1, \ldots, K$. The only assumption is that the data sites (x_i, y_i) are in the union of the triangulation. Let T_{ijk} denote the trangle V_i, V_j, V_k, where the vertices are labeled in a counter-clockwise order. The local domain S_i is the union of all triangles that have V_i as a vertex. Figure 1a illustrates a triangulation that is apparently independent of the data sites, while Figs. 1b and 1c are plots of corresponding weight functions $W_i(x, y)$ (plotted on a slightly larger domain). In the following section, we discuss methods for selecting the vertices V_i and the triangulation, including an interesting situation where the vertices are identical to the data sites. For now, we assume they are fixed, but arbitrary.

The weight functions $W_i(x, y)$ are represented using the hybrid Bézier cubic patches in Foley and Opitz [14], although similar weight functions can be generated using the Clough-Tocher element discussed in Farin [10] or the quintic patch in Barnhill and Farin [4]. Instead of 10 control points that are used to define a cubic Bézier patch, the hydrid form uses 12 control points. The hybrid form differs from the standard cubic in that the inner cubic Bézier point is a rational quadratic function of three control points, with the nine boundary control points having the same influence as the standard cubic patch. The hybrid patch can then be evaluated as a standard cubic using the Bernstein or de Casteljau form. With a small perturbation of the de Casteljau algorithm, we can easily compute tangent planes, directional derivatives and surface normals. The hybrid patch also satisfies the convex hull property and it provides a closed form solution for the discrete side-vertex method in a scattered data environment.

For the Bézier cubic triangle patch (see Farin [11]), assume that we are given 10 control points $b_I = b_{ijk}$, where $|I| = i + j + k$ and $0 \le i, j, k \le 3$. If $U = (u_0, u_1, u_2)$ are the barycentric coordinates of a point in the domain triangle, then the cubic Bézier patch can be defined in Bernstein form by

$$B(U) = \sum_{|I|=3} b_I B_I(U), \quad \text{where} \quad B_I(U) = B_{ijk}(U) = \frac{3!}{i!j!k!} u_0^i u_1^j u_2^k.$$

For the hybrid cubic Bézier triangle patch, suppose that we are given 12 control points $b_{111}^0, b_{111}^1, b_{111}^2$ and $b_I = b_{ijk}$, where $|I| = 3$ and $|I| \ne (1, 1, 1)$. The hybrid patch is defined by

$$H(U) = \sum_{|I|=3} b_I B_I(U),$$

where

$$b_{111} = b_{111}(U) = w_0(U) b_{111}^0 + w_1(U) b_{111}^1 + w_2(U) b_{111}^2,$$

and

$$w_i(U) = \frac{u_j u_k}{u_0 u_1 + u_0 u_2 + u_1 u_2}, \quad i \ne j \ne k.$$

On the edge corresponding to $u_i = 0$, the position and first derivative of the hybrid cubic is identical to the standard cubic Bézier patch whose inner control point is $b_{111} = b_{111}^i$. Because of this, constructing a C^1 composite surface is straightforward. The hybrid patch can be considered as a rational Bézier patch of degree 5 over degree 2, but it is faster to evaluate and requires less storage if it is implemented in the hybrid cubic form. There are removable singularities at the vertices which simply take on the function values b_{300}, b_{030} and b_{003}. Related work was done in [20, 24] for triangular patches, while [6] used a similar approach for rectangular Gregory patches.

The local weight function $W_i(x, y)$ is defined as a composite hybrid cubic for (x, y) in the triangle T_{ijk} as $W_{ijk}(x, y)$, and it is zero for (x, y) outside of S_i. For the $W_i(x, y)$ to be C^1 local functions that sum to unity, 10 of the 12 control points of $W_{ijk}(x, y)$ are forced to be:

$$b_{300} = b_{201} = b_{210} = 1 \quad \text{and} \quad b_{111}^0 = b_{030} = b_{003} = b_{012} = b_{021} = b_{102} = b_{120} = 0,$$

where the barycentric coordinates u_0, u_1, u_2 correspond to the vertices V_i, V_j and V_k, respectively. The remaining two inner control points for $W_{ijk}(x, y)$ will be denoted by $p_{ijk} = b_{111}^1$ and $q_{ijk} = b_{111}^2$. The values of these two points are chosen in conjunction with other inner control points to satisfy C^1 conditions across boundaries and so that the weight functions sum to unity. Figure 2 illustrates this for the weight function $W_i(x, y)$ on the two triangles T_{ijk} and T_{ilj}. The solid black squares have value one, the open squares have values zero, and the solid triangles have values q_{ilj}, p_{ilj},

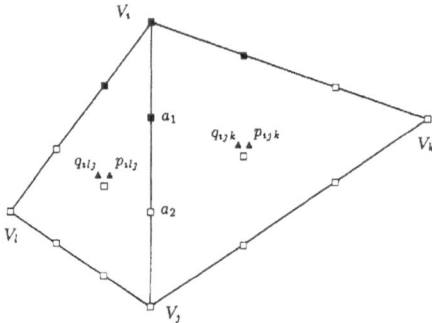

Figure 2. Weight function $W_i(x, y)$ on triangles T_{ijk} and T_{ilj}.

q_{ijk} and p_{ijk} from left to right in Fig. 2. Note that the weight function $W_i(x, y)$ equals one at V_i and that it equals zero at all other vertices V_j.

For C^1 continuity of $W_i(x, y)$ across the edge $V_i V_j$, assume that the neighboring triangles are T_{ijk} and T_{ilj}, as illustrated in Fig. 2. If we represent V_l using the barycentric coordinates of T_{ijk}, then

$$V_l = u_0 V_i + u_1 V_j + u_2 V_k. \tag{2}$$

For $W_{ijk}(x, y)$ and $W_{ilj}(x, y)$ to have a C^1 join on the edge $V_i V_j$, from Farin [10] we must have

$$p_{ilj} = u_0 a_1 + u_1 a_2 + u_2 q_{ijk}, \tag{3}$$

where a_1 and a_2 are the values of the two interior Bézier control points on the edge $V_i V_j$. For $W_i(x, y)$, we have that $a_1 = 1$ and $a_2 = 0$, thus

$$p_{ilj} = u_0 + u_2 q_{ijk}. \tag{4}$$

The weight function $W_j(x, y)$ (that is one at V_j) is defined using the same notation, and on the two triangles in Fig. 2, W_j is defined as the composite patch of $W_{jki}(x, y)$ and $W_{jil}(x, y)$. For $W_j(x, y)$ to be C^1 across the edge $V_i V_j$, we apply (3) with $a_1 = 0$ and $a_2 = 1$ which yields

$$q_{jil} = u_1 + u_2 p_{jki}, \tag{5}$$

where u_1 and u_2 also satisfy (2).

For a point (x, y) in the triangle T_{ijk}, only three weight functions are non-zero and

$$\sum_{m=1}^{K} W_m(x, y) = W_i(x, y) + W_j(x, y) + W_k(x, y) = W_{ijk}(x, y) + W_{jki}(x, y) + W_{kij}(x, y).$$

This sum is also a hybrid Bézier patch with the 9 boundary control points equal to one and the three inner control points are

$$b^0_{111} = q_{jki} + p_{kij}, \quad b^1_{111} = p_{ijk} + q_{kij}, \quad b^2_{111} = q_{ijk} + p_{jki}.$$

These must equal one on each triangle. For programming ease and for determining extra degrees of freedom, we will consider the inner control points that relate to each edge $V_i V_j$. For C^1 continuity across edge $V_i V_j$, we have (4) and (5), while the conditions for the weight functions to sum to unity are

$$q_{ijk} + p_{jki} = 1 \quad \text{and} \quad p_{ilj} + q_{jil} = 1. \tag{6}$$

These 4 equations and 4 unknowns include one dependent equation, thus there is still one degree of freedom available.

We considered several aproaches for specifying this extra degree of freedom. The simplest approach is to have the cross boundary derivatives vary linearly on the edge. This is equivalent to forcing the directional derivatives of $W_i(x, y)$ to be zero in the direction perpendicular to the edge becuse the weight function has zero derivatives at the vertices V_i and V_j. It is straightforward to show that

$$q_{ijk} = 1 - \frac{\langle V_j - V_i, V_k - V_i \rangle}{\| V_j - V_i \|^2}, \quad p_{ilj} = u_0 + u_2 q_{ijk}, \quad q_{jil} = 1 - p_{ilj}, \quad p_{jki} = 1 - q_{ijk},$$

where u_0, u_1 and u_2 satisfy (2), and $\langle *, * \rangle$ denotes scalar or inner product. An example of this is given in Fig. 1b (plotted on a larger domain) using the triangulation in Fig. 1a. As observed in Farin [1] and Foley and Opitz [14], smoother surfaces are possible using methods other than linearly varying cross boundary derivatives. We also used the same approach as Farin [10] in that we minimized the jump in the C^2 discontinuity across the edges, subject to satisfying (4), (5) and (6). This approach was effective on some triangulations, but it yielded unacceptable results on some of the triangulations tested in Zeckzer [36].

The method that yielded the most stable and visually pleasing results minimized

$$(q_{ijk} - 0.5)^2 + (p_{jki} - 0.5)^2 + (p_{ilj} - 0.5)^2 + (q_{jil} - 0.5)^2, \tag{7}$$

subject to satisfying (4), (5) and (6). This method was motivated by the fact that the control points on the edge $V_i V_j$ for W_i are $1, 1, 0, 0$, while they are $0, 0, 1, 1$ for W_j. By trying to get these four control points close to 0.5, we are choosing a tangent plane that is approximately balanced. Figure 1c illustrates the weight function using this approach and it is observed to be visually smoother than the linearly varying method in Fig. 1b. Minimizing (7), subject to satisfying (4), (5) and (6), has the following simple closed form solution:

$$p_{ilj} = \frac{u_2^2 + u_0 - u_1 + 1}{2u_2^2 + 2}, \quad q_{jil} = 1 - p_{ilj}, \quad q_{ijk} = \frac{p_{ilj} - u_0}{u_2}, \quad p_{jki} = 1 - q_{ijk},$$

where u_0, u_1 and u_2 satisfy (2). If the triangles are all equilateral, then all of the above methods yield the same results with the non-zero inner control points equal to 0.5.

In summary, for each interior edge $V_i V_j$, we compute the four values above. If $V_i V_j$ is a boundary edge in the triangulation, we set $q_{ijk} = p_{jki} = 0.5$. This is done in the initial stage of the interpolation method and this approach is used for all of the examples that follow. The initial stage also computes the necessary coefficients for each local interpolant $Q_i(x, y)$ that interpolates to the data sites in the region S_i. These linear systems of equations should be solved once for each local interpolant before any evaluations are made in the final stage. For programming efficiency, a list is computed for each i that contains the indices of all the data sites that are in the region S_i. The final stage involves evaluating the localized interpolant F at an arbitrary point (x, y) in the domain. To evaluate the localized interpolant $F(x, y)$, we first find the triangle T_{ijk} that contains (x, y) and compute the barycentric coordinates (u_0, u_1, u_2) of (x, y) with respect to V_i, V_j and V_k. The localized interpolant can then be computed as

$$F(x, y) = W_i(x, y)Q_i(x, y) + W_j(x, y)Q_j(x, y) + W_k(x, y)Q_k(x, y).$$

This is equivalent to first evaluating the three local interpolants at (x, y) and then computing $F(x, y)$ as the cubic Bézier triangle patch at (u_0, u_1, u_2), where

$$b_{300} = b_{210} = b_{201} = Q_i(x, y),$$

$$b_{030} = b_{120} = b_{021} = Q_j(x, y),$$

$$b_{003} = b_{102} = b_{012} = Q_k(x, y),$$

$$b_{111} = \frac{u_1 u_2 b_{111}^0 + u_0 u_2 b_{111}^1 + u_0 u_1 b_{111}^2}{u_0 u_1 + u_1 u_2 + u_0 u_2},$$

$$b_{111}^0 = q_{jki}Q_j(x, y) + p_{kij}Q_k(x, y),$$

$$b_{111}^1 = p_{ijk}Q_i(x, y) + q_{kij}Q_k(x, y),$$

$$b_{111}^2 = q_{ijk}Q_i(x, y) + p_{jki}Q_j(x, y).$$

The weight function $W_i(x, y)$ may be negative for some values of (x, y) depending on the method used to compute the inner control points and the structure of the triangulation. It can be shown that the weight function is non-negative if and only if the non-zero inner control points satisfy $p_{ijk} \geq -0.5$ and $q_{ijk} \geq -0.5$. Except for some extreme triangulations that we tested, the weight functions were non-negative when (7) was minimized or when linearly varying cross boundary derivatives were used. The approach that minimized the jump in the C^2 discontinuity was more prone to yield weight functions that were negative.

5. Local Regions and Examples

We have investigated several strategies for selecting the arbitrary triangulation of the arbitrary set of points V_i and each approach has yielded effective results on the test functions in Franke[16]. The local domain S_i is the union of all triangles that have V_i as a vertex. A simple approach is to create a triangle mesh consisting of equilateral triangles, such as the one shown in Fig. 3a. Figure 3b is the localized interpolant that uses MQ as the local interpolant on each region applied to the $N = 100$ point data in Franke [16] using his primary test function. This triangulation approach is effective if the data sites are uniformly distributed, but it faces the same limitations as the grid structure used in Franke [17]. Several examples of this approach are given in Zeckzer [36]. Another approach that we experimented with involves fixing the number of vertices V_i to K, and then applying the "cluster" algorithm in Schreiber [33] that computes V_i so that these points best approximate

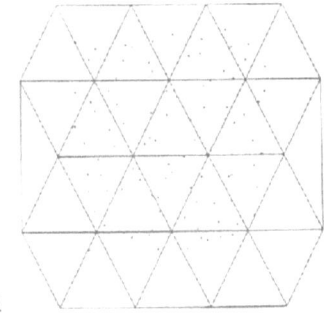

a

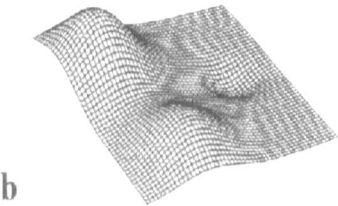

b

Figure 3. a Local regions from an equalateral triangle mesh and **b** the localized MQ interpolant to $N = 100$ points

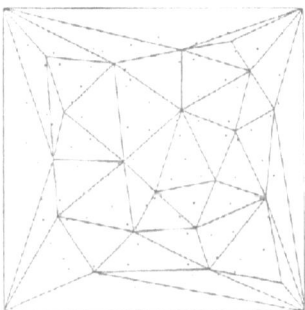

Figure 4. Local triangle regions based on a cluster algorithm

the N data sites (x_j, y_j) relative to a weighted distance measure. This approach needs to be modified because some data sites will fall outside of the convex hull of the V_i. A simple modification is shown in Fig. 4 where the 4 corner points are added to the set of V_i and the vertices are connected using the Delaunay triangulation. We have also experimented with adding additional points to define the boundary of the convex hull which improves the shape of the triangles near the boundary. However, this may not be necesary because the localized interpolant appears to be well behaved in the boundary regions in Fig. 4. A somewhat costly method currently under investigation involves finding vertices so that the Delaunay triangulation of these points has approximately the same number of points in each triangle. An

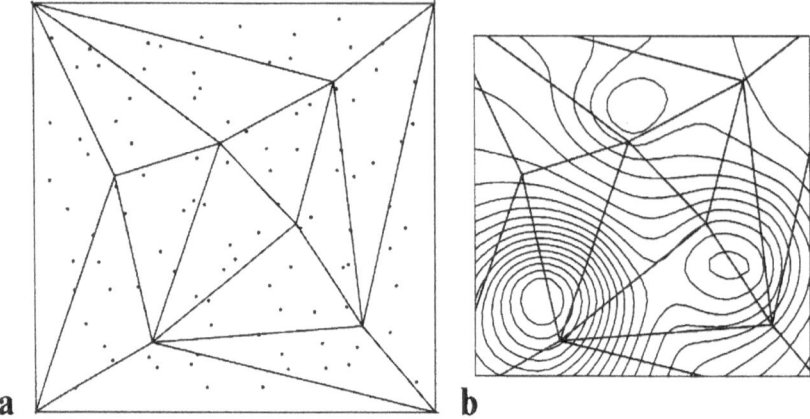

Figure 5. a Local regions with approximately the same number of points per triangle and **b** the localized
TPS interpolant

example of this is shown in Fig. 5a, while Fig. 5b displays contour plots of the
localized interpolant, restricted to the unit squre applied to the $N = 100$ point set in
Franke[16] on his primary test function. This example uses thin plate splines as the
local interpolants. One motivating factor for this approach is that if the localized
interpolant is implemented on a massively parallel computer, the initial stage that
computes all of the coefficients for the local interpolants will have a balanced load
for each processor.

An interesting method results when we choose the local region vertices V_i to be equal
to the given data sites (x_i, y_i), for $i = 1, \ldots, N$, and then triangulate these points using
any acceptable method. We will refer to this method as a *local triangle* method (as
opposed to a *localized* method) because it is just as "local" as other standard C^1
triangle based interpolants. For interpolation of bivariate scattered data using
a triangle based method, the following steps are generally involved. The points
$V_i = (x_i, y_i)$ in the plane are first triangulated using some local or global optimal
criteria, such as those discussed in [8, 9, 27, 35]. The next step involves estimating
partial derivatives of the underlying function at the vertices V_i based upon some or
all of the data (x_i, y_i, f_i). Several of these techniques are given in [16, 27, 29, 30].
Finally, the interpolating function is defined over each triangle using the position
and derivative information at the vertices of the triangle. Some methods for this step
are given in the previous references and in [1, 2, 4, 10, 14, 28]. To estimate the partial
derivatives at V_i, these methods at least use the data at all vertices V_j, where $V_i V_j$ is an
edge in the triangulation. This is precisely the same information used in our local
triangle method that is used to compute the local interpolant $Q_i(x, y)$. The new local
triangle method is then defined over the triangle T_{ijk} as the rational polynomial
combination of the three local interpolants $Q_i(x, y)$, $Q_j(x, y)$ and $Q_k(x, y)$. Observe
that this local triangle approach uses crossboundary derivatives that are computed
together with the derivatives at the vertices, whereas most other local triangle
methods separate these two steps.

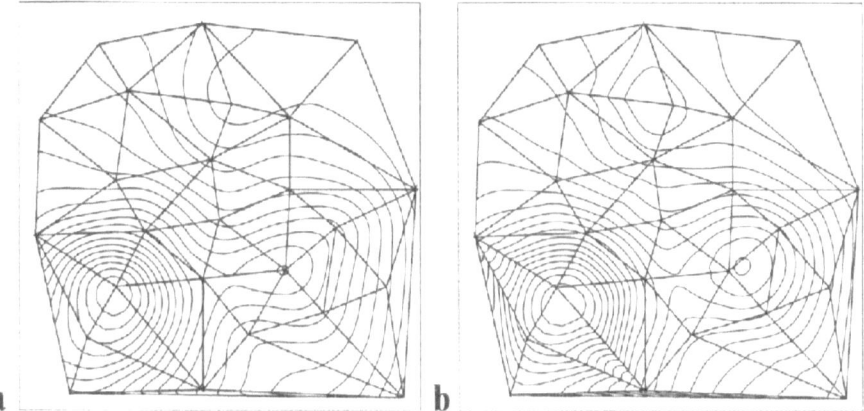

Figure 6. Contour plots of **a** the local triangle interpolant to 25 points and **b** the cubic side-vertex interpolant

Figure 6a is a contour plot of the local triangle method using local thin plate spline interpolants applied to the $N = 25$ point set in Franke [15, 16]. Figure 6b is the cubic blended side-vertex interpolant (see Nielson [28]) to the same data that uses exact partial derivatives at the vertices and linear varying cross boundary derivatives across the edges. Figures 7a and b are surface renderings of the same interpolants shown in Fig 6. The difference in the visual smoothness is evident in many locations in Figs. 6 and 7, but the difference is more obvious in the isophote plots in Fig. 8a and 8b. Isophotes (see [21, 30]) are points on the surface that have equal diffuse light intensity and they can be computed by contouring the function whose values are the angle between the surface normal and the light direction. The isophote surface

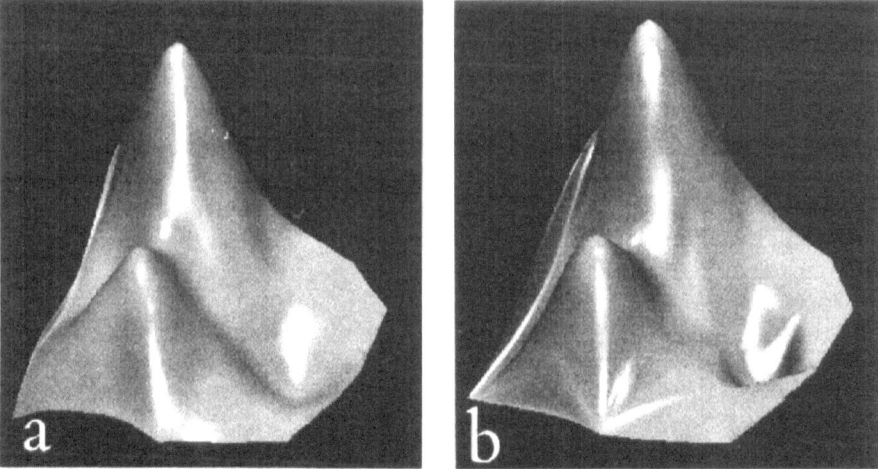

Figure 7. Surface rendering of **a** the local triangle TPS interpolant to 25 points and **b** the cubic side-vertex interpolant

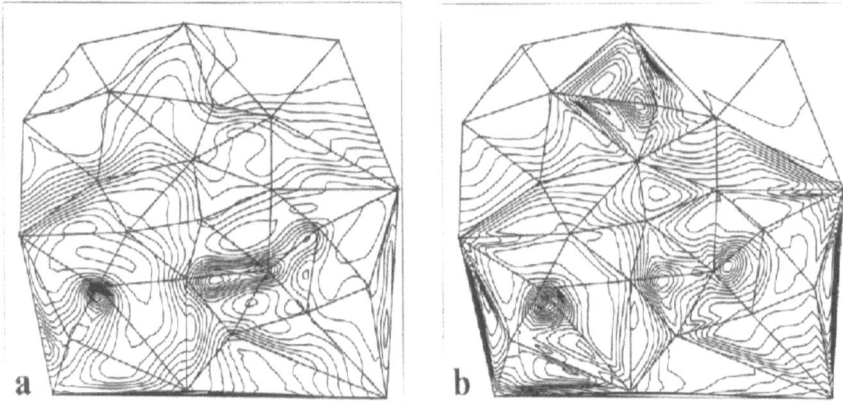

Figure 8. Isophotes of **a** the local triangle TPS interpolant to 25 points and **b** the cubic sidevertex interpolant

interrogation technique is similar to the reflection lines used in Farin [10], except the strip lights can be considered to be concentric circles. If the surface is G^k, then the isophotes are G^{k-1} curves. The visual smoothness of the local triangle method presented here compares favorably with the smoothness of the effective C^1 triangle methods in [10, 14]. Additional examples are given in Zeckzer [36], including a data set with $N = 1995$ points generated using the primary test function in Franke [16]. This example is not shown here because it is visually indistinguishable from the exact test function.

6. Concluding Remarks

The problem of selecting the local regions is currently under investigation, although most of the approaches presented here for region selection yield effective results. When the triangular regions form a triangulation of the scattered data locations, the resulting interpolant can be considered as a new local triangle patch method that generally has effective derivatives at the vertices and cross-boundary derivatives along the edges of the triangulation. This local triangle approach uses cross-boundary derivatives that are computed together with the derivatives at the vertices, whereas most other local triangle methods separate these two steps. Determining the optimal number of data sites for each region is dependent on what the objectives are. In the example where there are $N = 100$ points off of a smooth function, such as those tested Franke[15, 16], then the global MQ and TPS methods yield more accurate results than the localized approaches. However, the global methods require more time and storage, plus any local change in a single function value will change the entire surface. Another advantage of the localized approach is that they have the precision of the local interpolants in local regions. For example, if each local interpolant has linear precision and the data is from a linear function on the three regions S_i, S_j and S_k, then the localized interpolant will be linear in the triangle T_{ijk}

because the weight functions sum to unity. Initial testing indicates that having 10 to 25 points in each region will yield fairly accurate results without being costly. The localized interpolant is not very sensitive to the number of points per region provided that there are at least three points per region. One approach to dealing with the situation where there are no data points in a region (or in several regions) is given in Zeckzer[36]. A simple approach is to simply use data from neighboring regions to define the local "interpolant", although any approximation method can be used because there are no data sites in the region.

As noted earlier, we could have used the Clough-Tocher patch or the quintic patch in Barnhill and Farin [4] for the local weight functions. Other methods that could also be easily used are triangular B-splines or the Sibson type basis functions in Farin [12]. Extending this to trivariate methods is also straightforward.

Acknowledgements

This work was supported in part by the Deutsche Forschungsgemeinschaft (DFG) contract Ka477/12-1 awarded to the University of Kaiserslautern and the NSF and AFOSR grant DMS-9116930 at Arizona State University. The second author was supported by the ONR grant N00014-91-J-1456 awarded to W.R. Uttal at Arizona State University.

References

[1] Barnhill, R. E.: Representation and approximation of surfaces. In: Mathematical software III (Rice, J. R., ed.), pp. 69–120. New York: Academic Press.

[2] Barnhill, R. E.: Surfaces in computer aided geometric design: A survey with new results. Comput. Aided Geom. Des. 2, 1–17 (1985).

[3] Barnhill, R. E., Birkhoff, G., Gordon, W. J.: Smooth interpolation in triangles. J. Approx. Theor. 8, 114–128 (1973).

[4] Barnhill, R. E., Farin, G.: C^1 quintic interpolation over triangles: two explicit representations. Int. J. Num. Methods Eng. 17, 1763–1778 (1981).

[5] Carlson, R. E., Foley, T. A.: The paramter R^2 in multiquadric interpolation. Comput. Math. Appl. 21, 29–42 (1991).

[6] Chiyokura, H., Kimura, F.: Design of solids with free-form surfaces. Comput. Graphics 17, 289–298 (1983).

[7] Duchon J.: Splines minimizing rotation invariant semi-norms in Sobelev spaces In: Multivariate approximation theory (Schempp, W. Zeller, K., eds.), pp. 85–100. Basel: Birkhauser 1975.

[8] Dyn, N., Levin, D., Rippa, S.: Algorithms for the construction of data dependent triangulations. In: Algorithms for approximation II (Mason, J. C., Cox, M. G., eds.), pp. 185–192. London: Chapman and Hall 1990.

[9] N. Dyn, Levin D., Rippa, S.: Data dependent triangulations for piecewise linear interpolation. IMA J. Numer. Anal. 10, 137–154 (1990).

[10] Farin, G.: A modified Clough-Tocher interpolant. Comput. Aided Geom. Des. 2, 19–27 (1985).

[11] Farin, G.: Curves and surfaces for computer aided geometric design. San Diego: Academic Press 1990.

[12] Farin, G.: Surfaces over Dirichlet tessellations. Comput. Aided Geom. Des. 7, 281–292 (1990).

[13] Foley, T. A.: Interpolation and approximation of 3-D and 4-D scattered data. Comput. Math. Appl. 13, 711–740 (1987).

[14] Foley, T. A., Opitz, K.: Hybrid cubic Bézier triangle patches. In: Mathematical methods in computer aided geometric design II (Lyche, T. Schumaker, L. L., eds.), pp. 275–286. New York: Academic Press 1992.

[15] Franke, R.: A critical comparison of some methods for interpolation of scattered data. Naval Postgraduate School, Technical Report NPS-53-79-003, 1979.

[16] Franke, R.: Scattered data interpolation: tests of some methods. Math. Comp. 38, 181–200 (1982).

[17] Franke, R.: Smooth interpolation of scattered data by local thin plate splines Comput. Math. Appl. 8, 273–281 (1982b).

[18] Franke, R.: Recent advances in the approximation of surfaces from scattered data. In: Topics in multivariate approximation (Schumaker, L. L., Chui, C. C., Utreras, F., eds.), pp. 175–184. New York: Academic Press.

[19] Franke, R., Nielson, G. M.: Scattered data interpolation: a tutorial and survey. In: Geometric modeling: methods and their applications (Hagen, H., Roller, D., eds.), pp. 131–160. Heidelberg: Springer.

[20] Goodman, T.N.T., Said, H.: A C^1 triangular interpolant suitable for scattered data interpolation.

[21] Hagen, H., Schreiber, T., Gschwind, E.: Methods for surface interrogation. In: Visualization '90 (Kaufman, A., ed.), pp. 187–193. Los Alamitos: IEEE Press 1990.

[22] Hardy, R. L., Multiquadric equations of topography and other irregular surfaces, J. Geophys. Res. 76, 1905–1915 (1971).

[23] Hardy, R. L.: Theory and applications of the multiquadric-biharmonic method. Comput. Math. Appl. 19, 163–208 (1990).

[24] Herron, G. F.: A characterization of certain C^1 discrete triangular interpolants. SIAM J. Numer. Anal. 22, 811–819 (1985).

[25] Hoschek, J., Lasser, D.: Grundlagen der geometrischen Datenverarbeitung. Leipzig: Teubner 1989.

[26] Landcaster, P., Salkauskas, K.: Curve and surface fitting: an introduction. New York: Academic Press 1986.

[27] Lawson, C. L.: Software for C^1 surface interpolation. In: Mathematical software III (Rice, J. R., ed.), pp. 161–194. New York: Academic Press 1977.

[28] Nielson, G. M.: The side-vertex method for interpolation in triangles. J. Approx. Theor. 14, 318–336 (1979).

[29] Nielson, G. M.: A method for interpolation of scattered data based upon a minimum norm network. Math. Comp. 40, 253–271 (1983).

[30] Pottmann, H.: Scattered data interpolation of scattered data based upon generalized minimum norm net works. Constr. Approx. 247–256 (1991).

[31] Powell, M. J. D.: Radial basis functions for multivariate interpolation: a review. In: Algorithms for approximation (Mason, J. C., Cox, M. G., eds.), pp. 143–167 Oxford: Oxford University Press 1987.

[32] Powell, M. J. D.: The theory of radial basis function approximation in 1990. In: Advances in numerical analysis II: wavelets, subdivision algorithms and radial functions (Light, W., ed.), pp. 105–210. Oxford: Oxford University Press 1991.

[33] Schreiber, T.: A Voronoi diagram based adaptive K-means-type clustering algorithm for multi-dimensional weighted data. In: Computational geometry—methods, algorithms and applications (Bieri, H., Noltemeier, H., eds.), pp. 265–275. Berlin: 1991.

[34] Schumaker, L. L.: Fitting surfaces to scattered data. In: Approximation theory (Lorentz, G. G., Chui, C. K., Schumaker, L. L., eds.) pp. 203–268. New York: Academic Press 1976.

[35] Schumaker, L. L.: Triangulation methods. In: Topics in multivariate approximation (Chui, C., Schumaker, L. L., Utreras, F., eds.), pp. 219–232. New York: Academic Press 1987.

[36] Zeckzer, D.: Dreiecks-basierte lokale Scattered Data Interpolation unter Verwendung radialer Basismethoden. Diplomarbeit, Fachbereich Informatik, Universität Kaiserslautern, 1992.

Dr. T. A. Foley
Dr. S. Dayanand
Department of Computer Science
and Engineering
Arizona State University
Tempe, AZ 85287-5406
U.S.A.

Dr. D. Zeckzer
Informatik Department
Universitat Kaiserslautern
D-67653 Kaiserslautern
Federal Republic of Germany

Computing Suppl. 10, 177–187 (1995)

© Springer-Verlag 1995

Repeated Knots in Least Squares Multiquadric Functions*

R. Franke, Monterey, **H. Hagen**, Kaiserslautern, and **G. M. Nielson**, Tempe

Abstract. A previous paper by the authors [2] noted that there was a strong tendency to obtain near-repeated knots in their algorithm for least squares approximation of scattered data by multiquadric functions. In this paper we observe that this leads naturally to the inclusion of derivatives of the multiquadric basis function in the approximation, and give an algorithm for accomplishing this. A comparison of the results obtained with this algorithm and the previous one is made. While the multiple knot algorithm usually has the advantage in terms of accuracy and computational stability, there are datasets for which this is reversed.

Key words: Least squares, multiquadrics, radial basis functions, repeated knots, scattered data.

1. Introduction

This paper continues an investigation into the use of multiquadric functions to approximate scattered data. An initial report on our efforts appears in [2], where we reported that a phenomenon that seemed particularly interesting was the tendency of the optimization process to result in near-repeated knots. The implication is that the method was attempting to build a directional derivative of the basis function into the approximation. This paper documents our experiences with an algorithm that detects the occurrence of near-repeated knots, and when they occur, replaces basis functions corresponding to near-repeated knots with a single multiquadric basis function plus appropriate derivatives of the multiquadric basis function, at the pertinent knot.

While it is useful to have the previous paper at hand, for completeness we will briefly review the background necessary to make this report somewhat self-contained. We restrict our discussion to functions of two independent variables, the methods are easily extendible to arbitrary dimensions, and we expect that many of the conclusions will carry over.

The scattered data approximation problem is easily described and occurs frequently in many branches of science. The problem occurs in any discipline where measurements are taken at irregularly spaced values of two or more independent variables, and is especially prevalent in environmental sciences. We will suppose that triples of data, $(x_j, y_j, z_j), j = 1, \dots, N$ are given, assumed to be measurements (perhaps with error) of an underlying function $z = f(x, y)$. The function f is to be approximated

---

* This work was performed in part while the first and last authors were visiting Universität Kaiserslautern and were supported by a DFG-grant Ka 477/13-1.

by a function $F(x, y)$ from the given data. A recent survey of such methods is given in [3].

Multiquadric functions were introduced for interpolation of scattered data by Hardy [6]; also see [7] for a historical survey and many references. The method is one of a class of methods known now as "radial basis function methods" that includes other attractive schemes such as thin plate splines ([1, 5], and others). The basic idea of such methods is quite simple, and we describe it in some generality; for purposes of being definite it is pertinent to note that for the multiquadric method the radial function is $h(d) = \sqrt{(d^2 + r^2)}$. In general, suppose a function of one variable, $h(d)$, where d denotes distance, is given.

For interpolation (that is, exact matching of the given data), a basis function, $B_j(x, y) = h(d_j)$ is associated with each data point. Here $d_j = \sqrt{((x - x_j)^2 + (y - y_j)^2)}$, the distance from (x, y) to (x_j, y_j). Thus each basis function is a translate of the radial function, h. The approximation is a linear combination of the basis functions, along with some polynomial terms that may be necessary in some cases, or may be used to assure that the approximation method has polynomial precision. Thus,

$$F(x, y) = \sum_{j=1}^{N} a_j B_j(x, y) + \sum_{j=1}^{M} b_j q_j(x, y) \tag{1}$$

where $\{q_j\}$ is a set of M polynomials forming a basis for polynomials of degree $< m$. The coefficients a_j and b_j are determined by the linear system of equations prescribing interpolation of the data, and exactness for polynomials of degree $< m$:

$$\sum_{j=1}^{N} a_j B_j(x, y_i) + \sum_{j=1}^{M} b_j q_j(x_i, y_i) = z_i, \quad i = 1, \ldots, N$$

$$\sum_{j=1}^{N} a_j q_i(x_j, y_j) = 0, \quad i = 1, \ldots, M. \tag{2}$$

For multiquadric basis functions, this system of equations is known to have a unique solution for distinct (x_j, y_j) data (see, for example, [8]); while m may be taken as zero (no polynomial terms), Micchelli's results show the multiquadric basis function is positive definite of order one, and thus a constant term should be included. We have done so in all our work. If higher degree polynomial precision is desired, inclusion of those terms imposes no particular burden.

While interpolation theory is important and indicates something about the suitability of the class of functions for approximation purposes, our emphasis here is on least squares approximation. This implies using fewer basis functions than there are data points. In analogy with univariate cubic splines, it is convenient to refer to the points at which the radial basis functions are centered as "knots", as was done in [9], and we do so here. If a set of knot points, (u_k, v_k), $k = 1, \ldots, K$, with $K < N$ have been specified, then the problem of fitting a multiquadric function by least squares is similar to that of solving the system of equations corresponding to those above in the least squares sense. We give the details. Now, let $B_k(x, y)$ denote the radial basis function associated with the point (u_k, v_k), $B_k(x, y) = \sqrt{((x - u_k)^2 + (y - v_k)^2 + r^2)}$.

The system of equations, specialized for our case, is now of the form

$$\sum_{k=1}^{K} a_k B_k(x_i, y_i) + c = z_i, \quad i = 1, \ldots, N$$

$$\sum_{k=1}^{K} a_k = 0. \tag{3}$$

There is a question of how to treat the last equation, which guarantees precision for constants. In [9] the corresponding constraint equations were imposed exactly, rather than approximately, because of physical considerations. While there is not the corresponding physical situation here, we have also imposed the last equation as a constraint. This constraint can be used to reduce the size of the system by solving for a_K in terms of the other a_k and substituting into the first set of equations.

If the knot points are a subset of the data points, then the system of equations (by virtue of containing, as a subset, interpolation equations) has full rank, and thus guarantees a unique solution of the least squares problem. When the knot locations may differ from the data points, the problem of whether the coefficient matrix is of full rank or not is unknown to us, although we feel certain that the matrix is of full rank when the knot points are distinct, and have encountered no situations that indicate otherwise.

The impetus behind our original investigation was to obtain surface approximations that would be efficient in subsequent applications. That is, we consider it to be acceptable to expend considerable computational resources to obtain the approximation in a preprocessing step. Once obtained, the approximation can be evaluated efficiently, so it would be feasible to use it numerous times in an application program.

In the previous paper [2] we found that there was a tendency of the algorithm to converge on solutions where the knot points were often close to each other, or near-repeated. This occurred both in clusters of two and three knots. At that time, no attempt was made to investigate this behavior further. The outstanding result of that reference was the finding that allowing the knots (and r value) to be determined as part of the minimization process yielded approximations that were much more efficient than those obtained either by determining the knots a priori or adaptively (we outlined a greedy algorithm for sequentially determining knot locations that worked well). While we found that using variable r values also yielded good improvement, we do not pursue that idea here.

Because the greedy algorithm we developed previously was generally used as the starting point for the work reported here, we give a brief outline of it.

a) Obtain the least squares fit by a constant function, the average of the data values. The two data points having maximum positive and maximum negative error are taken to be the first two knots, (u_1, v_1) and (u_2, v_2). The knot counter K is set to 2.
b) The least squares multiquadric fit with K knots is obtained, and the residuals are computed.

Table 1. Data sets used extensively in tests

n.m	This refers to point set *n* and function *m* from [4], for $n = 1, 2$, and 3, and $m = 1, \ldots, 6$. $n = 1$ is 25 points, $n = 2$ is 33 points, $n = 3$ is 100 points. $n = 4$ refers to the 200 point data set used in [9]. $m = 1$ is the humps and dip function, $m = 2$ is the cliff, $m = 3$ is the saddle, $m = 4$ is the gentle hill, $m = 5$ is the steep hill, and $m = 6$ is the sphere. In addition, $m = 7$ refers to the curved valley function from [10].
GT	This refers to the thinned glacier data consisting of 678 points, with certain contour lines removed, from [9].
GL	This refers to the thinned glacier data consisting of 873 points.
HF	This is the data set from [4] generated to have point density approximately proportional to curvature, consisting of 500 points.

c) The maximum absolute value of the residuals is found and the location of this residual, subject to the minimum knot separation value, is taken to be the next knot location (u_K, v_K). At this point the algorithm proceeds to step *b* unless the maximum number of knot locations to be computed has been reached.

As previously, we have used a QRP' decomposition of the coefficient matrix to solve the least squares problem for given knots and *r* value. This provides a stable and efficient means for solution of the problem with an indication if a matrix of less than full rank is encountered.

In order to test the algorithms we have used a number of data sets. Several of these are based on previously published and widely available (x, y) data sets and parent functions. We have also used a few less readily available data sets that we are willing to share with anyone interested in obtaining them. Table 1 gives a summary of most of the data sets.

As mentioned in the opening paragraph, the primary purpose of this paper is to discuss some further results we have obtained in allowing the coalescence of knots. A discussion of the algorithm we used is given in Section 2. Results of calculations obtained using several data sets and comparison with the previous method where only simple (but possibly near-repeated) knots are allowed, is given in Section 3. Section 4 summarizes our experiences.

2. Derivatives of Multiquadric Basis Functions and Multiple Knots

Once again we have used Matlab[1] and LEASTSQ (a Levinburg-Marquardt method) to solve the nonlinear minimization problem

$$\min \sum_{i=1}^{N} \left[z_i - \sum_{k=1}^{K} a_k B_k(x_i, y_i) - c \right]^2 \tag{4}$$

where the minimization is taken over all (u_k, v_k), r, the a_k, and c (with the last equation of (3) imposed as a constraint) as an equivalent two step minimization. Thus, for each

---

[1] Math Works, 24 Prime Park Way, Natick. MA 01760.

given knot configuration and r value, the least squares solution of (3) is computed as a step toward (4). This results in the solution of a simpler, but equivalent problem since $2K$ parameters are eliminated from (4) by imposing the values of the a_j and c be always taken as obtained from the least squares solution of (3). Hence, our final process is more properly written as

$$\min \min_{i=1}^{N} \left[z_i - \sum_{k=1}^{N} a_k B_k(x_i, y_i) - c \right]^2 \qquad (5)$$

where the inner minimization is over the a_j and c (least squares solution of (3)), and the outer minimization is over the knot locations and the value of the parameter r. The global minimum of each of the two problems are clearly the same. Equation (5) is the more restrictive, but any minimum of (4) is a local minimum of (5), else a better solution is attainable for (4). This does not imply that the iterative methods employed to solve (5) would work equally well, nor find the same local minima, when applied to (4).

When two knots approach each other, the system (3) becomes ill-conditioned, and the coefficients for the two knots tend to become large and of opposite sign. This reminds one of derivatives, in this case a directional derivative. Because the two coefficients are not of the same magnitude, the appropriate replacement basis functions seem to be one of the basis functions, and a directional derivative of the basis function. In practice, the directional derivative is represented as a linear combination of the partial derivatives with respect to x and y. To prove this is the proper change requires showing that the coefficients grow at no greater a rate than the reciprocal of the distance between the knots. The work of Narcowich and Ward [11] shows that the norm of the inverse of the multiquadric interpolation matrix grows no more rapidly than linear with the reciprocal of the minimum distance between data points. It seems likely a similar result holds for the norm of the pseudoinverse of the system matrix for (3). We have verified this computationally for some cases.

The contribution of the two terms corresponding to the coalescent knots (say the jth and the kth) are of the form $a_j B_j(x, y) + a_k B_k(x, y)$. The six parameters, a_j, a_k, u_j, u_k, and v_k, in those terms are replaced by the five parameters in the terms $a_j B_j(x, y) + b_j(B_j(x, y))_x + c_j(B_j(x, y))_y$. In the previous expression the subscripts x and y refer to partial differentiation. We will refer to such a knot as a double knot.

Now, as it turns out, the coalescence of two knots, while the most prevalent, was by no means the only situation we encountered. The coalescence of three knots (or, once two knots have coalesced, the approach of a third knot to the location of a double knot) then results in the occurrence of a second derivative, a directional derivative of a directional derivative (the directions not necessarily being the same). In practice the replacement of the basis functions for the ith, jth, and kth knots is by the linear combination $a_j B_j + b_j(B_j)_x + c_j(B_j)_y + d_j(B_j)_{xx} + e_j(B_j)_{xy} + f_j(B_j)_{yy}$. While we have instances of near-coalescence of four knots, our present code does not attempt to handle such cases properly.

The algorithm is required to change basis functions when it appears that knots are coalescing. We have used a simple idea that seems to be effective. We iterate the solution of (5) in stages having increasingly stringent convergence tolerances. As each tolerance level is met, a check for near-repeated knots is made, "near" being a changing tolerance consistent with the tolerance for the optimization. When pairs of knots are found within the tolerance, the new basis is generated, and the optimization process continued. When no new repeated knots are obtained, the tolerance is decreased by a factor of two. We have observed that the inclusion of the derivative terms, vice the basis functions at near-repeated knots, generally yields essentially the same approximation, as it should, for both double and triple knots. The sequence of tolerance levels we used is: 0.02, 0.01, 0.005 for problems posed on the $[0, 1]^2$ square.

3. Approximations with Repeated Knots

The results of our investigation concern the ability of the algorithm to obtain good approximations in comparison with the algorithm permitting only simple (but perhaps near-repeated) knots, the degree of dependence on the initial guess for three different datasets (two in some detail), and a comparison of condition numbers of the coefficient matrices for the simple and multiple knot problems.

We first describe the layout of Table 2, which gives a comparison of the results of the simple knot and multiple knot algorithms for fourteen datasets. The results are for the surface fits obtained by starting from the initial guess obtained by the greedy algorithm [2], with $r = 0.3$, and closeness tolerance, $ctol = 0.1$. The 3.n datasets were fit using twelve knot locations, while the 4.n datasets were fit using twenty knot locations. The columns of the table contain, in pairs, the values of r, the rms errors for the data points, the rms errors on a 21×21 grid, and information about the near-repeated or multiple knots obtained. The subscripts s and m denote the simple

Table 2. Comparison of results: near-repeated knots vs. multiple knots

dset	r_s	r_m	rms_s	rms_m	$grms_s$	$grms_m$	$knots_s$	$knots_m$
3.1	0.0976	0.1588	0.0081	0.0059	0.0115	0.0107	2nd, 1nt	1d, 1t, 1nt
3.2	7e-6	0.0002	0.0074	0.0073	0.0186	0.2477	3nd, 1nt	3d, 1t, 1nq
3.3	0.3116	0.2502	0.0014	0.0008	0.0016	0.0011	1nt	1t
3.4	0.2885	0.3118	0.0005	0.0005	0.0006	0.0005	2nd	3d
3.5	0.3720	0.3886	0.0008	0.0007	0.0010	0.0010	3nd	2d
3.6	2.3280	2.3044	0.0005	0.0005	0.0007	0.0007		
3.7	0.0990	0.3304	0.0403	0.0279	0.0733	0.0324	4nd	4d
4.1	0.0958	0.1588	0.0070	0.0017	0.0071	0.0024	1nt	1t
4.2	2e-5	2e-5	0.0035	0.0035	0.0083	0.0082	4nd	3d, 1nd
4.3	0.2577	0.2513	0.0002	0.0002	0.0002	0.0002	4nd	1d, 1nd
4.4	0.4163	0.4625	7e-6	6e-6	7e-6	6e-6	1nd	3d
4.5	0.3234	0.3151	0.0002	0.0003	0.0002	0.0003	4nd	2nd
4.6	0.5668	0.5680	5e-5	4e-5	5e-5	4e-5	5nd	2d, 2nt, 2nd
4.7	0.1111	0.1134	0.0115	0.0114	0.0202	0.0287	3nd, 1nt	5d, 1t

Table 3. Results with different initial guesses for case 3.2

ctol	r	r_0	rms_0	$grms_0$	multiple knots
0.1	0.3	0.0002	0.0073	0.2496	t@ (.09,.14), 3d@(.38,.38), (.60,.60), (.89,.88)
0.25	0.1	0.0239	0.0060	0.0085	3t@ (.31,.36), (.76,.79), (.77,.78), 1d@ (.12,.13)
0.1	0.1	0.0247	0.0071	0.0111	2d@ (.38,.40), (.67,.66)
0.2	0.1	0.0247	0.0071	0.0113	3d@ (.14,.14), (.44,.44), (.68,.67)
0.25	0.01	0.0739	0.0054	0.0095	4d@ (.13,.11), (.37,.39), (.69,.65), (.92,.91)
0.1	0.01	5e-5	0.0071	0.0259	d@ (.69,.69), 2nd@ (.14,.12), (.33,.40)
0.2	0.01	0.0001	0.0071	0.0114	d@ (.44,.44)
0.0	0.01	5e-5	0.0071	0.0260	d@ (.69,.69), 2nd@ (.14,.11), (.33,.40)

knot and multiple knot algorithms, respectively. The terms 2d and 1t, for example, refer to 2 double knots, and 1 triple knot. Likewise, 1nd refers to 1 near-double knot. In some cases two double knots may have been obtained, with another simple knot nearby one of the double knots – such a case might be denoted by 2d, 1nt.

For a given dataset, the values of r obtained are mostly comparable, with larger values generally being obtained by the multiple knot algorithm. In the instances where the value of r is not larger, the values are essentially the same, the exception being dataset 3.3. The cases showing the most difficulty are the cases 3.2 and 4.2 (the cliff function), which has a tendency toward small values of the parameter r. The effects of this tendency are discussed in a later paragraph.

The rms errors are mostly comparable, with the errors obtained by the multiple knot algorithm generally smaller, with some ties, and one instance where the errors are bigger. This occurs for dataset 4.5, but the errors for both approximations are quite small for this case.

The rms errors on the grid ($grms$) generally follow the same pattern, with one major exception. On the dataset 3.2, the errors for the grid are quite large. Closer inspection of this particular case reveals there is one grid point (near the triple, almost quadruple, knot) where the surface is badly ill-behaved. Coefficients of the first derivative terms are on the order of 5 to 20, while the coefficients of the two multiquadric basis functions are reasonably large, on the order of 10^3. However, the value of r is very small in this case, about 1.5e-4, thus the derivatives of the basis functions very rapidly approach the value one in one of the coordinate directions (the value is about 0.99 at a distance of 0.001 from the knot). Hence, very rapid changes in the value of the function occur. While otherwise similar situations occur in other cases, the key difference here is the small value of r that occurs.

In order to discover whether the initial guess was contributing to the problem, or whether the tendency toward a small r value and multiple knots was driven by the data, we ran a number initial guesses for the dataset 3.2. The initial conditions were obtained from the greedy algorithm with various values of r and closeness tolerance (ctol). The results of the computations are summarized in Table 3. The table gives the parameter ctol, the initial value of r, the final value of r, the rms errors at the data points and on the grid, and the number, multiplicity, and approximate location of the multiple knots. From this we see that while there is a tendency toward multiple

Table 4. Results with different initial guesses for case 3.1

ctol	r	sk	r_0	rms_0	$grms_0$	multiple knots
0.0	0.3	g	0.1593	0.0067	0.0113	t @ (.43, .76), 2nd @ (.86, .23), (.83, .42)
0.1	0.3	g	0.1588	0.0059	0.0107	t @ (.43, .75), d @ (.78, .43), d is nt
0.2	0.3	g	0.1712	0.0063	0.0107	t @ (.44, .74), d @ (.15, .26)
0.25	0.3	g	0.2013	0.0075	0.0107	t @ (.43, .79), d @ (.20, .33)
0.1	0.1	g	0.1450	0.0066	0.0111	t @ (.43, .77)
0.1	0.01	g	0.0737	0.0118	0.0157	d @ (.72, .36)
–	0.3	r	0.1454	0.0066	0.0112	t @ (.43, .77)
–	0.3	r	0.1566	0.0061	0.0077	t @ (.42, .75)
–	0.3	r	0.1600	0.0067	0.0113	t @ (.43, .76), 2nd @ (.83, .43), (.86, .23)
–	0.3	r	0.1527	0.0067	0.0111	nt @ (.44, .75)

knots, no particular locations are tended toward, although it seems natural that repeated knots invariably occur near the diagonal. The fact that the surface is a function of one variable (that is, $y - x$) with rapid changes occurring near the diagonal may account for this circumstance. The fact that the data is not regular may account for a large number of comparable local minima. One case is of particular note: The case in the second line involves three triple and one double knot, however two of the triple knots are close together, yielding a near-repeated knot of multiplicity six. This could be expected to lead to poor behavior of the surface. Closer inspection revealed that, even though the coefficients are rather large (order 10^7), the surface is well behaved and yielded one of the better approximations among this group. This good behavior is undoubtedly enhanced by the fact that the r value is large for this surface. In the three cases where the r value is smaller than 10^{-4} the resulting surface is poorly behaved near a multiple or near-multiple knot, this being reflected by the larger $grms_0$ values shown.

We ran a number of different initial guesses for case 3.1 to determine both the robustness of the optimization routine and to confirm the tendency of the method to converge on multiple knots from a variety of initial guesses. Nine different guesses were made, six resulting from the greedy algorithm with different closeness tolerances and r values, and three with random initial knot locations. As is seen in Table 4, the results for the various initial guesses are strikingly similar: except for two cases the r values are all close to 0.16, all but one result in a triple or near-triple knot near the dip at about $(0.45, 0.78)$, and all but one have comparable accuracies for rms_0 and $grms_0$. The one odd case had the starting value of 0.01 for r. A wide-ranging difference is that different double and near-double knots may occur for the various runs. Unlike dataset 3.2, this case appears to be quite stable.

We also ran some different initial conditions on dataset 3.7. The results in this case seem to indicate complications somewhere between the datasets 3.1 and 3.2. Dataset 3.7 has a tendency toward many multiple knots, perhaps because it has a more complex behavior of the surface than any of the other datasets.

The occurrence of near-multiple knots, as we noted previously, must have a unfavorable effect on the condition number of the coefficient matrix, and in order to

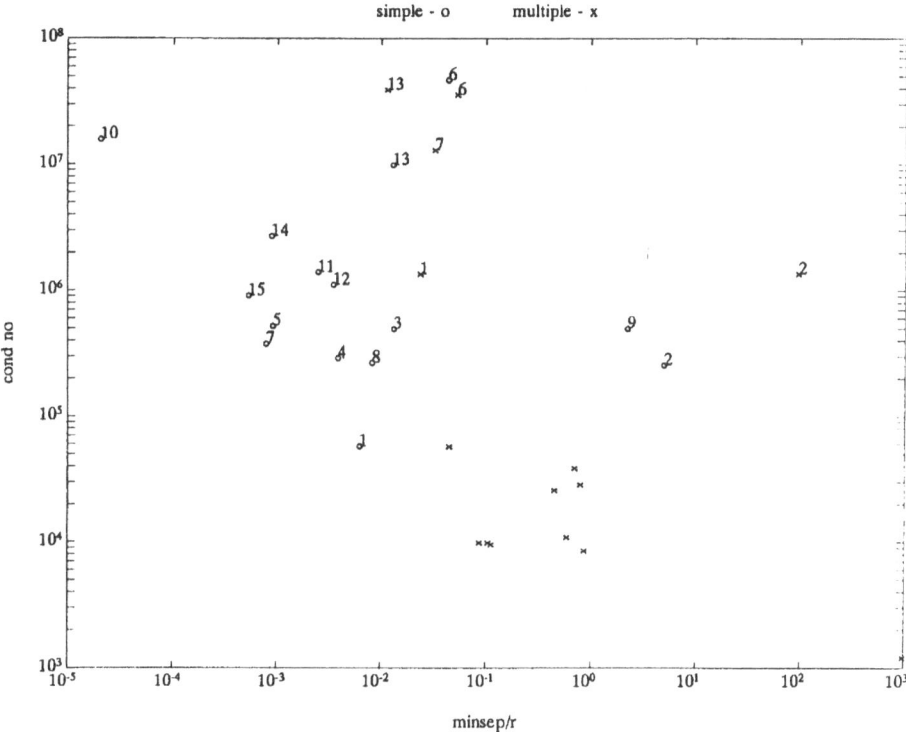

Figure 1. Minimum separation distance relative to r value vs. condition number for thirty problems. The o's denote simple knot cases, numbered from 1–15, the first 14 being as ordered in Table 2. Number 15 is the thinned glacier data, GL, with 25 knots. The x's denote the multiple knot cases – only those with condition number greater than 10^5 are labeled due to space considerations

investigate this we computed the condition number of the coefficient matrices for the fourteen cases of Table 2. A plot of the minimum separation distance divided by the r value vs. the condition number of the coefficient matrix is shown in Fig. 1. There are 15 points for each of the simple knot and multiple knot algorithms. The first fourteen points correspond to the data in Table 2, in the order given, while the fifteenth is for the thinned glacier dataset, GL, using 25 knots. For the latter data the underlying function is unknown. The simple knot algorithm found 9 near-double knots. The multiple knot algorithm found four double knots and three triple knots.

The points in Fig. 1 for the simple knot algorithm have been labeled with their number. Because of the clustering of many of the points for the multiple knot algorithm, only those with condition number great than 10^5 have been labeled.

What can be observed is that the condition numbers for the multiple knot algorithm are generally one to two orders of magnitude smaller than those for the simple knot algorithm. There are exceptions to this, most notably the datasets 3.1, 3.2, 3.7, and 4.5, where the behavior is reversed, and 3.6 for which the r value is relatively large in both cases, contributing to the size of the condition number.

parent surface

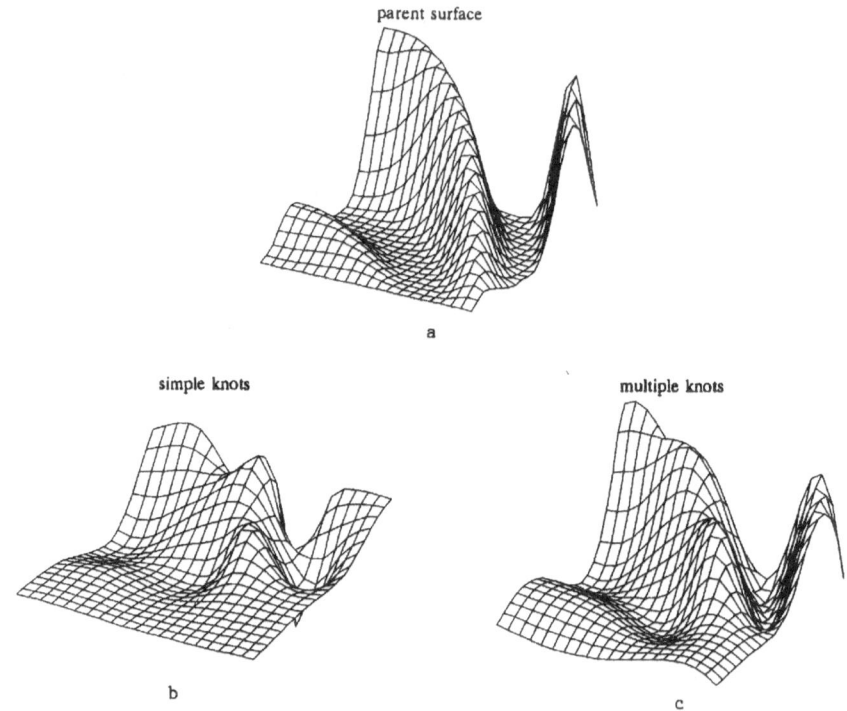

a

simple knots   multiple knots

b   c

Figure 2. a The parent surface from which the data was obtained. **b** The surface obtained with 12 knots using the simple knot algorithm. **c** The surface obtained with 12 knots using the multiple knot algorithm

Finally, for one of the parent surfaces, we include a plot of the underlying surface being approximated, along with the surfaces obtained from the simple knot algorithm and the multiple knot algorithm. The dataset is 3.7, the surface shown in Fig. 2a being sampled at 100 points. Figures 2b and 2c, respectively, show the surface reconstructed using the simple knot and multiple knot algorithms with twelve knot points. In this case it is clear that the multiple knot algorithm has a achieved a better fit.

4. Conclusion and Ideas for Further Work

We have formulated an algorithm to allow coalescence of near-repeated knots into multiple knots, bringing into the approximation the appropriate derivatives of the basis function in place of the basis functions for the nearby knots. As a general rule, this results in both better approximations, and more stable computations, although individual examples do not conform to this ideal. Our computations convince us that the occurrence of multiple knots is a natural phenomenon, and that algorithms for the problem need to account for the possibility.

The tendency toward repeated knots leads to one possible idea for further exploration. This would be to begin initially with some (or all) knots being double knots to take advantage of the apparently greater fitting power of the multiquadric plus its directional derivative. We have not yet had the opportunity to pursue this idea. While the advantage of starting with multiple knots seems clear, the disadvantage is that coalescence of these knots skips odd-multiplicity knots, especially triple knots, which occur regularly in our computations.

We have observed the tendency toward repeated knots with the LEASTSQ algorithm and another (FMINS – see [12]) in the previous paper and feel the repeated knots are not occurring as an artifact of the optimization process. Nonetheless, an algorithm that took advantage of the particular problem being solved could be more efficient and possibly avoid some of the local minima being found by the general minimization routines.

References

[1] Duchon, J.: Splines minimizing rotation-invariant seminorms in Sobolev spaces. In: Multivariate approximation theory (Schempp, W., Zeller, K., eds.), pp. 85–100. Basel Boston Stuttgart: Birkhäuser 1979.

[2] Franke, R., Hagen, H., Nielson, G. M.: Least squares surface approximation to scattered data using multiquadric functions. Adv. Comput. Math. *2*, 81–99 (1994).

[3] Franke, R., Nielson, G. M.: Scattered data interpolation and applications: a tutorial and survey. In: Geometric modelling: methods and their applications. (Hagen, H., Roller, D., eds.), pp. 131–160. Berlin Heidelberg New York Tokyo: Springer 1991.

[4] Franke, R.: Scattered data interpolation: tests of some methods. Math. Comp. *38*, 181–200 (1982).

[5] Harder, R. L., Desmarais, R. N.: Interpolation using surface splines. J. Aircraft *9*, 189–191 (1972).

[6] Hardy, R. L.: Multiquadric equations of topography and other irregular surfaces. J. Geophys. Res. *76*, 1905–1915 (1971).

[7] Hardy, R. L.: Theory and applications of the multiquadric-biharmonic method. Comput. Math. Appl. *19*, 163–208 (1990).

[8] Micchelli, C. A.: Interpolation of scattered data: distance matrices and conditionally positive definite functions. Constr. Approx. *2*, 11–22 (1986).

[9] McMahon, J., Franke, R.: Knot selection for least squares thin plate splines, SIAM J. Sci. Stat. Comput. *13*, 484–498 (1992).

[10] Nielson, G. M.: A first-order blending method for triangles based upon cubic interpolation. Int. J. Numer. Meth. Eng. *15*, 308–318 (1978).

[11] Narcowich, F. J., Ward, J. W.: Norms of inverses and condition numbers for matrices associated with scattered data. J. Approx. Theory *64*, 69–94 (1991).

[12] Woods, D. J.: Dept. Math. Sciences, Rice Univ., Report 85–5 (1985).

R. Franke
Department of Mathematics
Naval Postgraduate School
Monterey, CA 93943-5216
U.S.A.

H. Hagen
FB Informatik
Universität Kaiserslautern
D67653 Kaiserslautern
Federal Republic of Germany

G. M. Nielson
Department of Computer Science
Arizona State University
Tempe, AZ 85287
U.S.A.

Computing Suppl. 10, 189–198 (1995)

Stability Concept for Surfaces

H. Hagen and **St. Hahmann**, Kaiserslautern

Abstract. In CAD/CAM technologies the design of free form surfaces is the beginning of a chain of operation that ends with the numerically controlled (NC-) production of the designed object. Apart from the design process of parametric surfaces another important domain in geometric modeling exists: shape control. In this paper we present a stability concept for surfaces based on infinitesimal bendings.

Key words: Deformation, free form surfaces, infinitesimal bendings, stability.

1. Introduction

In CAD/CAM technologies the design of free-form surfaces is the beginning of a chain of operations, that ends with the numerically controlled (NC-) production of the designed object. The design of free form surfaces is often combined with certain conditions or constraints: some points that are output of a scanning process have to be interpolated or approximated. There are boundary conditions, e.g. given boundary curves or given tangents. Continuity conditions, such as C^1, C^2 or curvature continuities, are often necessary. Furthermore, another class of surface criteria exists, the so-called minimization criteria, to minimize some energy functionals.

After the construction of free form surfaces, quality control is an important step. Surface interrogation methods test certain properties. Most of them are related to the surface normal or the principle curvatures. Flatpoints, continuity, smoothness or simply aesthetic aspects are some of the properties to be tested, because they are not always guaranteed by the design process.

A new aspect of shape control is the stability of surfaces. The aim of this paper is to present a stability concept based on infinitesimal bendings. Infinitesimal bendings of a surface are deformations which keep the length of any arbitrary surface curve unchanged to first order. Why such deformations are interesting? The answer is given in physics, for example in the theory of thin shells [3]—see Fig. 1. This theory deals with curved bodies having one small dimension.

A first assumption which is often done is to regard the shell as a 2-dimensional body. This 2-dimensional theory is based on the fundamental assumption that the geometrical and statical quantities which determine the behavior of the shell can be assumed with sufficient accuracy to be functions of only two independent variables (namely the curvilinear coordinates of the middle surface). Another basic assumption consists of the investigation of only small displacement in comparison to transverse characteristic dimension, called infinitesimal theory. As a consequence,

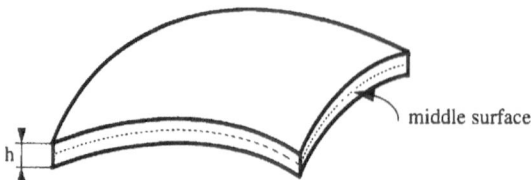

Figure 1. Section of a thin shell, middle surface

products of deformation parameters are neglected. And if everywhere on the middle surface, the length of all line elements remain unchanged, the displacement is said to be inextensional. The infinitesimal bendings we will talk about in this paper are also inextensional displacements. But regardless to the existence of many parallels to the theory of thin shells, the theory of infinitesimal bendings has only mathematical origins. Therefore the kind of stability defined here depends only on the shape of the surface and not on material coefficients.

A chapter about the fundamentals of infinitesimal bendings precedes a chapter where the concepts of stability and rigidity are introduced. Finally, the stability of parametric surfaces is tested and visualized with the so-called rotation vector fields.

2. Infinitesimal Bendings

In this chapter some fundamentals of infinitesimal bendings are given which will be needed later for the stability statements. A general survey is given in [1].

Parametric surfaces are represented as vector valued functions

$$\mathbf{X}: G \to \mathbb{R}^3, \quad (u, w) \mapsto \mathbf{X}(u, w) \tag{2.1}$$

where G is a connected domain of $\mathbb{R}^2$. $\mathbf{X}$ is assumed to be of class C^3 and $\mathbf{X}$ has to be regular. Let $\{\mathbf{X}_\varepsilon\}_{\varepsilon \in \mathbb{R}}$ be a continuous one-parameter family of surfaces, where $\mathbf{X}$ is contained in this family for $\varepsilon = 0$. If, for each ε, there is a continuous mapping between $\mathbf{X}$ and $\mathbf{X}_\varepsilon$ such that each point $\mathbf{P}$ of $\mathbf{X}$ corresponds to a unique point $\mathbf{P}_\varepsilon$ of $\mathbf{X}_\varepsilon$ then $\{\mathbf{X}_\varepsilon\}$ is called a *deformation* of the surface $\mathbf{X}$ (for illustration see Fig. 2).

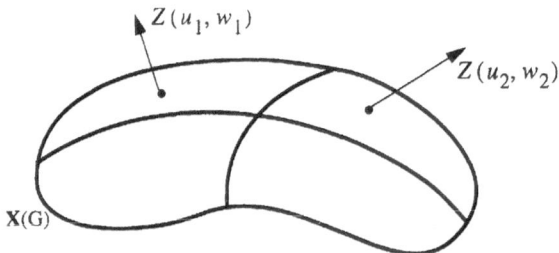

Figure 2. Deformation of the surface $\mathbf{X}$

A parametric representation is given by

$$\mathbf{X}_\varepsilon(u, w) = \mathbf{X}(u, w) + \varepsilon\mathbf{Z}(u, w) \tag{2.2}$$

where $\varepsilon \in \mathbb{R}$. $\mathbf{Z}(u, w)$ is a vector field, called *deformation vector field*. $\mathbf{Z}$ is assumed to be C^3.

The deformation (2.2) is called an *infinitesimal bending* of first order of the surface $\mathbf{X}$ if the length of any arbitrary curve on the surface $\mathbf{X}$ keeps unchanged to first order in ε, i.e., $L(\mathbf{c}_\varepsilon) = L(\mathbf{c}) + O(\varepsilon)$, where $\mathbf{c}(t) = \mathbf{X}(u(t), w(t))$ is an arbitrary surface curve. The vector field $\mathbf{Z}$ can be pictured as the *velocity field* of the surface points in the beginning of the deformation.

To give some properties of infinitesimal bendings, we need the following definition of the first variation of a function f:

Definition 1. Let $\{\mathbf{X}_t\}_{t\in I}$ ($I \subset \mathbb{R}$) be a deformation. Let $\{f_t\}_{t\in I}$ be a family of functions f_t defined on G such that $\begin{pmatrix} f : G \times I \to \mathbb{R} \\ (u, w, t) \mapsto f_t(u, w) \end{pmatrix}$ is continuous and has continuous partial derivatives.

The function δf defined on G by

$$\delta f := \frac{\partial f}{\partial t}\bigg|_{t=0} \tag{2.3}$$

is called the 1. variation of f.

This definition enables us to characterize the infinitesimal bendings with the following theorem:

Theorem 2. *Let $\mathbf{Z}$ be a deformation vector field of the surface $\mathbf{X}$. Let g_{ij} be the elements of the first fundamental form and denote the first partial derivatives of $\mathbf{X}$ and $\mathbf{Z}$ by $\mathbf{X}_u := \partial\mathbf{X}/\partial u$ and $\mathbf{Z}_u := \partial\mathbf{Z}/\partial u$. The following statements are equivalent:*

i) $\mathbf{Z}$ *defines an infinitesimal bending of first order of* $\mathbf{X}$
ii) $\delta g_{ij} = 0$ $\hfill (2.4)$
iii) $\langle \mathbf{Z}_u, \mathbf{X}_u \rangle = 0$
$\quad\quad \langle \mathbf{Z}_w, \mathbf{X}_w \rangle = 0$ $\hfill (2.5)$
$\quad\quad \langle \mathbf{Z}_u, \mathbf{X}_w \rangle + \langle \mathbf{Z}_w, \mathbf{X}_u \rangle = 0$

where $\langle \, , \, \rangle$ denotes the dot product.

Proof: (see [1])

Equation (2.4) means that the length of any surface curve does not change in first order in ε during the deformation. This follows because the first fundamental form can be used to measure distances on the surface [2].

The next theorem is at the same time a definition of the rotation vector field and plays an important role in the following discussion.

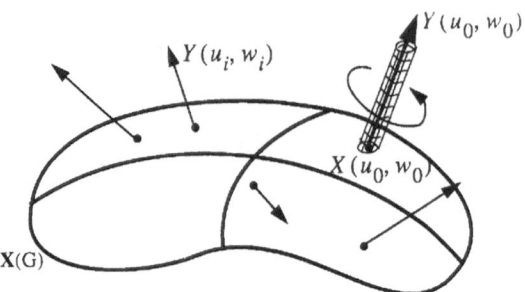

Figure 3. Rotation vector field

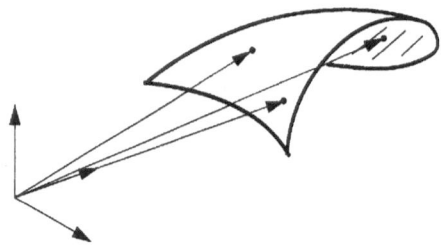

Figure 4. Instability surface

Theorem 3. (Existence Theorem). *If the deformation vector field* Z *verifies the three Eqs. (2.5), then there exists an unique vector field* Y *with the following properties:*

$$[Y, X_u] = Z_u \quad and \quad [Y, X_w] = Z_w. \tag{2.6}$$

where $[,]$ *denotes the vector product.* Y *is called rotation vector field.*

Proof: (see [1])

To give a geometric interpretation of Y: $Y(u_0, w_0)$ is the rotation vector of a rigid body attached to the surface $X(G)$ at the point $X(u_0, w_0)$ during the deformation (Fig. 3).

It is also possible to think of Y as a parametric surface. In this case Y is called the *instability surface* (also see Fig. 4) of the infinitesimal bending. Later we will see that a stable surface has an instability surface reduced to a single point.

3. Stable Rigid Surfaces

We now consider to a special case of infinitesimal bendings, those with a constant rotation vector field.

Definition 4. Infinitesimal bendings are called trivial (or infinitesimal motion) if and only if

$$Z(u, w) = [C, X(u, w)] + D \tag{3.1}$$

where C and C are constant vectors.

All bundles of line elements have the same momentary rotation if the rotation vector field is constant. This context leads to the following definition.

Definition 5. A surface which allows only trivial infinitesimal bendings is called infinitesimal rigid under infinitesimal bendings.

In other words, all infinitesimal bendings which are possible for the surface are trivial. The next theorem characterizes infinitesimal rigid surfaces.

Theorem 6. X is infinitesimal rigid if and only if for all deformations which satisfy $\delta g_{ij} = 0$ also holds $\delta h_{ij} = 0$. h_{ij} is the second fundamental form.

Therefore, $\delta g_{ij} = 0$ characterizes the infinitesimal bendings and $\delta h_{ij} = 0$ characterizes the infinitesimal rigid surfaces.

Another way to treat infinitesimal bendings is to apply the fundamental equations of this theory. We start with the relations $\mathbf{Z}_u = [\mathbf{Y}, \mathbf{X}_u]$ and $\mathbf{Z}_w = [\mathbf{Y}, \mathbf{X}_w]$ known from the Existence Theorem 3. To get a *jerk free deformation*, we need

$$\mathbf{Z}_{uw} = \mathbf{Z}_{wu}, \qquad (3.2)$$

and it follows that

$$[\mathbf{Y}_u, \mathbf{X}_w] = [\mathbf{Y}_w, \mathbf{X}_u]. \qquad (3.3)$$

This relation means that $\mathbf{Y}_u$, $\mathbf{Y}_w$, $\mathbf{X}_u$ and $\mathbf{X}_w$ (the partial derivatives of $\mathbf{X}$ and $\mathbf{Y}$) are coplanar. Therefore some real functions α, β, γ and δ exist such that

$$\exists \alpha, \beta, \delta : G \to \mathbb{R} : \mathbf{Y}_u = \alpha \mathbf{X}_u + \beta \mathbf{X}_w, \ \mathbf{Y}_w = \gamma \mathbf{X}_u + \delta \mathbf{X}_w.$$

From (3.3) follows

$$\mathbf{Y}_u = \alpha \mathbf{X}_u + \beta \mathbf{X}_w \qquad (3.4a)$$

$$\mathbf{Y}_w = \gamma \mathbf{X}_u - \alpha \mathbf{X}_w. \qquad (3.4b)$$

A further condition to impose is

$$\mathbf{Y}_{uw} = \mathbf{Y}_{wu} \qquad (3.5a)$$

which is equivalent to the equation:

$$(\alpha_w - \gamma_u)\mathbf{X}_u + (\beta_w + \alpha_u)\mathbf{X}_w - \gamma \mathbf{X}_{uu} + 2\alpha \mathbf{X}_{wu} + \beta \mathbf{X}_{ww} = \mathbf{O} \qquad (3.5b)$$

By expressing the derivatives of the vectors $\mathbf{X}_u$, $\mathbf{X}_v$ in the basis $\{\mathbf{X}_u, \mathbf{X}_w, \mathbf{N}\}$ ($\mathbf{N}$ unit normal vector of the surface), we obtain

$$\mathbf{X}_{uu} = \Gamma^1_{11}\mathbf{X}_u + \Gamma^2_{11}\mathbf{X}_v + h_{11}\mathbf{N}$$
$$\mathbf{X}_{uw} = \Gamma^1_{12}\mathbf{X}_u + \Gamma^2_{12}\mathbf{X}_v + h_{12}\mathbf{N} \qquad (3.6)$$
$$\mathbf{X}_{ww} = \Gamma^1_{22}\mathbf{X}_u + \Gamma^2_{22}\mathbf{X}_v + h_{22}\mathbf{N}$$

where the coefficients Γ^k_{ij}, $i, j, k = 1, 2$ are called the Christoffel symbols of $\mathbf{X}$ and where h_{ij} ($i, j = 1, 2$) is the second fundamental form of $\mathbf{X}$. (The reader who is not familiar with differential geometry can find the fundamentals in [2, 5]). Replacing Eqs. (3.6) in (3.5b) and decomposition of the resulting equation with respect to the basis $\{\mathbf{X}_u, \mathbf{X}_w, \mathbf{N}\}$, give the *fundamental equations*:

$$h_{11}\gamma - 2h_{12}\alpha - h_{22}\beta = 0$$
$$\Gamma^1_{11}\gamma - 2\Gamma^1_{12}\alpha - \Gamma^1_{12}\beta = \alpha_w - \gamma_u \qquad (3.7)$$
$$\Gamma^2_{11}\gamma - 2\Gamma^2_{12}\alpha - \Gamma^2_{22}\beta = \alpha_u + \beta_w.$$

If α, β, γ are solutions of the system (3.7), one gets the rotation vector field $\mathbf{Y}$ by integration and by a second integration the deformation vector field $\mathbf{Z}$. The surface $\mathbf{X}$ allows only the trivial infinitesimal bendings if and only if α, β, $\gamma \equiv 0$ is the unique solution of (3.7).

4. Visualization of the Rotation Vector Fields

The results of the previous chapter show that the determination of the infinitesimal bendings can be reduced to the solution of the fundamental equations (3.7). But the main interest of this paper is not to calculate the infinitesimal bendings of a surface. The aim is to present a concept for interrogating the stability. As it was shown in Sections 2 and 3, the rotation vector field is related to the definition of stable rigid surfaces.

Now there are two possible ways to accesses the rotation vector field. The first one is the fundamental system (3.7). The solutions α, β and γ have to be calculated in order to get $\mathbf{Y}_u$ and $\mathbf{Y}_w$. Then by integration one gets $\mathbf{Y}$ itself. The second possibility is based on the Existence Theorem 3 in combination with the jerk free deformation condition (3.2). We choose the second method, because $\mathbf{Y}$ can be calculated directly.

The problem can now be state as follows:

> Given a parametric surface $\mathbf{X}: G \rightarrow \mathbb{R}^3$, G a connected domain of $\mathbb{R}^2$. Wanted a vector valued function $\mathbf{Y}: G \rightarrow \mathbb{R}^3$ such that $[\mathbf{Y}_u, \mathbf{X}_w] = [\mathbf{Y}_w, \mathbf{X}_u]$.

To solve this system of partial differential equations, we choose a finite differences approach [4, 6]. The idea behind finite differences is to discretize of the problem by using a linear approximation to the partial derivatives. A linear system consisting of a non symmetric band matrix has to be solved. The solutions are discrete values of $\mathbf{Y}$ for certain parameter values. The advantage is that it is not necessary to integrate some formulas.

As mentioned above $\mathbf{Y} = constant$ is always a solution of this problem. The matrix is therefore singular and usually has more than one linearly independent solution. If there is at least one non constant rotation I vector field, then the surface is not stable because the corresponding infinitesimal bending is not trivial. But if all solutions are constant rotation vector fields, then the surface is stable. In this case the instability surface degenerates into a single point.

Parametric surfaces have different bending behaviors: there are surfaces which are more likely to bend and others are more likely to resist "pressure". With the help of the rotation vector fields, we want to visualize the bending property. Indeed the

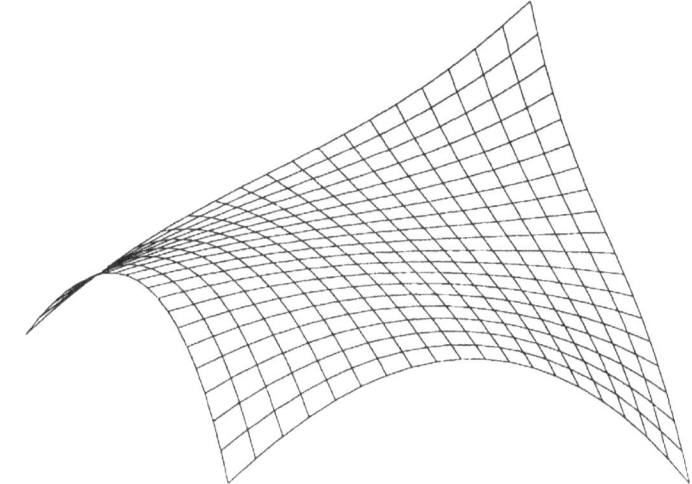

Figure 5a. Test surface

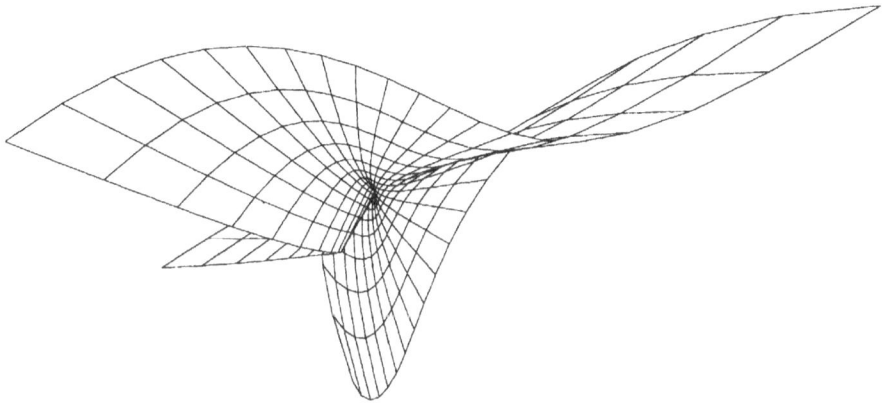

Figure 5b. Instability surface

notation of stability is closely related to the rotation vector field of infinitesimal bendings: A stable surface in the sense of infinitesimal bendings has a constant rotation vector field. Moreover, the more the rotation vectors vary in their directions the less the surface is stable. This behavior provides a visual test of stability as illustrated in the following examples.

In Fig. 5a we see a bicubic Bézier surface with its instability surface in Fig. 5b. The rotation vectors are attached on their corresponding parameter values in the parameter domain. The different directions of the rotation vectors are easily seen and indicate that they vary a lot (see Fig. 5c).

The second example is also a bicubic Bézier surface (see Fig. 6a). Its instability surface looks similar. We show here instability surfaces in Fig. 5b, 6b which are 'well

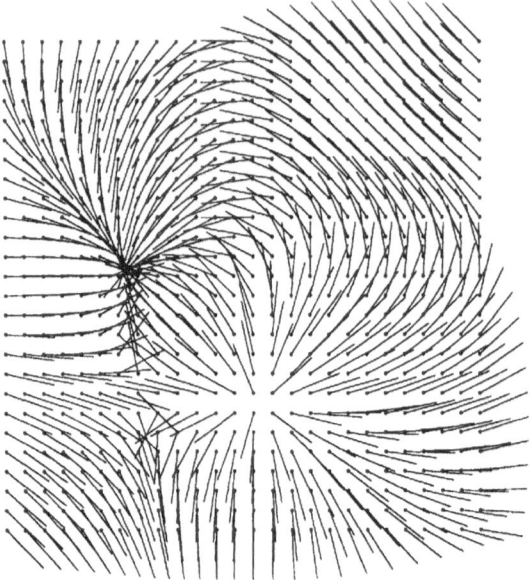

Figure 5c. Rotation vectors

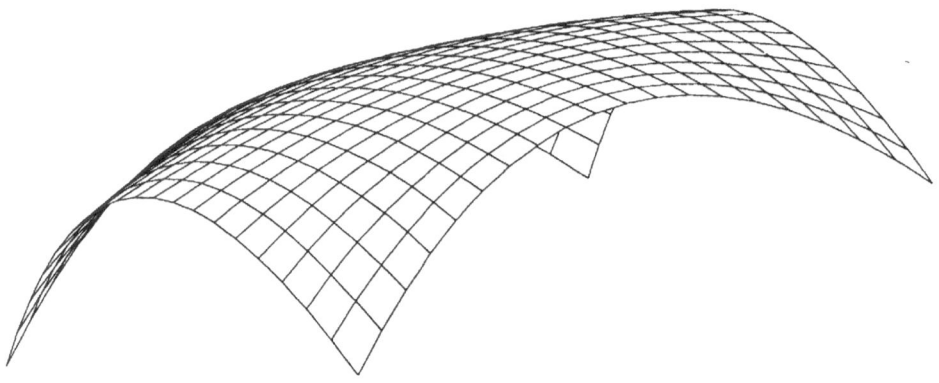

Figure 6a. Test surface

looking'. But it also happens that this surface can look total crazy, so that it is not possible to visualize it. More important are the rotation vectors. In Fig. 6c we can see that the rotation vectors seen from the same view point do not vary so much in their directions.

5. Conclusion and Remarks on Further Research

In this paper we presented a first approach for testing the stability of parametric surfaces in the sense of infinitesimal bendings. Stability is here a property of the surface itself. Only the shape and not the strength of the surface decides on the

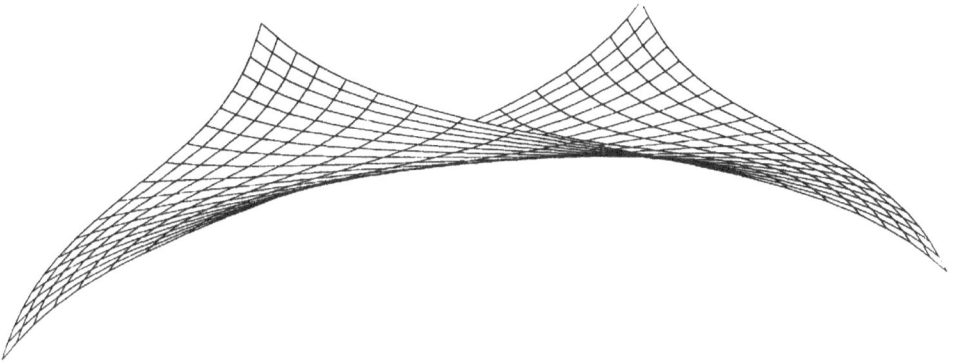

Figure 6b. Instability surface

Figure 6c. Rotation vectors

stability. We then introduced the notation of the rotation vector field. A kinetic meaning of these vectors was given and leaded to a visual stability test. The calculation of the rotation vectors was done discretely by a finite difference approach. We are working now on a polynomial approximation of these vector fields. With the help of this new representation we will develop additional visualization techniques. Further research will also include the development of new criteria of stability, such as the surface area of the instability surface, in order not to have only visual criteria.

References

[1] Efimov, N. W.: Flächenverbiegungen im Großen. Berlin: Akademie-Verlag 1957.
[2] Do Carmo, M. P.: Differential geometry of curves and surfaces, Englewood Cliffls: Prentice-Hall 1976.
[3] Gould, P. L.: Analysis of shells and plates. Berlin Heidelberg New York Tokyo: Springer 1988.
[4] Roache, P. J.: Computational fluid dynamics. Albuqueque: Hermosa Publishers 1972.
[5] Strang, G.: Introduction to applied mathematics. Cambridge Press: Welleseley 1986.
[6] Strubecker, K.: Differentialgeometrie III. Berlin: De Gryter 1959.

Hans Hagen
Stefanie Hahmann
Universität Kaiserslautern
FB Informatik
Postfach 3049
D-67653 Kaiserslautern
Federal Republic of Germany

Computing Suppl. 10, 199–210 (1995)

A Quartic Spline Based on a Variational Approach

B. Hamann, Mississippi State, and **T. A. Foley,** Tempe

Abstract. The C^2 continuous cubic spline can be viewed as the solution of a variational problem. The spline derived in this paper is obtained by solving a slightly different variational problem that depends on the input data. The goal is to obtain a spline that may have high second derivatives at the interpolated points and low second derivatives between two consecutive interpolated points. The solution is a C^2 continuous quartic spline.

Key words: Approximation, calculus of variation, interpolation, spline.

1. Introduction

Commonly, the C^2 continuous cubic spline interpolating the univariate data in $\{(x_i, F_i) | i = 0, \ldots, n\}$, where $x_{i+1} > x_i$, $i = 0, \ldots, (n-1)$, is derived using calculus of variation. The objective is to minimize an energy function, which has a piecewise cubic spline as solution. When dealing with parametric curves, one must interpolate the ordered points in $\{(x_i, y_i, z_i) | i = 0, \ldots, n\}$ considering an associated knot sequence $\{u_0 < \cdots < u_n\}$. Often, the solution of the univariate case is directly applied to the three coordinate functions of a parametric spline curve.

In the univariate case, the functional

$$\int (f''(x))^2 \, dx \qquad (1.1)$$

is minimized on $[x_0, x_n]$. In this paper, this functional is modified slightly, which leads to a C^2 continuous quartic spline. The modification of the functional (1.1) is motivated by the fact that the cubic spline tends to "overshoot" in regions where the given data changes rapidly. In the context of shape preservation, one wants to obtain a spline without (or at least very small) "overshoots". Thus, a different variational problem must be solved. One objective is to create a spline, which differs very little from the polygon obtained by connecting consecutive original data points. Intuitively, a spline that satisfies this criterion has high second derivatives at the knots x_i and low second derivatives between two knots x_i and x_{i+1}. Another objective is to construct a spline curve that has a second derivative that closely approximates a given function that may be dependent on the input data. It is shown in the following section how these objectives can be achieved by adding a single term to the functional (1.1).

It is not possible to list all references dealing with tension splines, monotone splines, or other shape preserving splines. The derivation of the C^2 continuous quartic uses a similar paradigm. Some of the related shape preserving curve and surface modeling concepts are discussed in [1,5–14, 16]. General references covering additional material include [3, 4, 15].

2. The Modified Variational Approach

In order to incorporate the "shape" of the polygon implied by the original data points, it is proposed to subtract a polynomial $\omega(x)$ from $f''(x)$ appearing in the functional (1.1). It is assumed that the polynomial $\omega(x)$ reflects the "desired" behavior of the second derivative of an interpolant to the given data. Therefore, the function $\omega(x)$ must be computed from a preprocessing step (Section 4) that depends on the input data.

The quartic spline is derived for the univariate case. The minimization problem to be solved considers the space $L^2[x_0, x_n]$ of functions $f''(x) - \omega(x)$ that are square-integrable on the interval $[x_0, x_n]$. The modified functional is

$$\int (f''(x) - \omega(x))^2 \, dx, \tag{2.1}$$

where $\omega(x)$ is a polynomial. Several methods for defining the polynomial $\omega(x)$ (for each interval $[x_i, x_{i+1}]$) are discussed in Section 4. This polynomial reflects the "desired" second derivative of a shape preserving interpolant.

A necessary condition for minimizing the expression (2.1) is given by Euler's equation (see [2]):

$$\frac{\partial G}{\partial f} - \frac{d}{dx}\left(\frac{\partial G}{\partial f'}\right) + \frac{d^2}{dx^2}\left(\frac{\partial G}{\partial f''}\right) - \frac{d^3}{dx^3}\left(\frac{\partial G}{\partial f'''}\right) + \cdots - \cdots = 0, \tag{2.2}$$

where, $G = G(f, f', f'', \ldots)$ is the integrand of the integral in (2.1). Substituting the integrand of (2.1) into (2.2) yields the differential equation to be solved. The integrand in (2.1) is

$$G(f, f', f'') = (f'')^2 - 2\omega f'' + \omega^2, \tag{2.3}$$

and the differential equation (2.2) becomes

$$\frac{\partial G}{\partial f} - \frac{d}{dx}\left(\frac{\partial G}{\partial f'}\right) + \frac{d^2}{dx^2}\left(\frac{\partial G}{\partial f''}\right) = \frac{d^2}{dx^2}(2f'' - 2\omega) = 2(f'''' - \omega'') = 0, \tag{2.4}$$

which is equivalent to

$$f''''(x) = \omega''(x). \tag{2.5}$$

Equation (2.5) implies that the polynomial $\omega(x)$ must be at least of degree two for $\omega''(x)$ not to vanish. Therefore, a quadratic polynomial $\omega(x)$ is used in the following. This is the lowest-degree polynomial that yields a spline different from the cubic spline. If a parametric spline curve $c(u) = ((x(u), y(u), z(u))$ is computed, the univariate approach is applied to the three coordinate functions. Examples will be given for both the non-parametric and the parametric case.

3. Computing the Quartic Spline Segments

In this section, it is assumed that a C^0 piecewise quadratic polynomial, denoted by $\omega_i(x)$, $x \in [x_i, x_{i+1}]$, is given. Integrating Eq. (2.5) implies that each spline segment is a quartic polynomial, which is written as

$$f_i(x) = \sum_{j=0}^{4} c_{i,j}(x - x_i)^j, \quad x \in [x_i, x_{i+1}], \quad i = 0, \ldots, (n-1). \tag{3.1}$$

Each quartic spline segment is determined by continuity conditions and the quadratic functions

$$\omega_i(x) = \sum_{j=0}^{2} \bar{c}_{i,j}(x - x_i)^j, \quad x \in [x_i, x_{i+1}], \quad i = 0, \ldots, (n-1). \tag{3.2}$$

Differentiating the quadratic functions (3.2), one obtains $\omega_i''(x) = 2\bar{c}_{i,2}$. Using Eq. (2.5) yields the coefficients for the quartic terms of each spline segment:

$$c_{i,4} = \frac{1}{12}\bar{c}_{i,2}, \quad i = 0, \ldots, (n-1). \tag{3.3}$$

Continuity conditions are imposed on the quartic spline segments; these are C^0, C^1, and C^2 continuity conditions. The quadratic polynomials $\omega_i(x)$ are required to be C^0 continuous at the interior knots, i.e., $\omega_{i-1}(x_i) = \omega_i(x_i)$, $i = 1, \ldots, (n-1)$. The continuity conditions for the spline segments are stated next.

The interpolation conditions imply

$$f_i(x_i) = c_{i,0} = F_i, \quad i = 0, \ldots, (n-1), \tag{3.4}$$

the conditions for C^0 continuity are

$$f_i(x_{i+1}) = f_{i+1}(x_{i+1}), \quad i = 0, \ldots, (n-2), \tag{3.5}$$

the conditions for C^1 continuity are

$$f_i'(x_{i+1}) = f_{i+1}'(x_{i+1}), \quad i = 0, \ldots, (n-2), \tag{3.6}$$

and, finally, the C^2 conditions are

$$f_i''(x_{i+1}) = f_{i+1}''(x_{i+1}), \quad i = 0, \ldots, (n-2), \tag{3.7}$$

Defining $h_i = (x_{i+1} - x_i)$, Eq. (3.7) can be written as

$$c_{i,3} = \frac{1}{3h_i}(c_{i+1,2} - c_{i,2} - 6c_{i,4}h_i^2), \quad i = 0, \ldots, (n-1). \tag{3.8}$$

The coefficients $c_{i,1}$ are derived from (3.5) and (3.8) as

$$c_{i,1} = \frac{1}{h_i}(c_{i+1,0} - c_{i,0}) - \frac{h_i}{3}(2c_{i,2} + c_{i+1,2}) + h_i^3 c_{i,4}, \quad i = 0, \ldots, (n-1). \tag{3.9}$$

A linear system of $(n-1)$ equations and $(n+1)$ unknowns is obtained for the coefficients $c_{i,2}$ using (3.6). This linear system is given by

$$h_{i-1}c_{i-1,2} + 2(h_{i-1}+h_i)c_{i,2} + h_ic_{i+1,2}$$

$$= 3\left(\frac{1}{h_{i-1}}c_{i-1,0} - \left(\frac{1}{h_{i-1}}+\frac{1}{h_i}\right)c_{i,0} + \frac{1}{h_i}c_{i+1,0} + h_{i-1}^3c_{i-1,4} + h_i^3c_{i,4}\right),$$

$$i = 1,\ldots,(n-1). \quad (3.10)$$

Assuming that values for $c_{0,2}$ and $c_{n,2}$ are known (using some end conditions for the spline), this linear system can be written in matrix form as

$$
\begin{pmatrix}
2(h_0+h_1) & h_1 & 0 & \vdots & \vdots & & \vdots \\
h_1 & 2(h_1+h_2) & h_2 & 0 & \vdots & & \vdots \\
0 & h_2 & 2(h_2+h_3) & h_3 & 0 & & \vdots \\
\vdots & 0 & \ddots & \ddots & \ddots & & 0 \\
\vdots & \vdots & 0 & h_{n-3} & 2(h_{n-3}+h_{n-2}) & & h_{n-2} \\
\vdots & \vdots & \vdots & 0 & h_{n-2} & & 2(h_{n-2}+h_{n-1})
\end{pmatrix}
\begin{pmatrix}
c_{1,2} \\
c_{2,2} \\
c_{3,2} \\
\vdots \\
c_{n-2,2} \\
c_{n-1,2}
\end{pmatrix}
$$

$$
= 3
\begin{pmatrix}
\dfrac{1}{h_0}c_{0,0} - \left(\dfrac{1}{h_0}+\dfrac{1}{h_1}\right)c_{1,0}+\dfrac{1}{h_1}c_{2,0}+h_0^3c_{0,4}+h_1^3c_{1,4}-\underline{h_0c_{0,2}} \\[3mm]
\dfrac{1}{h_1}c_{1,0} - \left(\dfrac{1}{h_1}+\dfrac{1}{h_2}\right)c_{2,0}+\dfrac{1}{h_2}c_{3,0}+h_1^3c_{1,4}+h_2^3c_{2,4} \\[3mm]
\vdots \\[2mm]
\vdots \\[3mm]
\dfrac{1}{h_{n-3}}c_{n-3,0} - \left(\dfrac{1}{h_{n-3}}+\dfrac{1}{h_{n-2}}\right)c_{n-2,0}+\dfrac{1}{h_{n-2}}c_{n-1,0}+h_{n-3}^3c_{n-3,4}+h_{n-2}^3c_{n-2,4} \\[3mm]
\dfrac{1}{h_{n-2}}c_{n-2,0} - \left(\dfrac{1}{h_{n-2}}+\dfrac{1}{h_{n-1}}\right)c_{n-1,0}+\dfrac{1}{h_{n-1}}c_{n,0}+h_{n-2}^3c_{n-2,4}+h_{n-1}^3c_{n-1,4}-\underline{h_{n-1}c_{n,2}}
\end{pmatrix}
$$

$$(3.11)$$

The two underlined terms in the first and last row of the right hand side of (3.11) indicate that the end conditions for the spline determine those values. The coefficients $c_{0,2}$ and $c_{n,2}$ can be derived from a local polynomial approximation of the given data at the ends. Obviously, the linear system of equations with its symmetric, tridiagonal, and diagonally dominant matrix can be solved in linear time.

If periodic (cyclic) interpolation of the given data is desired, end conditions need not to be specified, since the equations $f_0(x_0)=f_{n-1}(x_n)$, $f_0'(x_0)=f_{n-1}'(x_n)$, $f_0''(x_0)= f_{n-1}''(x_n)$, and $h_0=h_n$ hold. These periodicity constraints imply a linear system of equations with a cyclic matrix given by

$$
\begin{pmatrix}
2(h_{n-1}+h_0) & h_0 & 0 & \vdots & \vdots & h_{n-1} \\
h_0 & 2(h_0+h_1) & h_1 & 0 & \vdots & 0 \\
0 & h_1 & 2(h_1+h_2) & h_2 & 0 & \vdots \\
\vdots & 0 & \ddots & \ddots & \ddots & 0 \\
0 & \vdots & 0 & h_{n-3} & 2(h_{n-3}+h_{n-2}) & h_{n-2} \\
h_{n-1} & \vdots & \vdots & 0 & h_{n-2} & 2(h_{n-2}+h_{n-1})
\end{pmatrix}
\begin{pmatrix}
c_{0,2} \\
c_{1,2} \\
c_{2,2} \\
\vdots \\
c_{n-3,2} \\
c_{n-2,2} \\
c_{n-1,2}
\end{pmatrix}
$$

$$
= 3
\begin{pmatrix}
\dfrac{1}{h_{n-1}}c_{n-1,0} - \left(\dfrac{1}{h_{n-1}}+\dfrac{1}{h_0}\right)c_{0,0} + \dfrac{1}{h_0}c_{1,0} + h_{n-1}^3 c_{n-1,4} + h_0^3 c_{0,4} \\
\vdots \\
\vdots \\
\vdots \\
\dfrac{1}{h_{n-2}}c_{n-2,0} - \left(\dfrac{1}{h_{n-2}}+\dfrac{1}{h_{n-1}}\right)c_{n-1,0} + \dfrac{1}{h_{n-1}}c_{0,0} + h_{n-2}^3 c_{n-2,4} + h_{n-1}^3 c_{n-1,4}
\end{pmatrix}
. \quad (3.12)
$$

Solving the linear system provides the coefficients $c_{i,2}$. The coefficients $c_{i,3}$ and $c_{i,1}$ are then computed using (3.8) and (3.9).

4. Choosing the Quadratic Functions $\omega_i(x)$

The quadratic functions $\omega_i(x)$ are defined for each interval $[x_i, x_{i+1}]$, $i=0,\ldots,(n-1)$, interpolating second derivative estimates at x_i. They are C^0 continuous at the knots. These weight functions should have exactly one zero in the interior of $[x_i, x_{i+1}]$, if it is the desire to force the final interpolant to have second derivative extrema at the knots and small second derivatives between knots. In order to define the quadratic polynomials $\omega_i(x)$, it is necessary to derive estimates for the second derivatives at all knots. Several possibilities can be used to obtain these estimates.

In order to compute the quadratic weight function $\omega_i(x)$, a quadratic polynomial $q_i(x)$ is computed that interpolates the three data in

$$\{(x_{i-1}, F_{i-1}), (x_i, F_i), (x_{i+1}, F_{i+1})\}, \quad i=1,\ldots,(n-1). \quad (4.1)$$

Alternatively, a quartic polynomial $r_i(x)$ is computed that interpolates the five data in

$$\{(x_{i-2}, F_{i-2}), (x_{i-1}, F_{i-1}), (x_i, F_i), (x_{i+1}, F_{i+1}), (x_{i+2}, F_{i+2})\}, \quad i=2,\ldots,(n-2). \quad (4.2)$$

By differentiating these local approximants at $x = x_i$ one obtains values $q_i''(x_i)$ (or $r_i''(x_i)$). These values are used as second derivative estimates and are set equal to $\omega_{i-1}(x_i) = \omega_i(x_i)$, leaving one degree of freedom for defining the quadratic weight function. Obviously, special care is required at the ends. The values of $q_1''(x)(r_2''(x))$ and $q_{n-1}''(x)(r_{n-2}''(x))$ are used to obtain estimates at the ends. This, of course, is no problem in the periodic case, since the data are repeated cyclically.

The use of quartic polynomials $r_i(x)$ for second derivative estimation can be usd to generate a curve scheme with quartic precision. One approach is to define

$\omega_i(x)$ as the quadratic that interpolates the three values $r_i''(x_i)$, $r_{i+1}''(x_{i+1})$, and $\frac{1}{2}(r_i(\frac{1}{2}(x_i + x_{i+1})) + r_{i+1}(\frac{1}{2}(x_i + x_{i+1})))$, $i = 2, \ldots, (n-3)$. Special care is required for the end intervals when computing $\omega_0(x)$, $\omega_1(x)$, $\omega_{n-2}(x)$, and $\omega_{n-1}(x)$ (non-periodic case). A simple solution is the requirement $\omega_0(x) = \omega_1(x) = r_2''(x)$ and $\omega_{n-2}(x) = \omega_{n-1}(x) = r_{n-2}''(x)$.

Forcing quartic precision is effective if the data varies smoothly. Unfortunately, using quartic polynomials for second derivative estimation might not be "local enough" in certain cases and might lead to unwanted oscillations in the final spline if the data varies rapidly. For more local results, quadratic polynomials should be used for the estimation.

Having computed second derivative estimates at all knots, the quadratic polynomials $\omega_i(x)$ can be defined by specifying one more value for each interval. To generate a spline curve with small second derivative between the interpolated points, the weight function is computed using the following case distinctions:

- If the estimates at x_i and x_{i+1} are both zero set $\omega_i(x) = 0$.
- If the estimate at one knot is zero, and the estimate at the other knot is positive (negative) let $\omega_i(x)$ be the quadratic polynomial, which has a double zero at the knot where it must interpolate the zero estimate.
- If the estimates at both knots are greater (smaller) than zero let $\omega_i(x)$ be the quadratic polynomial that has a double zero in the open interval (x_i, x_{i+1}).
- If the estimate at one knot is negative, and the estimate at the other knot is positive force $\omega_i(x)$ to have a zero at $(x_i + x_{i+1})/2$.

One can introduce real-valued parameters α_i to scale the second derivative estimates at each knot x_i. Obviously, if $\alpha_i = 0$, $i = 0, \ldots, n$, the common cubic spline is obtained; if $\alpha_i = 1$, $i = 0, \ldots, n$, the "standard" quartic spline is obtained. Scaling the estimates of the second derivatives at the knots using increasing positive scaling factors α_i generally produces splines which approach the polygon implied by the original data. This justifies the view of these scaling factors as tension parameters. They can be used as an interactive shaping tool for the quartic spline.

Unfortunately, increasing these scaling factors beyond a certain threshold can lead to splines with unwanted inflection points or even "loops". It is not clear at this point how these factors must be chosen in order to get the most pleasing spline with maximum shape preservation. Once determined, the functions $\omega_i(x)$ define the coefficients $c_{i,4}$ according to (3.3). Examples of possible quadratic polynomials $\omega_i(x)$ are shown in Fig. 1.

In the parametric case, the method presented is applied to the three coordinate functions $x(u)$, $y(u)$, and $z(u)$ of the parametric spline curve $c(u)$. Three different sets of quadratic polynomials $\omega_i(u)$ are generated, one for each coordinate.

5. Conversion to Bernstein-Bézier Representation

The C^2 continuous quartic spline has been derived in monomial form. The transformation of a single polynomial $f_i(x)$ to its corresponding Bernstein-Bézier representation is given by

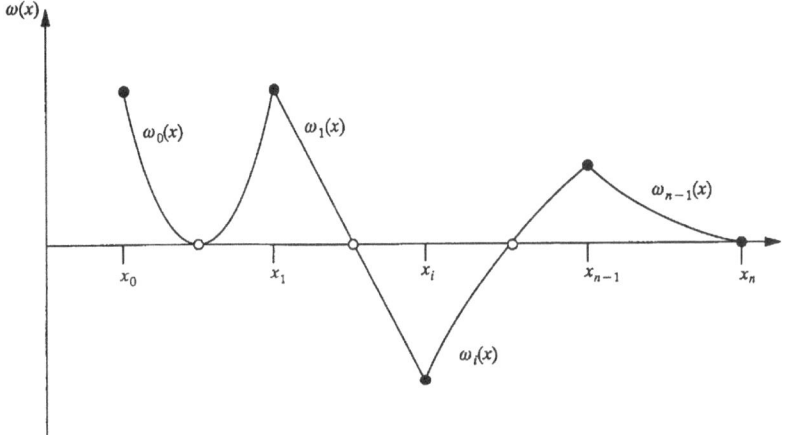

Figure 1. Possible quadratic polynomials $\omega_i(x)$

$$f_i(x) = \sum_{j=0}^{4} c_{i,j}(x - x_i)^j = \sum_{j=0}^{4} b_{i,j} B_j^4(x), \quad i = 0, \ldots, (n-1). \tag{5.1}$$

Here, x varies in $[x_i, x_{i+1}]$, $b_{i,j}$ are the Bézier ordinates for a single spline segment, and $B_j^4(x)$, $j = 0, \ldots, 4$, are the quartic Bernstein-Bézier polynomials defined as

$$B_j^4(x) = \frac{1}{h_i^4} \binom{4}{j}(x_{i+1} - x)^{4-j}(x - x_i)^j, \quad i = 0, \ldots, (n-1). \tag{5.2}$$

In matrix notation, the conversion between monomial and Bernstein-Bézier representation is written as

$$\begin{pmatrix} c_{i,0} & -c_{i,1}x_i & +c_{i,2}x_i^2 & -c_{i,3}x_i^3 & +c_{i,4}x_i^4 \\ 0 & c_{i,1} & -2c_{i,2}x_i & +3c_{i,3}x_i^2 & -4c_{i,4}x_i^3 \\ 0 & 0 & c_{i,2} & -3c_{i,3}x_i & +6c_{i,4}x_i^2 \\ 0 & 0 & 0 & c_{i,3} & -4c_{i,4}x_i \\ 0 & 0 & 0 & 0 & c_{i,4} \end{pmatrix} = \frac{1}{h_i^4}$$

$$\cdot \begin{pmatrix} x_{i+1}^4 & -4x_i x_{i+1}^3 & 6x_i^2 x_{i+1}^2 & -4x_i^3 x_{i+1} & x_i^4 \\ -4x_{i+1}^3 & 4(3x_i x_{i+1}^2 + x_{i+1}^3) & -12(x_i^2 x_{i+1} + x_i x_{i+1}^2) & 4(x_i^3 + 3x_i^2 x_{i+1}) & -4x_i^3 \\ 6x_{i+1}^2 & -12(x_i x_{i+1} + x_{i+1}^2) & 6(x_i^2 + 4x_i x_{i+1} + x_{i+1}^2) & -12(x_i^2 + x_i x_{i+1}) & 6x_i^2 \\ -4x_{i+1} & 4(x_i + 3x_{i+1}) & -12(x_i + x_{i+1}) & 4(3x_i + x_{i+1}) & -4x_i \\ 1 & -4 & 6 & -4 & 1 \end{pmatrix}$$

$$\cdot (b_{i,0} \quad b_{i,1} \quad b_{i,2} \quad b_{i,3} \quad b_{i,4})^T. \tag{5.3}$$

The Bézier ordinates $b_{i,j}$ can be computed directly from (5.3). Alternatively, it is possible to directly derive the quartic spline in Bernstein-Bézier representation.

6. Computing Biquartic Tensor Product Splines

In order to construct a tensor product surface method based on quartic splines, a set of cardinal basis functions is defined, where the quadratic weight functions depend on "cardinal data." More precisely, define the basis function $g_i(x)$ as the quartic spline interpolant that is zero at every knot except at x_i, where it is one. Thus, a C^2 continuous quartic spline can be written in cardinal form as

$$f(x) = \sum_{i=0}^{n} F_i g_i(x), \tag{6.1}$$

where the functions $g_i(x)$, $i = 0, \ldots, n$, are C^2 continuous quartic cardinal splines satisfying $g_i(x_k) = \delta_{i,k}$ (Kronecker delta). This linear combination interpolates the data, but it can be different from the method described earlier, where the quadratic weight functions depended on the given data. Using the cardinal form (6.1), the weight functions depend on the data $(x_k, \delta_{i,k})$.

Using the concept of cardinal spline bases, a C^2 continuous biquartic tensor product spline is written as

$$f(x, y) = \sum_{j=0}^{n} \sum_{i=0}^{m} F_{i,j} g_i(x) h_j(y) \tag{6.2}$$

interpolating the data in $\{(x_i, y_j, F_{i,j}) | x_{i+1} > x_i, y_{j+1} > y_j, i = 0, \ldots, m, j = 0, \ldots, n\}$. Again, the basis functions $g_i(x)$ and $h_j(y)$ satisfy the conditions $g_i(x_k) = \delta_{i,k}$ and $h_j(y_l) = \delta_{j,l}$. Both cardinal spline bases are computed using the univariate scheme.

In the parametric case, an interpolating parametric tensor product surface in three-dimensional space is written in cardinal form as

$$s(u, v) = \sum_{j=0}^{n} \sum_{i=0}^{m} x_{i,j} g_i(u) h_j(v), \tag{6.3}$$

where $x_{i,j} = (x_{i,j}, y_{i,j}, z_{i,j})$ is an interpolated point, and $g_i(u)$ and $h_j(v)$ are cardinal basis functons. Two increasing knot sequences are required in the parametric case, $\{u_0 < \cdots < u_m\}$ and $\{v_0 < \cdots < v_n\}$. A C^2 continuous quartic cardinal spline basis is shown in Fig. 2 using periodic end conditions, quadratic functions $\omega_i(x)$ for second derivative estimation, and tension parameters $\alpha_i = 1$.

7. Examples

The following examples demonstrate the new method for non-parametric and parametric cases. Figures 3 and 4 show C^2 continuous quartic splines/parametric spline curves obtained by increasing tension parameters α_i. The tension parameters used in Fig. 3 are 0 (= cubic spline, upper-left), 5 (upper-right), 10 (lower-left), and 15 (lower-right). For a particular spline/parametric spline curve, the tension parameters are the same at all knots. The second derivative estimates at the interior knots are based on local quadratic approximants q_i. In the non-periodic case, second derivative estimates at the end knots are zero, and natural end conditions are used

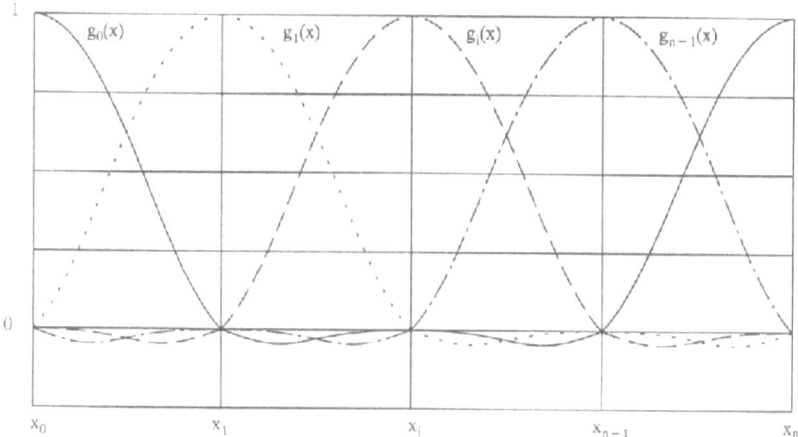

Figure 2. Cardinal spline basis functions using periodic end conditions

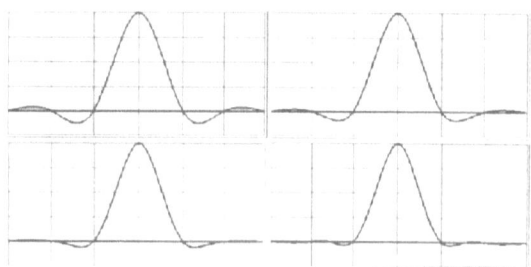

Figure 3. Quartic splines interpolating same data using increasing tension

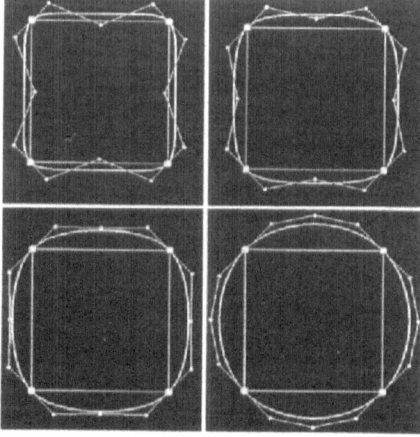

Figure 4. Quartic parametric spline curves interpolating same data using increasing tension

B. Hamann and T. A. Foley

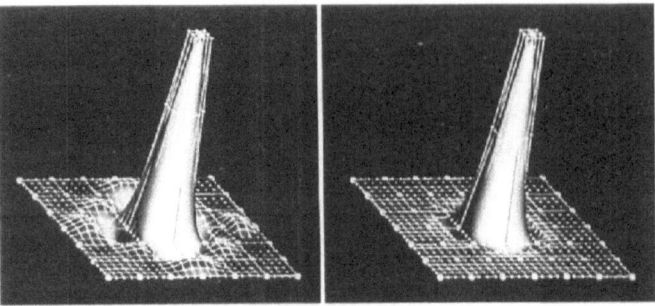

Figure 5. Product of two cubic and two quartic cardinal splines

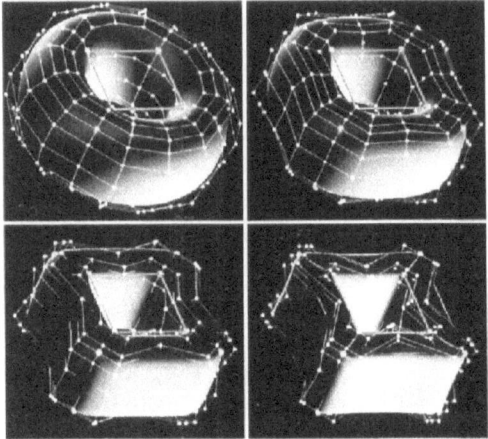

Figure 6. Modeling torus using increasing tension

for computing the quartic spline. Fig. 4 shows a periodic parametric quartic spline curve and its Bézier control polygon; the parametrization is based on chord length. The tension parameters used in Fig. 4 are 0 (= cubic spline, upper-left), 2.2 (upper-right), 4.4 (lower-left), and 6.6 (lower-right).

The next three examples show biquartic splines/parametric spline surfaces. Figure 5 shows the piecewise bicubic function (left)/piecewise biquartic function (right) defined as the product $f(x, y) = 6g_3(x)h_3(y)$ of the two cubic (left)/two quartic (right) cardinal splines $g_3(x)$ and $h_3(y)$. The tension parameters applied in x- and y-direction are 12 in the biquartic case. The knots are $x_i = i, i = 0, \ldots, 6$, and $y_j = j \; j = 0, \ldots, 6$. Figures 6 and 7 show the use of the method for interactively modeling parametric surfaces. The tension parameters used in Fig. 6 are 0 (= bicubic spline surface, upper-left), 6 (upper-right), 12 (lower-left), and 18 (lower-right). The tension parameters used in Fig. 7 are 0 (= bicubic spline surface, upper-left), 10 (upper-right), 15 (lower-left), and 20 (lower-right). All parametrizations are based on average chord

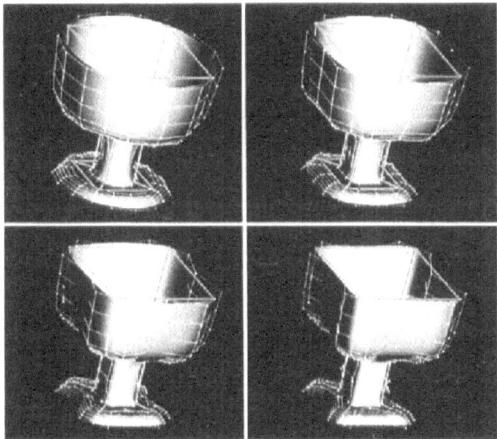

Figure 7. Modeling goblet using increasing tension

length, the same tension is applied in both parameter directions, and the associated Bézier control nets are shown.

8. Conclusions

A C^2 continuous quartic spline based on a minimization problem has been introduced. The spline is an alternative to the cubic spline. The quartic spline can be used as an interactive design tool by scaling second derivative estimates at the knots. This scaling parameter can be used to achieve interpolating curves of different shape with high second derivatives at the interpolated points. Since the quadratic polynomials ω_i depend on the knots, it is investigated how the quartic spline reacts to changes in the parametrization.

Acknowledgements

This research was supported by Department of Energy grant DE-FG02-87ER25041 and National Science Foundation grant DMS-9116930, both to Arizona State University, and by National Science Foundation grant ASC-9210439, to Mississippi State University. Algorithms were implemented on a Silicon Graphics workstation 4D/220 GTX. The authors wish to thank all members of the Computer Aided Geometric Design research group in the Computer Science Department at Arizona State University and all members of the National Grid Project team at the NSF Engineering Research Center for Computational Field Simulation at Mississippi State University for their advice and support.

References

[1] Akima, H.: A new method of interpolation and smooth curve fitting based on local procedures. J. Assoc. Comp. Mach. *17*, 589–602 (1970).

[2] Clegg, J. C.: Calculus of variations, Edinburgh: Oliver & Boyd 1968.

[3] de Boor, C.: A practical guide to splines. New York: Springer 1978.

[4] Farin, G.: Curves and surfaces for computer aided geometric design, 3rd edn. San Diego: Academic Press 1993.

[5] Foley, T. A.: Local control of interval tension using weighted splines. Comput. Aided Geom. Des. 3, 281–294 (1986).

[6] Foley, T. A.: Interpolation with interval and point tension controls using cubic weighted ν-splines. ACM Trans. Math. Software 13, 68–96 (1987).

[7] Foley, T. A.: A shape preserving interpolant with tension controls. Comput. Aided Geom. Des. 5, 105–118 (1987).

[8] Fritsch, F. N., Carlson, R. E.: Monotone piecewise cubic interpolation. SIAM J. Numer. Anal. 17, 238–246 (1980).

[9] Goodman, T. N. T.: Shape preserving representations. In: Mathematical methods in computer aided geometric design (Lyche, T., Schumaker, L. L., eds.), pp. 333–351. San Diego: Academic Press 1989.

[10] Goodman, T. N. T. Unsworth, K.: Shape preserving interpolation by curvature continuous parametric curves. Comput. Aided Geom. Des. 5, 323–340 (1988).

[11] Hagen, H., Schulze, G.: Automatic smoothing with geometric surface patches. Comput. Aided Geom. Des. 4, 231–235 (1987).

[12] Hamann B., Farin, G., Nielson, G. M.: A parametric triangular patch based on generalized conics. In: NURBS for curve and surface design (Farin, G., ed.), pp 75–85. Philadelphia: SIAM 1991.

[13] Nielson, G. M.: Some piecewise alternatives to splines under tension. In: Computer aided geometric design (Barnhill, R. E., Riesenfeld, R. F., eds.), pp. 209–235. San Diego: Acadamic Press 1974.

[14] Salkauskas, K.: C^1 splines for interpolation of rapidly varying data. Rocky Mountain J. Math. 14, 239–250 (1984).

[15] Schumaker, L. L.: Spline functions: basic theory. New York J. Wiley: 1961.

[16] Schweikert, D. G.: An interpolation curve using a spline in tension. J. Math. Phys. 45, 312–317 (1996).

Dr. B. Hamann
NSF Engineering Research Center
for Computational Field Simulation
Mississippi State University
P.O. Box 6176,
Mississippi State
MS 39762, U.S.A.
or
Department of Computer Science
Mississippi State University
P.O. Drawer CS
Mississippi State
MS 39762, U.S.A.

Dr. T. A. Foley
Computer Science Department
Arizona State University
Tempe
AZ 85287-5406, U.S.A.

Computing Suppl. 10, 211–226 (1995)

A Knowledge-Based System for Geometric Design

U. **Langbecker** and H. **Nowacki,** Berlin

Abstract. A knowledge base for geometric design has been developed including mainly two types of knowledge for problem formulation resp. problem solution. The problem formulation knowledge serves to classify the problem type according to statements given by the user about geometric data, mathematical representation, criterion function and discrete as well as integral constraints. Problem solution knowledge pertains to the choice of adequate problem solvers for each problem type, all of them regarded as constrained optimization problems. A prototype system was implemented for the design of curves using the declarative programming language PROLOG embedded in a hybrid environment with MATHEMATICA, X11 and C, enabling the user to choose multitudinous combinations of criteria functions and constraints at session time. The presented approach lends itself to achieving great flexibility within a large class of shape generation problems.

Key words: Knowledge-based geometric design, problem formulation knowledge, problem solution knowledge, PROLOG.

1. Introduction

The generation and processing of curves and surfaces is a field of major interest within Computer Aided Geometric Design (CAGD). Well known to a few experts there is a large number of methods, representation and data types used to solve special problems. But for most of the people faced with concrete design problems familiarity with that large amount of details cannot be assumed. Making this huge body of presently amenable knowledge for geometric construction problems or at least a considerable part of it accessible to a larger community would open a big potential for new solutions. Therefore the following task was proposed.

Develop a knowledge based system with expert knowledge on freeform geometric design, related to both problem formulation and problem solution. This knowledge should be defined explicitly in terms of facts and rules and dynamic updating of the knowledge base should be possible using an inference mechanism. An implementation of a prototype should refer to modeling of curves.

To achieve that goal many questions have to be answered, including the following:

- How can the problem description be organized and in which way it is possible to structure and classify the knowledge?
- How can this knowledge be formalized and fed into a program and which kind of program is the most appropriate for these goals?

To answer these question the rest of this paper is organized in the following way. In Section 2 we consider some basic structures and techniques for knowledge based systems. Then, within Sections 3 and 4, we focus our attention on the two major types of knowledge used within our approach, namely the problem formulation knowledge and the problem solution knowledge. The components and the function-ality of the current prototype are the focus of interest in Section 5 and after the treatment of some examples finally a short summary is given.

2. Knowledge-Based Techniques

Within this section we outline only a few ideas about knowledge-based systems (KBS), that are needed in the course of this paper.

In general KBSs are used typically in situations characterized by

- a large scope,
- a very complex behaviour, and
- many interdependent parameters and constraints.

This also applies to a large class of CAGD problems. There are many types of curves, Bezier splines, Hermite splines, B-Splines, β-, γ-, ν-, τ-splines, splines in tension, geometric splines, natural splines, etc., occurring more or less frequently. And this enumeration is far from complete, because new methods are continuously being developed to fit new needs. Some of them are new combinations of already known ones.

Therefore the knowledge-based approach seem to be an appropriate means for getting a unified perspective on geometric modelling methods.

Every KBS is structured in the following way. First of all it has a central component, the *inference engine*, which is responsible for the administration and processing of the knowledge. This knowledge is contained in the *knowledge base*. The *knowledge acquisition component* serves to feed the experts knowledge into the systems, while the *user interface* enables the user to pose queries to the system and to receive the inferred answers and other information via the *message board* (see Fig. 1). Within our prototype we did not include a knowledge acquisition component.

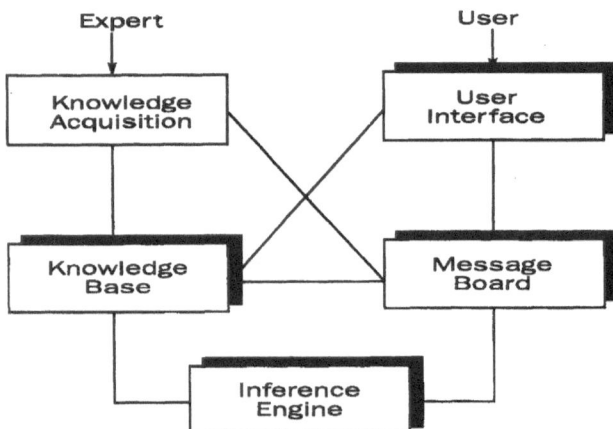

Figure 1. Principle structure of a knowledge-based system; shadowed components were realized within the developed prototype

One of the most important capabilities of a KBS is to make implicit knowledge explicit, that is to prove, to identify, to derive new representations from the inherited knowledge.

Turning now to design applications, one can identify two major phases of design, namely *problem formulation* and *problem solution*. While usually for the latter software is being used, the former is the user's responsibility, both by mind and by hand.

The developed prototype supports the user also in declaratively setting up his problem in mathematical terms, completing the *problem formulation knowledge* and mapping the mathematical representation directly onto a logical one. The problem can now being tackled with the help of the *problem solution knowledge*.

The following definitions state in short terms, what we understand by the afore mentioned concepts, while this is treated in more depth in the next two sections.

Problem formulation knowledge *is aimed at structuring and classifying all requirements of a given problem. It serves as a basis for constructing a unique mathematical problem. This type of knowledge is mainly stated declaratively.*

Problem solution knowledge *is responsible for analysing the given problem formulation, for selecting one or more methods and for determining the sequence in which they have to be applied to get a numerical result and to transform this into a graphical representation. This type of knowledge is mainly stated procedurally.*

3. Problem Formulation Knowledge

In the following we outline what is meant by problem formulation knowledge and how it is structured. First we describe which kind of formulation knowledge could be part of our system, but because only a part of the functionality was implemented only the relevant constraints are shown in more detail.

Our approach is based heavily on the assumption that it is possible to use a unified problem formulation as basis for a unified problem solution.

Most problems in shape generation can be stated in terms of

- the given data elements for shape properties,
- the mathematical shape representation form, and
- the generation process described by a criterion function and constraints.

Hence the shape generation can be viewed as a constrained optimization process [1, 17].

Based on this assumption we analyzed the mathematical basis of a number of methods and generalized as far as possible, based on systematic overviews, e.g. [7, 5].

For the given data it is possible to distinguish between discrete and integral shape requirements. While the former include coordinate vectors for points, tangents, curvatures, higher derivatives and combinations thereof, the latter cover area under the curve, centroid location and higher order moments.

The mathematical representation (1) for a parametric curve $Q(t)$ is characterized by the basis functions, i.e., polynomial, rational, exponential or trigonometric, the type of coefficients, i.e. Bezier, Hermite or monomial, the parametrization, i.e., equidistant, chordal or centripetal, but also by the segmentation including statements about number, boundaries and weights of segments.

$$Q(t) = (x(t), y(t), z(t))^T = \sum_{j=1}^{N} \sum_{k=0}^{n_j} B_{j,k}(t, \underline{\tau}) \underline{C}_{j,k} \tag{1}$$

The notation is according to the following:

N	number of segments
n_j	polynomial degree in segment j
$\underline{C}_{j,k}$	coefficients in matrix form
$B_{j,k}(t)$	k'th basis function in segment j
$t \in [a, b]$	global parameter
$\underline{\tau} = \{\tau_0, \ldots, \tau_N\}$	segment knot vector $a = \tau_0 < \ldots < \tau_N = b$

Here only parametric, polynomial representations are taken into account and we choose the Bezier representation because of its good behaviour as far as numerical stability, recursive computability (de Casteljau's algorithms), easy scaling for non-standard parameter intervals and good geometric interpretation are concerned, but other representations could be chosen as well.

The generation process is characterized by an objective function and constraints, which are either discrete or integral. We considered the following constraints of discrete type:

- continuity of certain order and type (geometric, parametric),
- interpolation of coordinates for points, directions and derivatives,
- prescribed end conditions, e.g. of type I (prescribed tangent), of type II (prescribed curvature), of periodic or natural type.

$$\text{I: } E = D^1 \underline{Q}(t_i) - \underline{T}_i = 0 \quad \text{or} \quad \text{II: } E = D^2 \underline{Q}(t_i) - \underline{K}_i = 0 \quad \text{for} \quad i = 1 \text{ or } N$$

- *single point* as well as *weighted sum* approximations

$$A = \sum_{i=1}^{N} \{w_i(\underline{Q}(t_i) - \underline{R}_i)\}^2 - \varepsilon \leq 0 \quad \text{or} \quad A_i = (\underline{Q}(t_i) - \underline{R}_i)^2 - \varepsilon_i) \leq 0$$

Please note that the approximation conditions are not subjected to a given tolerance, rather the RMS error is minimized. This notion has been adopted to avoid nonlinear inequality constraints and possibly empty feasible domains for small tolerances while the number of segments is fixed.

Area conditions of the form $F = S - S_0 = 0$ belong to the group of integral constraints, but are not elaborated upon here.

The objective function J is formulated as a weighted sum of several discrete and integral criteria with respect to arbitrary order of derivatives. Note that the used criteria can differ from one segment to another.

$$J = \underbrace{\sum_{k=1}^{N} p_k \sum_{i=1}^{m_k} w_{k,i} \int_{\tau_{k-1}}^{\tau_k} \left\| \frac{d^i \underline{Q}(t)}{dt^i} \right\|^2 dt}_{T_1} + \underbrace{\sum_{k=0}^{N} \sum_{l=1}^{m_k - 1} v_{k,l} \left\| \frac{d^l \underline{Q}(\tau_k)}{dt^l} \right\|^2}_{T_2} + \underbrace{\sum_{k=0}^{M} \theta_k \| \underline{Q}(t_k) - \underline{P}_k \|^2}_{T_3}$$

where the following notation is adopted:

$\{m_1, \ldots, m_N\}$	max. order of integral criterion
$\{w_{k,1}, \ldots, w_{k,m_k}\}$	weights of integral criteria in segment k
$\{p_1, \ldots, p_N\}$	weights of segments
$v_{k,l}; \theta_k$	weights of knots or points, resp.
$\{\underline{P}_0, \ldots, \underline{P}_M\}$	given points

Now the problem has a *constrained optimization format*:

"Minimize J subject to $A \leq 0$, $E = 0$, $F = 0$ etc."

This gives a wide ranging class of already well known methods but also certain new combinations of fairness criteria, constraints and representations, characterizing new methods not developed yet. It can be seen that among others the conventional $C^{(2)}$ cubic splines (with minimal strain energy), the change of curvature minimal splines [13], the segment weighted cubic splines [6], the point weighted cubic splines (v-splines, [14]), the geometric splines [9] and the quintic τ-splines [8] are comprised as special cases proving the great flexibility of this approach.

For implementation purposes we restricted our prototype to those cases in which all constraints are linear expressions and the criterion is a quadratic polynomial.

The presented characterization for shape generation problems is applicable to surfaces, too, despite the fact that at least higher complexity of the criterion function must be expected, but also it is difficult to formulate appropriate fairness measures. Though the developed prototype is restricted to curves only, it can be extended to model even complex surfaces in the same way.

This might be the right place for giving some examples for facts containing the problem formulation. All user requirements given either by file input or graphical input devices are stored in the database in form of facts tagged by the predicate *given*. If one wants to design, e.g., a composite polynomial curve of 3 segments, each of degree 3 and joined twice differentiable at the break points, then this could be expressed in the following way:

```
given(degree_per_segment([3,3,3])).
given(base_per_segment([bezier, bezier, bezier])).
given(order_per_break([2,2])).
```

Further statements could specify the parameter to lie within the range of $[0, 3]$ and the use of natural boundary conditions at the left boundary.

```
given (parameter_domain ([0, 2])).
given (boundary ([natural, none])).
```

The curve design problem intended to be solved may contain some more specifications, e.g., related to given points and the placement of breakpoints. The point data include an identification number, a characterization whether this is a point (indicated by 0) or a n-th order derivative at this point (indicated by n), the coordinate vector and a flag to indicate if this point or derivative shall be interpolated or approximated.

```
default (parameter_domain ([0, 3])).
given (points_for_parametrization ([1, 2, 3, 4])).
given (points_for_segmentation ([1, 2, 3, 4])).
given (point (1, 0, [1.0, 9.0, 0.0], interpolation)).
given (point (1, 1, [1.0, 0.0, 0.0], approximation)).
```

But not all information must be given by the user, default values are taken otherwise and the related facts are tagged by the predicate *default* instead of *given*, e.g.

```
default (type_of_parametrization ([equidistant])).
```

To give an actual value afterwards overrides the default value of the same predicate used before. Finally there is a third classifying predicate *derived*, pointing to the fact that this information was generated by the inference mechanism of the system using the given facts. This keeps intermediate results and prevents from starting from scratch each time anew, if only slight modifications in the problem formulation occur. Examples for that are:

```
derived (number_of_segments (3)).
derived (segmentation ([0, 1, 2, 3])).
```

Obviously the number of segments can be derived from the list containing the points for segmentation while the segmentation itself depends on the parameter domain, the type of parametrization and the points for segmentation also. Please notice that the two facts shown above must be compatible.

The three classification predicates *default*, *given* and *derived* can be seen in a hierachical order with increasing priority. This means that an attribute value of a certain priority can be overwritten only by a higher priority, but not by a lower one.

Even when formatted, this notation is relatively easy to read and close to a verbal formulation and though all examples are shown only in textual fashion, nearly all settings for attribute values or more generally for the definition of requirements can be performed using graphical input devices very conveniently.

From the examples given above it also can be seen that for the representation of knowledge nested lists are being used very often.

A preprocessing of all user defined data is the last step of the problem formulation in order to get a completed and consistent formal description of the curve design problem.

4. Problem Solution Knowledge

To get a representation of a geometric object fulfilling our requirements we have to match several subgoals in a defined order, starting with the formalization and completion of user specifications as described in the section before. After that all terms involved have to be evaluated to the simplest possible form and as an intermediate result we get a unified *formal problem description* stored in the data base and corresponding directly to an optimization format. Now this description has to be analyzed and a number of appropriate methods, numerical and/or symbolical, have to be selected and applied. This is done according to certain heuristic measures, explained in detail somewhat later. The final numerical result serves as a basis for generating a representation of the geometric object.

While a considerable part of the problem formulation knowledge has the form of facts depending on the users goals, the problem solution knowledge mainly contains rules describing processes and heuristics for the problem solution. The methods are solving algorithms as well as transformations, including elimination and evaluation procedures which may be applied successively.

Testing if a method has been applied successfully leads to an end of the iteration, and the same applies if there are no methods left which can be used.

There are several types of knowledge included, showing the hybrid character the system. For instance we have rules for

- analysis and validation of user's input
- determining the sequence of and interdependence between rules, i.e. the meta-knowledge, in PROLOG
- numerical algorithms, in C
- supervision, guidance and recording of the solution process
- initializing and managing data structures, in C
- transformation of symbolic expression, in MATHEMATICA

One of the heuristics used within our approach is a simple but also a powerful one. Given a pool of several methods which may be applied to the actual problem description, initially the most specific and least time-consuming method will be tried, followed by the next more general one in the case of failure etc. The process stops if one method has reached successful completion.

The current implementation includes three different linear equation solvers:

- QR-decomposition (for quadratic, symmetric and positive definite matrices)
- Gauss' elimination method (for quadratic matrices)
- Householders method (for arbitrary matrices)

The case of arbitrary matrices is the most general case solvable by linear solvers.

The choice of the method is performed on the basis of the properties of the matrix. If, for example, a quadratic matrix proves to be singular during Gauss' elimination, then the more general Householders method is applied.

As we have restricted ourselves to linear conditions and a quadratic objective function there is enough potential to expand the knowledge base in order to enlarge the facilities of our prototype by adding further methods. To that end the usage of a numerical toolkit offering several nonlinear solvers and also quadratic programming solvers should be very helpful.

Using the backtracking facilities of PROLOG the system is able to produce all potential solutions according to the given problem descriptions. The choice of a "best" solution remains up to the user. Another implemented strategy is to reduce the dimension of the problem step by step eliminating a certain number of variables in each. So one can see this solution process as a series of successive transformations.

While making use of both numerical and symbolic capabilities to get a result, pure numeric algorithms are used whenever possible to be more efficient. For the same goal if a symbolic procedure is called the first time, the gained result, i.e. a formula, is put into the knowledge base as a fact. This can be interpreted as making implicit knowledge explicit.

5. Components and Functionality of the Prototype

A prototype of a knowledge based system containing geometric knowledge for problem formulation and solution has been implemented as a hybrid environment with a PROLOG kernel and several other components with special facilities (Fig. 2).

The developed prototype adopts the principle structure of a KBS as outlined in Section 2, including all parts except the knowledge acquisition component (cf. shadowed boxes in Fig. 1).

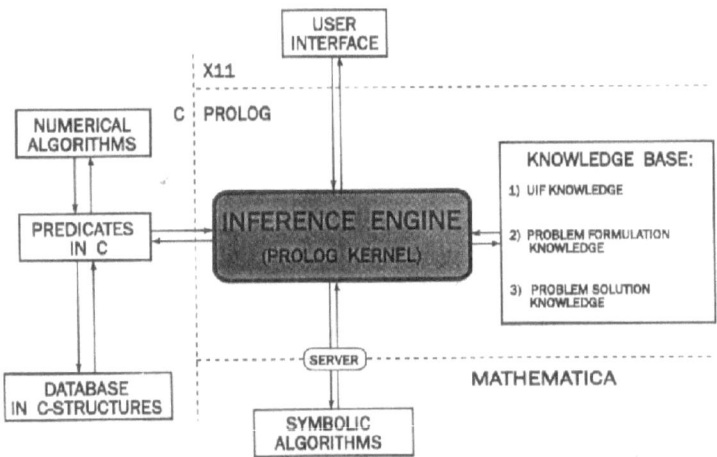

Figure 2. Components of the developed hybrid knowledge-based system

The central component of the system is the inference engine which administers and controls the other components. For that purpose a PROLOG system with a well supported programming interface in C language and a X11 graphical interface was chosen. It seemed to be very natural to take a declarative, logic programming language for our purposes, i.e., to formulate a problem in terms of requirements. On the other hand PROLOG is also a good tool to express dependencies between different parameters, e.g., degree, differentiation order for fairness measures and continuity conditions at breakpoints, enabling the system to perform implicit dependency checks to keep the database consistent.

The knowledge base consists of facts and rules formulated in PROLOG and, indirectly, in C and MATHEMATICA. This is realized via the programming interface mentioned before. Our graphical user interface (X11 and PROLOG) differs from most others in that it is very close to mathematical problem formulations due to the nature of declarative problem definition.

As a message board in general has to explain the status, agenda and systems configuration and should offer some error diagnostics and help functions, the respective component of our system comprises

- a protocol of the inference path, i.e., a collection of all conditions and constraints and their types, steps of elimination, resulting dimension of the problem, number of parameters etc.
- a window, which shows selected parts of the current knowledge base, i.e., facts for default, given and derived facts contained in the knowledge base.

6. Examples

In this section we present several examples in order to give an idea of the flexibility our prototype offers, despite the fact, that the mathematical form of the results is known for particular cases and that numerical results can be computed in principle by other software, too. But the main advantage of our approach is to get a unified description for a great variety of problems. This allows the user very flexibly to choose the problem type at session time.

In the first example we consider 5 points on a circle to be interpolated by a cubic spline curve with 4 segments. The first point is identical to the last one to ensure the closure of the resulting curve. It can be seen how the curve reflects the different end conditions, e.g., if we use natural or periodical end conditions (Fig. 3).

If we fix the periodic end conditions for a moment and choose different parametrization for first two segments (1:1 resp. 1:3) this again gives two different shaped curves (Fig. 4).

We now modify the continuity constraints interactively, i.e., starting with a C^1 continuous curve we decrease the order of continuity by one at each breakpoint to get a C^0 curve, as you can see in Fig. 5. The fairing was done with a second order criterion and no end conditions were specified.

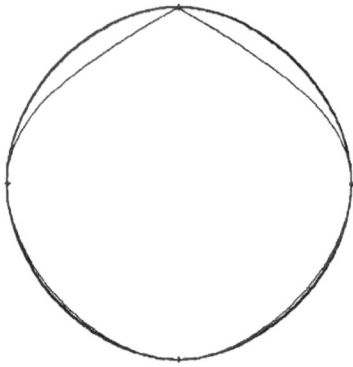

Figure 3. Interpolating curves with periodical or without end conditions

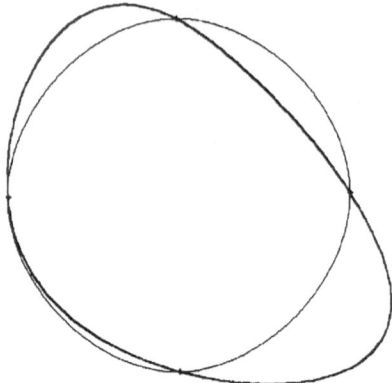

Figure 4. Different parametrizations

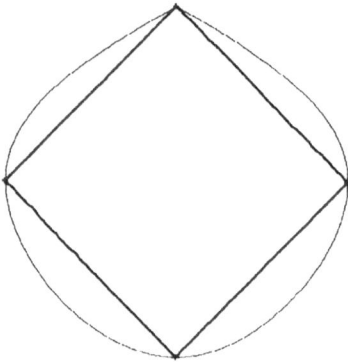

Figure 5. Different continuity conditions

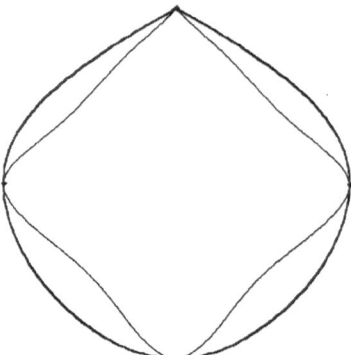

Figure 6. Different fairness criteria

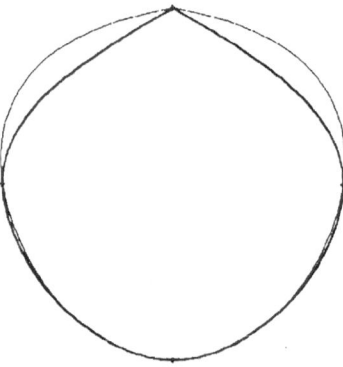

Figure 7. Different fairness criteria

The last two figures within this example show that different fairness criteria yield different curve shapes, e.g., in Fig. 6 the thin curve minimizes the first order derivative, while the fat curve corresponds to the second order minimum. The curve is required to be of fifth degree in all its segments, no end conditions are considered, and to be C^2 continuous at all breakpoints. Replacing the order of derivative of order 1 by order 3 instead gives Fig. 7.

In the second example points lying on a quarter circle are to be interpolated by a quintic spline curve with 3 segments. The tangent vectors at the endpoints shall be interpolated, too. The remaining degrees of freedom are chosen to minimize a weighted second order fairness criterion. If the weight of the middle segment changes from 1 to 5, the fat curve turns into the thin one (Fig. 8). In Fig. 9 again the results of first and second order fairing have been compared, giving the thin and the bold curve respectively. Both curves are C^2 continuous and of degree 5. The effect of changing the length of the tangent vector has been demonstrated in Fig. 10. A cubic spline is generated according to the tangential end conditions. If the prescribed tangents are unit vectors we get the thin curve while a scaling by factor 10 (and leaving the direction unchanged) gives the bold curve.

In the third example given points grouped in form of a staircase are approximated by a piecewise quintic curve of 4 segments. In this example C^2 continuity conditions at all break points are required and no end conditions are given. If this approximation curve is computed without any fairing, we get the thin shaped curve in Fig. 11, while using the first order fairness criterion the bold one will be obtained.

A variation of the order of derivative from 2 to 3, used in the fairness criterion gives slight modification in the curve (Fig. 12). One can observe, the higher the order of derivative, the more flexible is the behaviour of the curve. To finish this example we change some constraints. The endpoints (marked by crosses) are interpolated yielding the bold curve and setting the weights for two points equal to 10 while all others remain equal to 1 gives thin one (Fig. 13).

Figure 14 shows digitized data points of a ship section. As the different markers indicate, these points should be interpolated or approximated. No end conditions are defined. According to our specification we get a cubic spline curve minimizing the functional based on the second derivative (Fig. 15).

In the next example we are given a number of digitized points from a profile from NACA series (Fig. 16). These points are interpolated by a cubic spline consisting of 8 segments, while no end conditions are specified. The cubic spline using natural end conditions has been computed for equidistant and chordal parametrization giving the thin and the bold curve, respectively, as a result (cf. Fig. 17).

The final example demonstrates what can happen, if there is no solution to the problem statement given by the user. The given points are lying on a quarter circle and should be interpolated by a cubic spline with 3 segments. There is a tangential end condition at the left boundary and a natural at the other boundary. By adding a further point to be interpolated we get an overconstrained problem. Hence the resulting curve is discontinuous at the breakpoints due to the effect of RMS fitting (Fig. 19).

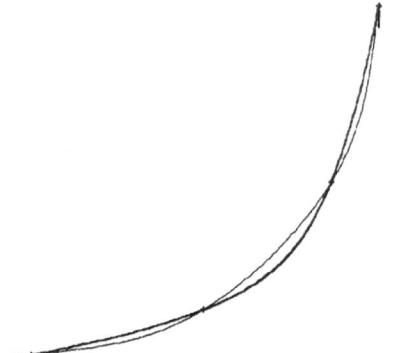

Figure 8. Interpolation curve due to second order fairing using different segment weights

Figure 9. First order versus second order fairing

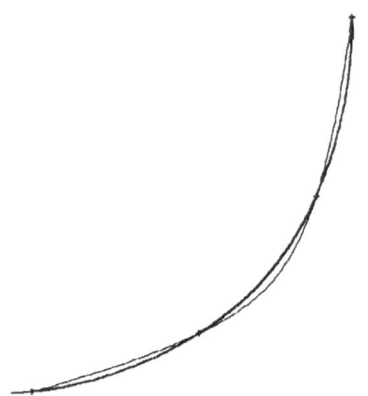

Figure 10. Interpolating curve for different length of tangent vectors

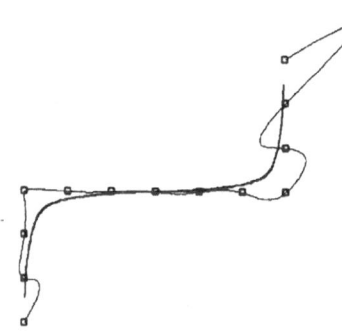

Figure 11. Approximation with and without fairing

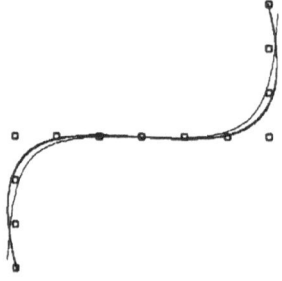

Figure 12. Different orders of fairness criteria

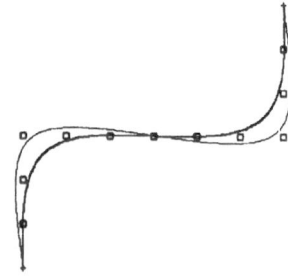

Figure 13. Changing some constraints

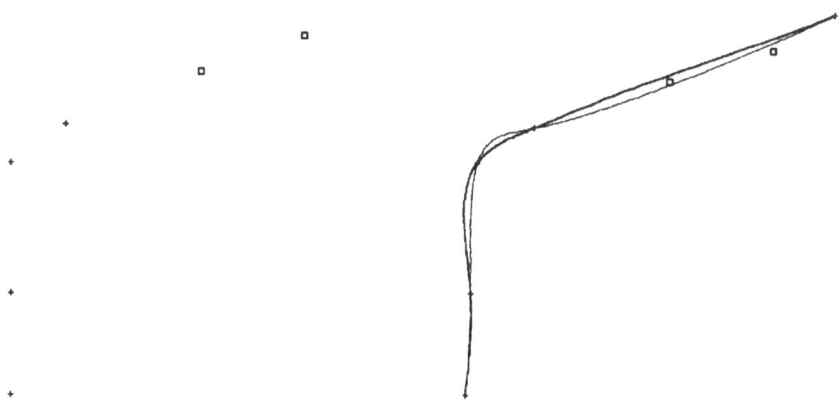

Figure 14. Given points to interpolate or approximate

Figure 15. Different parametrization types

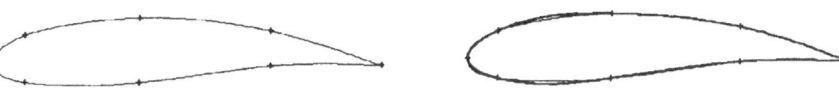

Figure 16. Interpolating given points

Figure 17. Different parametrization types

Figure 18. Curve with control polygon

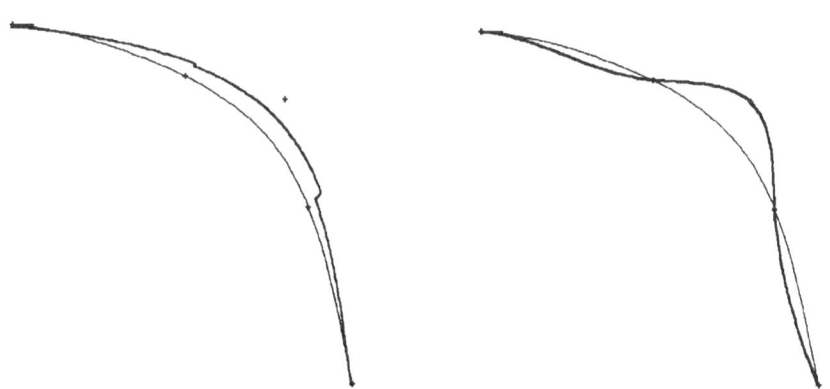

Figure 19. More constraints than unknowns

Figure 20. Elevate the degree for one segment

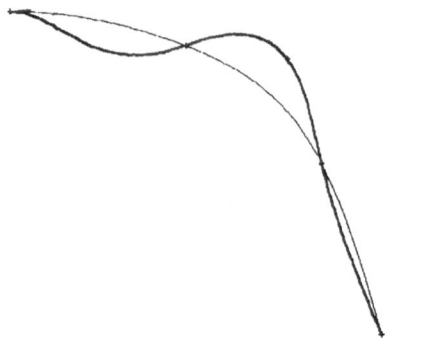

Figure 21. Reduce the order of continuity at one break

To eliminate these unwanted discontinuities, there are at least 3 different possibilities. Either we could

- elevate the degree of one segment from 3 to 4 (Fig. 20), or
- reduce the order of continuity at one break from 2 to 1 (Fig. 21), or
- drop the natural end condition,

and in all cases the resulting curve is continuous again!

7. Summary

We have presented a new concept of a knowledge based system for geometric design purposes based on a unified problem description format for shape generation problems in an explicit and declarative form. It was shown that this problem formulation can serve as a basis for a unified problem solution.

To solve a given problem one has to state this problem in mathematical terms. After that the problem description is mapped onto a logical format inherent to a declarative language, getting the problem formulation knowledge as a result. An inference mechanism controls the invocation of flexible numerical solvers for linear constraints and quadratic criterion functions in order to get a solution. All knowledge involved in that process is therefore called problem solution knowledge. The numerical result is finally transformed into a graphical representation, typically accompanied by information about the inference path and the uniqueness of the solution.

A prototype implementation was developed and tested using PROLOG as inference engine embedded in a hybrid programming environment with MATHEMATICA, C and X11 components.

So it is possible to construct a curve out of requirements and without hardcoding the solution for each single case in advance.

Though most problems are considered in 2-space because graphical input and output devices are most convenient for this case, the system can handle problems in n-space as well.

The developed prototype can serve as a tool both for students and experienced persons.

In the first case, it can be called *Teaching Environment*, the students can learn about the underlying theory of geometric design just by playing with the parameters. This seems a good way of becoming acquainted with the qualitative behaviour of a special feature or method so that a good geometric feeling can be gained.

In the second case, it can be called *Constraint Design Environment*, a more experienced person, maybe an engineer, can work on constraint design problems or can do some parametric design presuming that a more application oriented user interface is developed.

Some examples in curve design have been shown, illustrating the openness and flexibility of our approach. Especially worth mentioning is the fact that the system allows making choices about the problem type at session time, demonstrating great flexibility in that way.

Nevertheless there are some open issues requiring further investigation. For instance including solvers for inequality constraints would increase the power of the system considerably. Therefore it will be very helpful to use the capabilities of a symbolic system to a larger extent to simplify systems of inequalities and to identify redundant terms or possibly empty feasible domains.

Acknowledgements

The results presented in this paper originated from the project "Object Oriented and Knowledge Based Geometric Modelling" within the Sonderforschungsbereich 203, located at the Technical University of Berlin until the end of 1992 (cf. [15]). This research was funded by the Deutsche Forschungsgemeinschaft (DFG). We are grateful to L. Schumaker for pointing out some related papers also considering logic-based numerical problem solving environments [11, 12], but in a different context. We also want to thank Uwe Umlauf, Feng Jin and Uwe Friedrich for their helpful discussions and their contributions to both ideas and implementation of our system.

References

[1] Bercovier, M., Jacobi, A.: Approximation and/or construction of curves by minimization methods with or without constraints. Math. Model. Numer. Anal. *26*, 211–232 (1992).
[2] Bjorkenstam, U., Westberg, S.: General cubic curve fitting algorithm using stiffness coefficients. CAD *19*, 58–64 (1987).
[3] Cheng, F., Barsky, B. A.: Interproximation: interpolation and approximation using cubic spline curves. CAD *23*, 700–706 (1991).
[4] Clocksin, W. F., Mellish, C. S.: Programming in Prolog. Berlin Heidelberg New York: Springer 1984.
[5] Farin, G.: Curves and surfaces in CAGD—A practical guide. Boston, New York: Addison Wesley 1993.
[6] Foley, T. A.: Local control of interval tension using weighted spline. CAGD *3*, 281–294 (1986).
[7] Hoschek, J., Lasser, D.: Grundlagen der geometrischen Datenverarbeitung. Stuttgart: B. G. Teubner 1989.
[8] Lasser, D.: Visually continuous quartics and quintics. Computing *45*, 135–141 (1990).
[9] Hagen, H.: Geometric spline curves. CAGD *2*, 223–227 (1985).
[10] Kitzmiller, C. T., Kowalik, J. S.: Symbolic and numerical computing in knowledge based systems. In: Coupling symbolic and numeric computing (Kowalik, J. S., ed.), pp. 3–15. Amsterdam: North-Holland 1986.

[11] Mason, J. C., Reid, I.: Numerical problem-solving environments—current and future trends. In: Scientific software systems (Mason, J. C., Cox, M. G., eds.), pp. 202–209. London New York: Chapman and Hall 1988.

[12] Iles, R. M. J., Reid, I., Mason, J. C.: What do we mean by expert systems? In: Scientific software systems (Mason, J. C., Cox, M. G., eds.), pp. 202–209. London New York: Chapman and Hall 1988.

[13] Meier, H., Nowacki, H.: Interpolating curves with gradual changes in curvature. CAGD 4, 297–305 (1990).

[14] Nielson, G. M.: Some piecewise polynomial alternatives to splines under tension. CAD, 209–235 (1974).

[15] Nowacki, H., et al.: Objektorientierte und wissenbasierte Verfahren der Flächenmodellierung, final report, SFB 203, TU Berlin 1992.

[16] Nowacki, H., Liu, D., Lü, X.: Fairing Bezier curves with constraints. CAGD 7, 43–55 (1990).

[17] Nowacki, H.: Mathematische Verfahren zum Glätten von Kurven und Flächen, In: Geometrische Verfahren der Graphischen Datenverarbeitung (Encarnacao, J. L., Hoschek, J., Rix, J., eds.), pp. 22–45. Berlin Heidelberg New York Tokyo: Springer 1990.

[18] Ohsuga, S.: Toward intelligent CAD systems. CAD 21, 315–337 (1989).

[19] Rogers, D. F., Fog, N. G.: Constrained B-Spline curve and surface fitting. CAD 21, 641–648 (1989).

[20] Wolfram, S.: Mathematica—a system for doing mathematics by computer. Redwood City: Addison-Wesley 1991.

Uwe Langbecker
Horst Nowacki
Institut für Schiffs- und Meerestechnik
Sekr. SG 10, Salzufer 17–19
Technische Universität Berlin
D-10587 Berlin
Federal Republic of Germany

Computing Suppl. 10, 227–242 (1995)

Bézier Representation of Trim Curves

D. Lasser and **G. P. Bonneau,** Kaiserslautern

Abstract. The composition of Bézier curves and tensor product Bézier surfaces, polynomial as well as rational, is applied to exactly and explicitly represent trim curves of tensor product Bézier surfaces. Trimming curves are assumed to be defined as Bézier curves in surface parameter domain. A Bézier spline approximation of lower polynomial degree is built up as well, which is based on the exact trim curve representation in coordinate space.

Key words: Bézier representation, trim curves, curves on surfaces, composition, approximation, free form deformation.

1. Introduction

Very few three-dimensional objects are representable by a single (free form) surface patch. Generally surfaces have to be *glued* together smoothly as in a patch work to form spline surfaces, and (spline) surfaces have to be processed further within CAD/CAM solid modelling systems. Some of these interrogation techniques, such as intersecting and blending, trim surfaces to an area bounded by border curves (see Fig. 1) commonly referred to as *trim curves*. In context of blending they are also called *contact* or *link curves*. This is to highlight the fact that in blending, trim curves are curves of contact along which primary surface and blend surface are linked together (see Fig. 2).

Literature on the subject of trim curves and trimmed surfaces can be grouped into three categories. First, there is the solid modelling, second, the computer graphic, and third, the CAGD oriented literature.

Undoubtedly, *solid modelling* is the area with most contributions. Issues discussed are the ones most important for Boolean set operation and for boundary evaluation: First, the intersection problem of trimmed surfaces and of sculptured solids, i.e. solids with a free form outer (trimmed) surface, and second, set membership classification, which classifies a point as being interior to, on the boundary of, or exterior to a set [3, 4, 7, 12]. Engineering analysis interrogation algorithms for solids bounded by trimmed surfaces, i.e. computation of surface area, volume, center of gravity, moments of intertia and other mechanical or mass properties are subject of [12] and [5]. [29] triangulates trimmed surfaces for the purpose of data exchange between CAD systems and a stereolithography apparatus to generate a solid hard copy directly from a 3D CAD model. The faceting algorithm described in [29] is suitable for rendering of parametrically defined surfaces.

Rendering is also the main issue of the *computer graphic* oriented literature on trimmed surface. Even though, very different methods are applied: [26] creates

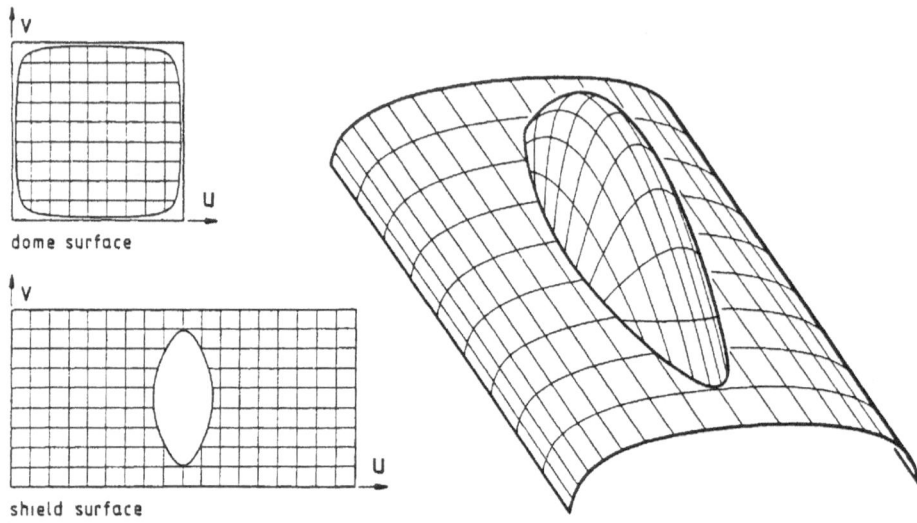

dome surface

shield surface

Figure 1. Surface–surface intersection creates trimmed surfaces

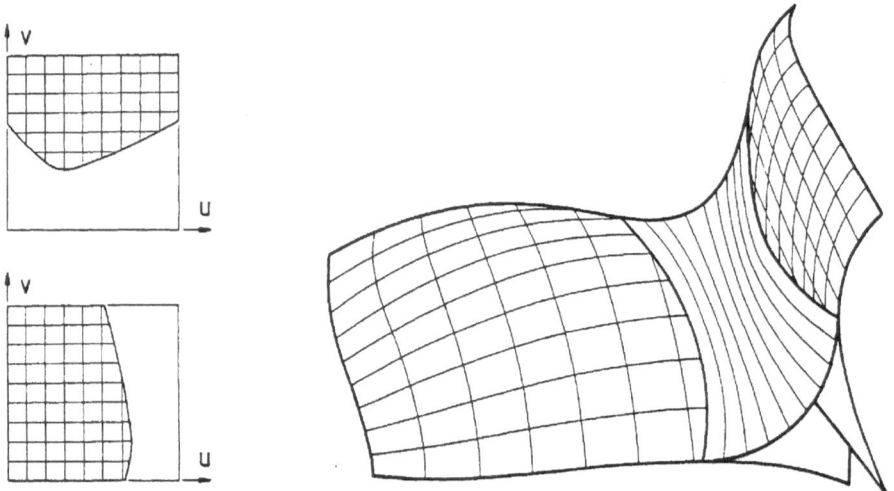

Figure 2. Blending of surfaces creates trimmed surfaces

a faceting of trimmed rational tensor product surfaces. Facets are lighted, smooth shaded and z-buffered using standard 3D polygon rendering techniques. [28] performs a scan line based rendering of trimmed NURBS surfaces using adaptive forward differencing together with a Hermite shading function approximation along surface curves. [25] renders trimmed rational tensor product Bézier surfaces using the ray tracing methology. While [26] and [28] have to calculate intersection points of isoparametric lines and trim curves to actually perform trimming, [25] realizes trimming by point classification, i.e. by determining if a point on a patch lies inside or outside a trimmed region.

In CAGD contributions on trimmed surfaces concentrate on two issues: data exchange between CAD systems and blending of surfaces. In blending, determination of trim curves is one of the crucial points. It can be done interactively or automatically, in coordinate space or in parameter space [1, 6, 14, 15, 20, 22, 27]. Data exchange is subject of [18], describing an approximate conversion of NURBS surfaces trimmed by NURBS curves by lower degree B-spline representations, and of [30, 31], exactly converting a polynomial tensor product Bézier surface trimmed by polynomial Bézier curves into a composite Bézier surface. The later algorithm is based on subdividing the trimmed parameter space region of a surface in planar three- and four-gonal regions, and on the substitution of a bilinear Coons representation of these parameter space areas into the surface representation.

Most algorithms assume trim curves $K(t)$ to be given in polynomial or rational representation in parameter space of surface $F(u, v)$: $\mathbb{R}^2 \to \mathbb{R}^3$, i.e. $K(t) = (u(t), v(t))$: $\mathbb{R} \to \mathbb{R}^2$, even if they have been calculated approximatively or in discrete points only, by surface intersection or projection, for example. Further trim curve processing then usually involves one or more approximation procedures at different levels of processing.

While it is well understood that the mathematical nature of trimming is functional composition, $F(K(t))$, see Fig. 3, evaluation of $F(K(t))$ is done throughout the literature pointwise and up to very recently no exact closed form representation of $F(K(t))$ had been given, except for the triviale case of monomial representations of $K(t)$ and of $F(u, v)$, which can be found in [2] already (in context of free form deformation application). [30, 31] also gives an explicit and exact representation of compositions, in context of trim curves and trimmed tensor product surfaces. Though the work is primarily done for Bézier representations, [30, 31] unnecessarily

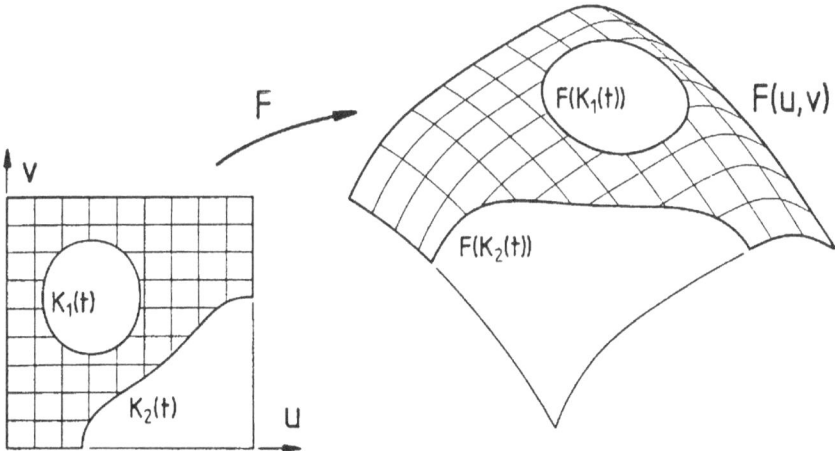

Figure 3. The mathematical nature of trimming is composition

performs conversions at several stages of the algorithm which makes the method less accurate, [13]. Composition especially is done in the monomial bases only, which requires conversion to and from this bases.

To have an exact and explicit Bézier representation of compositions of Bézier representations would be of great value for many CAD/CAM, computer graphics and CAGD applications. T. DeRose [8] has been the first who actually performed functional composition in case of simplicial Bézier representations. DeRose also pointed out a whole selection of applications of functional composition such as evaluation, subdivision, (non-linear) reparametrization and geometric continuity. A more interesting example of application might be given by the idea of curve and surface modeling in the sense of free form deformation. A second, very important application of functional composition concerns the subject of curves on surfaces, e.g. trimming of surfaces by surface curves, with applications in blending and in solid modeling, i.e. trimmed surfaces. [10] looks at Bézier curves and tensor product Bézier surfaces, both polynomial and rational. But there are no explicit Bézier-like equations given for $\mathbf{F}(\mathbf{K}(t))$ and furthermore, Table 6.1 of [10] listing polynomial degrees of $\mathbf{F}(\mathbf{K}(t))$ is erroneous. [23] provided exact Bézier-representations of $\mathbf{F}(\mathbf{K}(t))$ for polynomial and rational Bézier curves and surfaces, multivariate composition is treated as well. [9] deals with the same topic, but is highlighting the blossoming nature of algorithms much stronger than [23] does.

In this paper we apply (and discuss a special case of) the theoretical results given in [23] to the problem of exactly representing curves on surfaces, here trim curves, in Bézier form. An approximative representation is given as well. The paper is structured as follows: Section 2 reviews definitions of Bézier curves and surfaces. In Sections 3 and 5 Theorem 1 (polynomial case) and Theorem 2 (rational case) are given which are fundamental for the exact representation of Bézier curves on tensor product Bézier surfaces. Sections 4 and 6 apply Theorem 1 and Theorem 2 to exactly and explicitly represent trim curves of surfaces. Section 7 notes one more application of functional composition and of theorems given in Sections 3 and 5, planar curve design via free form deformation. Section 8 is concerned with the task of approximating trim curves by lower degree polynomials.

2. Bézier Representations

A planar *Bézier curve* $\mathbf{K}(t)$ of degree N in t is defined by

$$\mathbf{K}(t) = \sum_{I=0}^{N} \mathbf{K}_I B_I^N(t), \quad t \in [0, 1], \tag{1}$$

where $K_I \in \mathbb{R}^2$ and

$$B_I^N(t) = \binom{N}{I} t^I (1-t)^{N-I}$$

are the (ordinary) *Bernstein polynomials* of degree N in t.

A *tensor product Bézier surface* $\mathbf{F}(u, v)$—briefly *TBP-surface*—of degree (l, m) in (u, v) is defined by

$$\mathbf{F}(u, v) = \sum_{i=0}^{l} \sum_{j=0}^{m} \mathbf{F}_{i,j} B_i^l(u) B_j^m(v), \quad u, v \in [0, 1], \tag{2}$$

where $\mathbf{F}_{i,j} \in \mathbb{R}^3$ and with Bernstein polynomials $B_i^l(u)$ and $B_j^m(v)$. By reason of the tensor product definition algorithms in u and in v commute, and the result is independent of the order.

A planar *rational Bézier curve* $\mathbf{K}(t)$ of degree N in t is defined by

$$\mathbf{K}(t) = \frac{\displaystyle\sum_{I=0}^{N} \beta_I \mathbf{K}_I B_I^N(t)}{\displaystyle\sum_{I=0}^{I} \beta_I B_I^N(t)}, \quad t \in [0, 1], \tag{3}$$

and a *rational tensor product Bézier surfaces* $\mathbf{F}(u, v)$—briefly *rational TBP-surface*—of degree (l, m) in (u, v) is defined by

$$\mathbf{F}(u, v) = \frac{\displaystyle\sum_{i=0}^{l} \sum_{j=0}^{m} \omega_{i,j} \mathbf{F}_{i,j} B_i^l(u) B_j^m(v)}{\displaystyle\sum_{i=0}^{l} \sum_{j=0}^{m} \omega_{i,j} B_i^l(u) B_j^m(v)}, \quad u, v \in [0, 1]. \tag{4}$$

Coefficients $\mathbf{K}_I$ and $\mathbf{F}_{i,j}$ are called *Bézier points*. They form in their natural ordering, given by their subscripts, the vertices of the *Bézier polygon* and of the *Bézier net*. Scalars $\beta_I \in \mathbb{R}$ and $\omega_{i,j} \in \mathbb{R}$ are called *weights*. Positive weights result in curves and surfaces which have all the properties and algorithms which we do have for polynomial representations.

The Bézier description is a very powerful tool because the expansion in terms of Bernstein polynomials yields, first, a numerically very stable behavior of all algorithms. And, second, a geometric meaning of Bézier points (and weights). For an extensive coverage of properties of Bernstein polynomials and Bézier representations see e.g. [11, 19].

3. Composition of Polynomial Curves and Surfaces

Theorem 1 describes the composition $\mathbf{F}(t) = \mathbf{F}(\mathbf{K}(t)) = \mathbf{F}(u(t), v(t))$ of a planar Bézier-curve $\mathbf{K}(t)$ and a TPB-surface $\mathbf{F}(u, v)$. There are several problems in curve and surface modeling pointed out by DeRose [8] that can be solved using functional composition. Examples are evaluation, subdivision, (nonlinear) reparametrization and geometric continuity of Bézier representations. Two more examples of practical interest are curve and surface modeling via free-form deformation and the description of curves on surfaces. The later can be thought of as trim curves with applications in blending as well as in connection with trimmed surfaces as the occur in solid modeling. In that context Theorem 1 is fundamental for the exact representation of trim curves.

Theorem 1. *Let* $\mathbf{K}(t) = (u(t), v(t)): \mathbb{R} \to \mathbb{R}^2$ *be a planar polynomial Bézier curve of degree* N, (1), *with Bézier points* $\mathbf{K}_I = (u_I, v_I)$, *and let* $\mathbf{F}(u, v) = (x(u, v), y(u, v), z(u, v))$: $\mathbb{R}^2 \to \mathbb{R}^3$, *by a polynomial TPB-surface of degree* (l, m), (2), *with Bézier points* $\mathbf{F}_{i,j} = (x_{i,j}, y_{i,j}, z_{i,j})$. $\mathbf{F}(t) = \mathbf{F}(\mathbf{K}(t)) = \mathbf{F}(u(t), v(t))$ *is polynomial and can be represented as Bézier curve of degree* rN, *where* $r = l + m$. *We have*

$$\mathbf{F}(t) = \mathbf{F}(\mathbf{K}(t)) = \sum_{R=0}^{rN} \mathbf{R}_R B_R^{rN}(t), \tag{5}$$

with Bézier points

$$\mathbf{B}_R = \sum_{|\mathbf{I}| = R} C_R^{l,m}(N, \mathbf{I}) \mathbf{F}_{0,0}^{l,m}(u_{\mathbf{I}^u}^l, v_{\mathbf{I}^v}^m), \tag{6}$$

and constants

$$C_R^{l,m}(N, \mathbf{I}) = \frac{\prod\limits_{Q^u=1}^{l} \binom{N}{I_{Q^u}^u} \prod\limits_{Q^v=1}^{m} \binom{N}{I_{Q^v}^v}}{\binom{rN}{R}}. \tag{7}$$

Proof of Theorem 1 is done analogously to [8] and has been given in detail in [23] (cf. [9]).

$\sum_{|\mathbf{I}| = R}$ has the meaning of summation over all $\mathbf{I} = (\mathbf{I}^u, \mathbf{I}^v)$, where $\mathbf{I}^u = (I_1^u, \ldots, I_l^u)$, $\mathbf{I}^v = (I_1^v, \ldots, I_m^v)$ and where $0 \le I_1^u, \ldots, I_l^u \le N$ and $0 \le I_1^v, \ldots, I_m^v \le N$ and $|\mathbf{I}| = |\mathbf{I}^u| + |\mathbf{I}^v| = I_1^u + \cdots + I_l^u + I_1^v + \cdots + I_m^v = R$.

Note that construction points $\mathbf{F}_{0,0}^{l,m}(u_{\mathbf{I}^u}^l, v_{\mathbf{I}^v}^m)$ arise in the calculation of the polar form (see e.g. [19]) of $\mathbf{F}(u, v)$. They can be computed recursively using the de Casteljau construction, i.e. for the u parameter direction by

$$\mathbf{F}_{i,j}^{i+\alpha, j+\beta}(u_{\mathbf{I}^u}^\alpha, v_{\mathbf{I}^v}^\beta) = (1 - u_{I_\alpha^u}) \mathbf{F}_{i,j}^{i+\alpha-1, j+\beta}(u_{\mathbf{I}^u}^{\alpha-1}, v_{\mathbf{I}^v}^\beta) + u_{I_\alpha^u} \mathbf{F}_{i+1,j}^{i+\alpha, j+\beta}(u_{\mathbf{I}^u}^{\alpha-1}, v_{\mathbf{I}^v}^\beta),$$

and for the v parameter direction by

$$\mathbf{F}_{i,j}^{i+\alpha, j+\beta}(u_{\mathbf{I}^u}^\alpha, v_{\mathbf{I}^v}^\beta) = (1 - v_{I_\beta^v}) \mathbf{F}_{i,j}^{i+\alpha, j+\beta-1}(u_{\mathbf{I}^u}^\alpha, v_{\mathbf{I}^v}^{\beta-1}) + v_{I_\beta^v} \mathbf{F}_{i,j+1}^{i+\alpha, j+\beta}(u_{\mathbf{I}^u}^\alpha, v_{\mathbf{I}^v}^{\beta-1}),$$

where $\mathbf{F}_{i,j}^{i,j} \equiv \mathbf{F}_{i,j}$. The argument $(u_{\mathbf{I}^u}^\alpha, v_{\mathbf{I}^v}^\beta)$ has the meaning that $\mathbf{F}_{i,j}^{i+\alpha, j+\beta}$ has to be calculated by performing α de Casteljau constructions in u direction for the u parameter values given by the indices $\mathbf{I}^u = (I_1^u, \ldots, I_\alpha^u)$, i.e. for the parameter values $u_{I_1^u}, \ldots, u_{I_\alpha^u}$ and β de Casteljau constructions in v direction for the v parameter values given by the indices $\mathbf{I}^v = (I_1^v, \ldots, I_\beta^v)$, i.e. for the parameter values $v_{I_1^v}, \ldots, v_{I_\beta^v}$. Calculations for different parameter values commute, and the order of performed calculations does not effect the final result.

The special case $N = 1$ deserves some extra attention: While it is well known that for a TPB-surface of degree (l, m) isoparameter lines map to Bézier curves of degree l or m, respectively, which can be calculated by applying the de Casteljau (subdivision) algorithm, Theorem 1 generalizes this statement in the sense that now we know,

lines of parameter space in general position map to Bézier curves of degree $l+m$[1].
Actually, Theorem 1 does not really differentiate between isoparametric lines and
lines in general position. Theorem 1 represents both of them as Bézier curves of
degree $l+m$. This sounds like Theorem 1 is wrong. Indeed, the only way it can work
is that,

Statement 1. In case of isoparametric lines Theorem 1 yields degree raised Bézier
curves.

Proof: To prove Statement 1, we go with the following strategy: First, we assume
$K(t)$ be isoparametric and specialize Theorem 1 to $N = 1$, second, we show that $F(t)$ is
degree raised, and third, we perform degree reduction and compare the result with
the Bézier representation of isoparametric lines, calculated using the de Casteljau
(subdivision) algorithm:

First, we specialize Theorem 1 to $N = 1$ and assume (because of the tensor product
structure of $F(u, v)$) w.l.o.g. $K(t)$, $t \in [0, 1]$, being given as part $u \in [u_0, u_1] \subseteq [0, 1]$ of
a u parameter line, i.e. $v_0 = v_1 \equiv v_*$ (calculations and arguments are analogous in
case of a v parameter line $u_0 = u_1 \equiv u_*$). This should result in a parametric curve
$F(K(t)) = F(u, v_*)$, $u \in [u_0, u_1]$, of degree l in u:

Because of $N = 1$, i.e. $I^u_{Q^u} \in \{0, 1\}$, and $I^v_{Q^v} \in \{0, 1\}$ (5) becomes

$$C^{l,m}_R(N, I) = \frac{1}{\dbinom{l+m}{R}},\tag{8}$$

and (4) simplifies to

$$B_R = \frac{1}{\dbinom{l+m}{R}} \sum_{|I|=R} F^{l,m}_{0,0}(u^l_{I^u}, v^m_{I^v}).$$

Because de Casteljau constructions commute, and the polar form of $F(u, v)$ is
symmetric w.r.t. permutations of the argument, i.e. $F^{l,m}_{0,0}(u^l_{I^u}, v^m_{I^v})$ is so, the sum
$\sum_{|I|=R} F^{l,m}_{0,0}(u^l_{I^u}, v^m_{I^v})$ is in the case of $v_0 = v_1 = v_*$ equivalent to $\sum_{a=0}^{R} \dbinom{l}{a}\dbinom{m}{R-a} F_a$,
using $\dbinom{n}{k} \equiv 0$ for $k < 0$ and for $k > n$, and where F_a stands for $F^{l,m}_{0,0}(u^l_{I^u}, v^m_*)$ with
a being the number of indices $I^u_{Q^u}$ of I^u of value 1, i.e. $a = |I^u| \le l$. Thus far, the
representation of $F(K(t))$ according to Theorem 1 simplifies to

$$F(t) = F(K(t)) = \sum_{R=0}^{l+m} B_R B^{l+m}_R(t),\tag{9}$$

---

[1] This is, by the way, in contrast to the situation for triangle Bézier surfaces. In case of a triangle Bézier
surface of degree n isoparametric lines as well as lines of parameter space in general position map to Bézier
curves of degree n.

with

$$B_R = \sum_{a=0}^{R} \frac{\binom{l}{a}\binom{m}{R-a}}{\binom{l+m}{R}} F_a. \tag{10}$$

Now, $F(K(t))$ is supposed to be of degree l instead of degree $l+m$. To prove the degree l of $F(K(t))$ we calculate forward differences

$$\Delta^k B_R = \sum_{j=0}^{k} (-1)^j \binom{k}{j} B_{R+k-j},$$

with B_{R+k-j} given by (10), and then prove validity of

$$\bigwedge_{i=1,\dots,m} \bigwedge_{R=0,\dots,m-i} \Delta^{l+i} B_R = 0. \tag{11}$$

Bézier points of the degree l representation finally result by degree reduction of (9), (10) from degree $l+m$ to degree l.

The last two steps, proof of (11) and degree reduction, are quite cumbersome. That is why we prefer to go a *short-cut* and are doing both steps at the same time:

First, we note that (10) is equivalent to

$$B_R = \sum_{\substack{a=R-m \\ a \geq 0}}^{R} \frac{\binom{l}{a}\binom{m}{R-a}}{\binom{l+m}{R}} F_a, \tag{12}$$

and second, we realize that (12) is the degree raising formula for raising the polynomial degree of a Bézier curve of degree l defined by Bézier points F_a from l to $l+m$. This completes the proof, because F_a stands for $F_{0,0}^{l,m}(u_{[u}^l, v_{*}^m)$, which means that F_a has been calculated using the de Casteljau algorithm exactly the precise number of times and for the correct parameter values. ∎

Remark 1. Bézier splines can be treated easily, too. First, we determine in the domain space of $F(u, v)$ the intersections of the spline curve $K(t)$ and the patch edges of spline surface $F(u, v)$ and split $K(t)$ at these intersection points by adding new knots applying de Casteljau's subdivision algorithm. For all segments of $F(u, v)$, all spline curve segments of $K(t)$ situated in the (u, v) domain of the specific spline segment of $F(u, v)$ can now be treated directly by Theorem 1. If $K(t)$ is C^a-continuous in $t = t^*$, and $F(u, v)$ is C^b-continuous in the corresponding point $K(t) = K(t^*)$ of (u, v) space, $F(K(t))$ is C^e-continuous in $F(K(t^*))$, with $e = min\{a, b\}$, 23].

4. Polynomial Trim Curves and Surfaces

Apply Theorem 1, curves on surfaces (e.g. trim curves) can be represented exactly, provided both curve as well as surface are of polynomial nature and, in that case, w.l.o.g. are given in Bézier representation.

Assuming $\mathbf{K}(t)$ not being degree raised, we first check in the event of $N = 1$, by comparing $\mathbf{K}_0$ with $\mathbf{K}_1$, if $\mathbf{K}(t)$ is isoparametric. If so, $\mathbf{F}(\mathbf{K}(t))$ is not of degree $l + m$ and, we do not have to calculate Bézier points $\mathbf{B}_R$ of $\mathbf{F}(\mathbf{K}(t))$ using Eqs. (6) and (7). Following Statement 1, $\mathbf{F}(\mathbf{K}(t))$ is of degree l or m, respectively, instead of degree $l + m$, and Bézier points $\mathbf{B}_R$ are defined by $\mathbf{B}_R \equiv \mathbf{F}_{0,0}^{l,m}(u_{\mathrm{I}u}^l, v_*^m)$, where $|\mathbf{I}^u| = R$, in case of $\mathbf{K}(t)$ being u parameter line $v_0 = v_1 \equiv v_*$, or by $\mathbf{B}_R = \mathbf{F}_{0,0}^{l,m}(u_*^l, v_{\mathrm{I}v}^m)$, where $|\mathbf{I}^v| = R$, in case of $\mathbf{K}(t)$ being v parameter line $u_0 = u_1 = u_*$, respectively. Otherwise, $\mathbf{F}(\mathbf{K}(t))$ is of degree $N(l + m)$, and it's Bézier representation is given by Theorem 1.

Figures 3–6 illustrate Theorem 1 and Statement 1 for the examples of trimming Bézier (spline) curves defined in the domains of TPB-surfaces. In case of Fig. 3 a closed cubic C^1 subspline and a quartic curve are mapped on a TPB-surface of degree (2, 4).

5. Composition of Rational Curves and Surfaces

Theorem 2 describes the composition of a planar rational Bézier curve of degree N, $\mathbf{K}(t)$, defined in domain space of a rational TPB-surface of degree (l, m), $\mathbf{F}(u, v)$.

Theorem 2. *Let* $\mathbf{K}(t) = (u(t), v(t))$: $\mathbb{R} \to \mathbb{R}^2$ *be a planar rational Bézier curve of degree* N, (3), *with Bézier points* $\mathbf{K}_I = (u_I, v_I)$ *and weights* β_I. *And let* $\mathbf{F}(u, v) = (x(u, v), y(u, v), z(u, v))$: $\mathbb{R}^2 \to \mathbb{R}^3$, *be a rational TPB-surface of degree* (l, m), (4), *with Bézier points* $\mathbf{F}_{i,j} = (x_{i,j}, y_{i,j}, z_{i,j})$ *and weights* $\omega_{i,j}$. $\mathbf{F}(t) = \mathbf{F}(\mathbf{K}(t)) = \mathbf{F}(u(t), v(t))$ *is rational and can be represented as rational Bézier curve of degree* rN, *where* $r = l + m$.

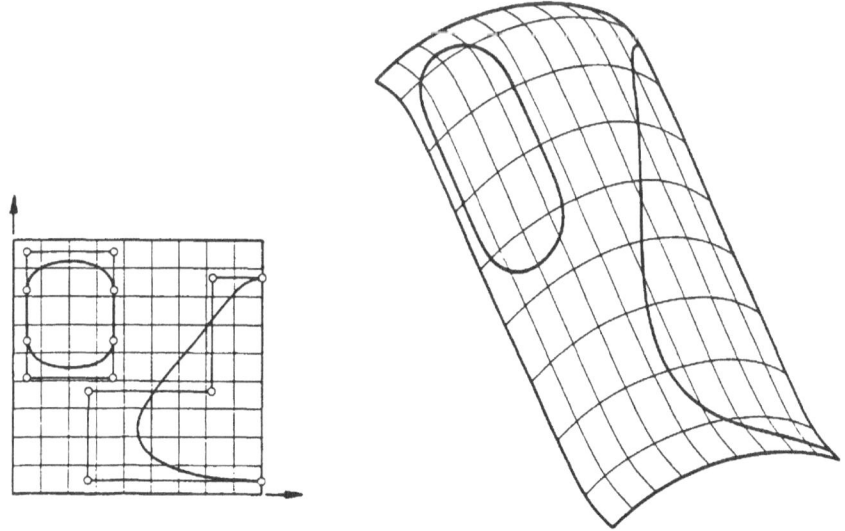

Figure 4. Mapping of a quintic and of a closed GC^1 linear/cubic trimming curve on a TPB-surface of degree (5, 3)

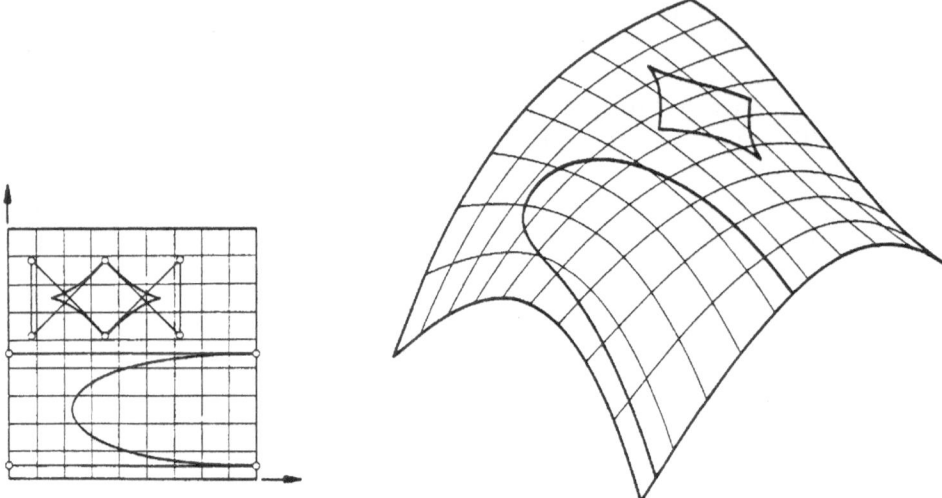

Figure 5. Mapping of cubic trimming curves on a bicubic TPB-surface

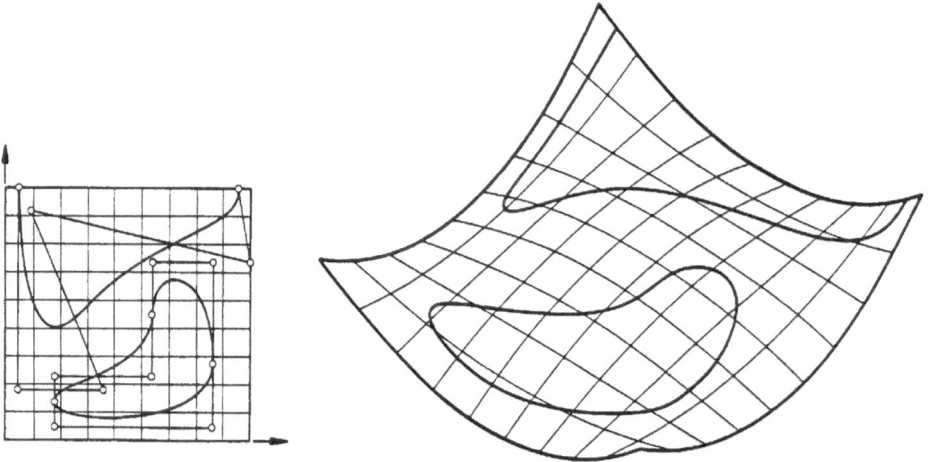

Figure 6. Mapping of a closed cubic GC^1 subspline and of a quintic trimming curve on a biquadratic TPB-surface

We have

$$\mathbf{F}(t) = \mathbf{F}(\mathbf{K}(t)) = \frac{\displaystyle\sum_{R=0}^{rN} \Omega_R \mathbf{B}_R B_R^{rN}(t)}{\displaystyle\sum_{R=0}^{rN} \Omega_R B_R^{rN}(t)}. \tag{13}$$

Weights are given by

$$\Omega_R = \sum_{|I|=R} B_R^{l,m}(N, \mathbf{I})\omega_{0,0}^{l,m}(u_{I^u}^l, v_{I^v}^m),\tag{14}$$

Bézier points are given by $\mathbf{B}_R = \Omega_R \mathbf{B}_R / \Omega_R$, *where*

$$\Omega_R \mathbf{B}_R = \sum_{|I|=R} B_R^{l,m}(N, \mathbf{I})\omega_{0,0}^{l,m}(u_{I^u}^l, v_{I^v}^m)\,\mathbf{F}_{0,0}^{l,m}(u_{I^u}^l, v_{I^v}^m),\tag{15}$$

and with constants

$$B_R^{l,m}(N, \mathbf{I}) = \frac{\displaystyle\prod_{Q^u=1}^{l}\beta_{I_{Q^u}^u}\binom{N}{I_{Q^u}^u}\prod_{Q^v=1}^{m}\beta_{I_{Q^v}^v}\binom{N}{I_{Q^v}^v}}{\displaystyle\binom{rN}{R}}.\tag{16}$$

Proof of Theorem 2 is essentially like the one of Theorem 1 with the difference, that rational representations are now involved. Proof of Theorem 2 has been given in [23]. $\sum_{|I|=R}$ and $\mathbf{I} = (\mathbf{I}^u, \mathbf{I}^v)$ have the same meaning as in Theorem 1. $\omega_{0,0}^{l,m}\,\mathbf{F}_{0,0}^{l,m}$ and $\omega_{0,0}^{l,m}$ are defined recursively by de Casteljau's construction, analogously to Theorem 1.

Theorem 2 maps straight lines of parameter space to rational Bézier curves of degree $l + m$. We have to show consistency of Theorem 2 with the fact that isoparametric lines actually map to rational Bézier curves of degree l or m, respectively. Therefore, we formulate.

Statement 2. In case of isoparametric lines Theorem 2 yields degree raised rational Bézier curves.

Proof: Statement 2 can be proven exactly the same way Statement 1 has been: Let $N = 1$ and $\mathbf{K}(t), t \in [0,1]$, being givdn as part $u \in [u_0, u_1] \subseteq [0,1]$ of a u parameter line, i.e. $v_0 = v_1 \equiv v_*$ (calculations and arguments are analogous in case of a v parameter line $u_0 = u_1 \equiv u_*$). W.l.o.g. β_0 and β_N can be chosen equal to one, $\beta_0 = \beta_N = 1$. Now, because of $N = 1$, these are the only weights defining $\mathbf{K}(t)$ and furthermore, we have $I_{Q^u}^u \in \{0,1\}$ and $I_{Q^v}^v \in \{0,1\}$. Equation (16) which defines $B_R^{l,m}(N, \mathbf{I})$ therefore simplifies to (8) and Eqs. (14) and (15) become (cf. Statement 1)

$$\Omega_R = \sum_{\substack{a=R-m \\ a\geq 0}}^{R}\frac{\binom{l}{a}\binom{m}{R-a}}{\binom{l+m}{R}}\omega_a,\tag{17}$$

$$\Omega_R\mathbf{B}_R = \sum_{\substack{a=R-m \\ a\geq 0}}^{R}\frac{\binom{l}{a}\binom{m}{R-a}}{\binom{l+m}{R}}\omega_a\mathbf{F}_a,\tag{18}$$

where ω_a and $\omega_a F_a$ are short for $\omega_{0,0}^{l,m}(u_{\mathbf{I}^u}^l, v_*^m)$ and for $\omega_{0,0}^{l,m}(u_{\mathbf{I}^u}^l, v_*^m)F_{0,0}^{l,m}(u_{\mathbf{I}^u}^l, v_*^m)$ with $a = |\mathbf{I}^u| \leq l$. Realizing that $(\omega_a F_a, \omega_a)$ as well as $(\Omega_R B_R, \Omega_R)$ are 4D-Bézier points of a 4D homogeneous coordinate representation of rational Bézier curves we see that (17) and (18) are degree raising formulas for rational Bézier curves for raising the degree from l to $l + m$, and we are done.   ∎

Remark 2. Rational Bézier splines can be treated as described in Remark 1.

Remark 3. Theorem 2 includes special cases $\mathbf{K}(t)$ and $\mathbf{F}(u, v)$ both being polynomial, $\mathbf{K}(t)$ being polynomial and $\mathbf{F}(u, v)$ being rational, and $\mathbf{K}(t)$ being polynomial and $\mathbf{F}(u, v)$ being rational. In the first case statement of Theorem 1 results. Special cases two and three both yield rational curves [23].

6. Rational Trim Curves and Surfaces

According to Theorem 2 and Statement 2 we go with the following strategy (assuming $\mathbf{K}(t)$ not being degree raised):

If $N = 1$ and $\mathbf{K}_0$ and $\mathbf{K}_1$ are isoparametric, $\mathbf{F}(\mathbf{K}(t))$ is of degree l defined by weights $\Omega_R \equiv \omega_{0,0}^{l,m}(u_{\mathbf{I}^u}^l, v_*^m)$ and by Bézier points $\mathbf{B}_R$ defined via $\Omega_R \mathbf{B}_R \equiv \omega_{0,0}^{l,m}(u_{\mathbf{I}^u}^l, v_*^m)F_{0,0}^{l,m}(u_{\mathbf{I}^u}^l, v_*^m)$, where $|\mathbf{I}^u| = R$, in case of $\mathbf{K}(t)$ being u parameter line $v_0 = v_1 \equiv v_*$, and $\mathbf{F}(\mathbf{K}(t))$ is of degree m defined by weights $\Omega_R \equiv \omega_{0,0}^{l,m}(u_*^l, v_{\mathbf{I}^v}^m)$ and by Bézier points $\mathbf{B}_R$ defined via $\Omega_R \mathbf{B}_R \equiv \omega_{0,0}^{l,m}(u_*^l, v_{\mathbf{I}^v}^m)F_{0,0}^{l,m}(u_*^l, v_{\mathbf{I}^v}^m)$ where $|\mathbf{I}^v| = R$, in case of $\mathbf{K}(t)$ being v parameter line $u_0 = u_1 \equiv u_*$. Otherwise $\mathbf{F}(\mathbf{K}(t))$ is of degree $N(l + m)$ with Bézier representation according to Theorem 2.

Figures 7 and 8 illustrate Theorem 2 and Statement 2 for the examples of rational trimming Bézier (spline) curves defined in the domains of rational TPB-surfaces.

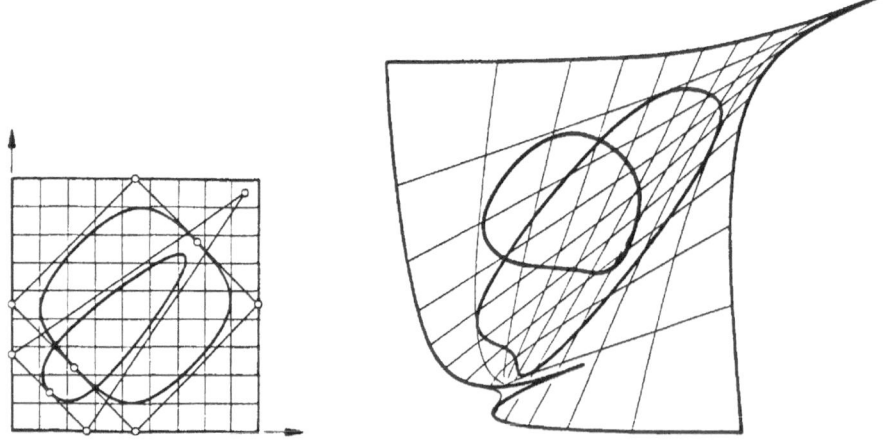

Figure 7. Mapping of a rational quartic and of a closed rational C^1 cubic on a rational biquadratic TPB-surface

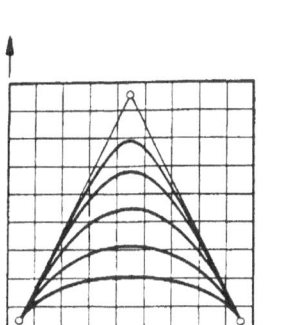
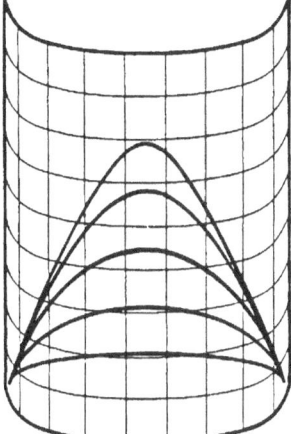

Figure 8. Mapping of rational quadratic curves ($\beta_1 = \frac{1}{4}, \frac{1}{2}, 1, 2, 4$) on a rational bicubic TPB-surface
($\omega_{i,1} = \omega_{i,2} = 2$)

7. Free Form Deformation

Theorems 1 and 2 give the exact and explicit description of surface curves. These curves can be thought of as the boundary curves of trimmed surfaces, resulting from intersecting or blending operations, for instance. Surface curves described by Theorems 1 and 2 also can be interpreted as deformations of planar curves under the mapping of the equation which defines the associated free form surface. Therefore, they provide the mathematical foundation for the free form deformation (briefly FFD) approach to planar curve design via surface modelling.[2]

Figure 9 illustrates the example of modelling a planar shape which is bounded by linear and quadratic (Bézier) curves. First, the form is embedded in the domain $[0,1] \times [0,1]$ of a TPB-surface (upper left illustration of Fig. 9). The deformation procedure is chosen to be polynomial biquintic, resulting in a corresponding control point net which covers the deformation domain (lower left illustration of Fig. 9). The Bézier points of the grid are given by $\mathbf{F}_{i,j} = (i/5,\ j/5,\ 0)$, $0 \le i, j \le 5$. The actual (biquintic) deformation of the form involves the translation of the Bézier points $\mathbf{F}_{i,j} \rightarrow \tilde{\mathbf{F}}_{i,j}$ (right illustration of Fig. 9) and evaluation of the FFD defining (biquintic) equation with coefficients $\tilde{\mathbf{F}}_{i,j}$. While the position of every point in the interior as well as those on the boundary curves of the object are altered by the deformation equation, Theorem 1 and Statement 1 actually provide an exact and closed form representation of all boundary curves of the form. According to Statement 1, linear isoparametric boundary curves map to quintic Bézier curves, and according to Theorem 1 linear non-isoparametric boundary curves map to Bézier curves of degree 10 while quadratic boundary curves map to Bézier curves of degree 20.

---

[2] An introduction into the EFD idea, i.e. curve and surface modelling by the way of volume design, has been given in [19].

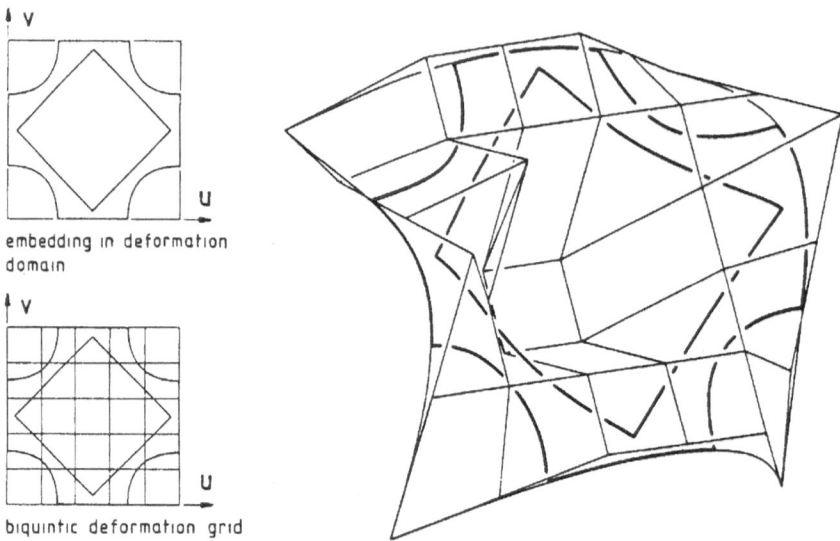

embedding in deformation
domain

biquintic deformation grid

Figure 9. Biquintic deformation of a planar form bounded by linear and quadratic curves

8. Approximation of Trim Curves

Major disadvantage of exactly representing trim curves is a rather high degree of
$F(K(t))$ which might cause problems in following interrogation actions such as
intersection operations and which also might not be supported by certain CAD
systems. Therefore, an approximation of $F(t) = F(K(t))$ by a low polynomial degree
Bézier (spline) curve $X(t)$ might be desirable. An easily implemented quite powerful
technique to do so is the one described in [16] based on parameter optimization and
the concept of geometric continuity:

We resume reduction to cubic (spline) curves only, i.e. we assume the approximating
curve $X(t)$ to be given in Bézier form,

$$X(t) = \sum_{k=0}^{3} \mathbf{b}_k B_k^3(t), \quad t \in [0, 1], \tag{19}$$

where $\mathbf{b}_k$ are unknown Bézier points of $X(t)$.

Requiring that the endpoints of $X(t)$ and of $F(t)$ coincide, and that the two curves
meet each other with first order smoothness at these points, it follows that Bézier
points of $X(t)$ are determined by

$$\begin{aligned}
\mathbf{b}_0 &= \mathbf{B}_0, \quad \mathbf{b}_1 = \mathbf{B}_0 + \lambda_1(\mathbf{B}_1 - \mathbf{B}_0) \\
\mathbf{b}_3 &= \mathbf{B}_{rN}, \quad \mathbf{b}_2 = \mathbf{B}_{rN} + \lambda_2(\mathbf{B}_{rN-1} - \mathbf{B}_{rN}).
\end{aligned} \tag{20}$$

In order to find the best λ_1, λ_2 we choose $L + 1 > rN$ points $\mathbf{P}_l$ on $F(t)$ with respect to
the equidistant parameter values $t_l = l/L$, $\mathbf{P}_l \equiv F(t_l)$, $l = 0, \ldots, L$, and minimize the
absolute value $d = \sum_{l=0}^{L} |\mathbf{d}_l|^2$ of the error vectors $\mathbf{d}_l = \mathbf{P}_l - X(t_l)$. Substituting (20) in

the expression for $\mathbf{d}_l$ yields

$$d = \sum_{l=0}^{L} [\mathbf{D}_l - \lambda_1(\mathbf{B}_1 - \mathbf{B}_0)B_1^3(t_l) - \lambda_2(\mathbf{B}_{rN-1} - \mathbf{B}_{rN})B_2^3(t_l)]^2, \tag{21}$$

where

$$\mathbf{D}_l = \mathbf{P}_l - \mathbf{B}_0 B_0^3(t_l) - \mathbf{B}_0 B_1^3(t_l) - \mathbf{B}_{rN} B_2^3(t_l) - \mathbf{B}_{rN} B_3^3(t_l). \tag{22}$$

λ_1 and λ_2 can be found by solving the linear system of equations resulting from the necessary conditions $\partial d/\partial \lambda_1 = 0$ and $\partial d/\partial \lambda_2 = 0$ for the minimum of d.

The result depends on the parametrization of the points $\mathbf{P}_l$, of course. In addition, error vectors $\mathbf{d}_l$ do not necessarily give the shortest distance between points $\mathbf{P}_l$ and approximating curve $\mathbf{X}(t)$. Hence, after solving the system we apply a parameter correction $t_l \rightarrow t_l^*$ i.e. $\mathbf{P}_l \rightarrow \mathbf{P}_l^* = \mathbf{F}(t_l^*)$, thus, $\mathbf{D}_l \rightarrow \mathbf{D}_l^*, d \rightarrow d^*$, as described in [17] and then solve the new system, which results from $\partial d^*/\partial \lambda_1 = 0$ and $\partial d^*/\partial \lambda_2 = 0$. This process is iterated in order to force nearly all error vectors to be perpendicular to $\mathbf{X}(t)$ in $\mathbf{X}(t_l)$. If the given error tolerance cannot be satisfied, $\mathbf{F}(t)$ is split into two segments at the point where the error is maximal and then the algorithm is applied to each of the two pieces. The algorithm results in a spline curve with a minimal number of cubic Bézier curve segments.

Please note that the Bézier approximation can be built up using the exact point and derivative information of the originally given surface curve in Bézier form, $\mathbf{F}(\mathbf{K}(t))$. At no point of the algorithm estimates of derivatives have to be done, neither conversions from or to any other representation have to be performed!

References

[1] Bardis, L., Patrikalakis N. M.: Blending rational B-spline surfaces. EUROGRAPHICS '89, 453–462 (1989).
[2] Bézier, P.: General distortion of an ensemble of biparametric surfaces. Comput. Aided Des. 10, 116–120 (1987).
[3] Casale, M. S.: Free-form solid modeling with trimmed surface patches. IEEE Comput. Graphics Appl. 7, 33–43 (1987).
[4] Casale, M. S., Bobrow, J. E.: A set operation algorithm for sculptured solids modeled with trimmed patches. Comput. Aided Geom. Des. 6, 235–248 (1989).
[5] Casale, M. S., Bobrow, J. E., Underwood, R.: Trimmed-patch boundary elements: bridging the gap between solid modeling and engineering analysis. Comput. Aided Des. 24, 193–199 (1992).
[6] Choi, B. K., Ju, S. Y.: Constant-radius blending in surface modelling. Comput. Aided Des. 21, 213–220 (1989).
[7] Crocker, G. A., Reinke, W. F.: Boundary evaluation of non-convex primitives to produce parameteric trimmed surfaces. SIGGRAPH '87, ACM Computer Graphics, pp. 129–136 (1987).
[8] DeRose, T. D.: Composing Bézier simplices. ACM Trans. Graphics 7, 198–221 (1988).
[9] DeRose, T. D., Goldman, R. N., Hagen, H. Mann, St.: Functional composition algorithms via blossoming. ACM Trans. Graphics 12, 113–135 (1993).
[10] Elber, G.: Free form surface analysis using a hybrid of symbolic and numeric computation. PhD Thesis, University of Utah 1992.
[11] Farin, G.: Curves and surfaces for computer aided geometric design. A practical guide. 2nd edn. New York: Academic Press 1990.
[12] Farouki, R. T.: Trimmed-surface algorithms for the evaluation and interrogation of solid boundary representations. IBM J. Res. Dev. 31, 314–334 (1987).

[13] Farouki, R. T.: On the stability of transformations between power and Bernstein polynomial forms. Comput. Aided Geom. Des. *8*, 29–36 (1991).
[14] Filip, D. J.: Blending parametric surfaces. ACM Trans. Graphics *8*, 164–173 (1989).
[15] Harada, T., Toriya, H., Chiyokura, H.: An enhanced rounding operation between curved surfaces in solid modeling. In: CG International '90 (Chua, T. S., Kunii, T. L., eds.), pp. 563–588. Berlin Heidelberg New York Tokyo: Springer 1990.
[16] Hoschek, J.: Approximate conversion of spline curves. Comput. Aided Geom. Des. *4*, 59–66 (1987).
[17] Hoschek, J., Schneider, F. J., Wassum, P.: Optimal approximate conversion of spline surfaces. Comput. Aided Geom. Des. *6*, 293–306 (1989).
[18] Hoschek, J., Schneider, F. J.: Spline conversion for trimmed rational Bézier- and B-spline surfaces. Comput. Aided Des. *22*, 580–590 (1990).
[19] Hoschek, J., Lasser, D.: Grundlagen der Geometrischen Datenverarbeitung, 2nd edn. Stuttgart: Teubner 1992.
[20] Klass, R., Kuhn, B.: Fillet and surface intersections defined by rolling balls. Comput. Aided. Geom. Des. *9*, 185–193 (1992).
[21] Kobori, K., Iwazu, M., Jones, K. M.: Polygonal subdivision of parametric surfaces. In: Advanced computer graphics (Kunii, T. L., ed.), pp. 50–59. Berlin Heidelberg New York Tokyo: Springer 1986.
[22] Koparkar, P. A.: Designing parametric blends: surface model and geometric correspondence. Visual Comput. *7*, 39–58 (1991).
[23] Lasser, D.: Composition of tensor product Bézier representations. Interner Bericht 213/91, Fachbereich Informatik, Universität Kaiserslautern 1991.
[24] Miller, J. R.: Sculptured surfaces in solid models: Issues and alternative approaches. IEEE Comput. Graphics Appl. *6*, 37–48 (1986).
[25] Nishita, T., Sederberg, Th. W., Kakimoto, M.: Ray tracing trimmed rational surface patches. ACM Comput. Graphics *24*, 337–345 (1990).
[26] Rockwood, A. P., Heaton, K., Davis, T.: Real time rendering of trimmed surfaces. SIGGRAPH '89, ACM Comput. Graphics *23*, 107–116 (1989).
[27] Pegna, J., Wolter, F. E.: Designing and mapping trimming curves on surfaces using orthogonal projection. Adv. Des. Auto. *1*, 235–245 (1990).
[28] Shantz, M., Chang, S. L.: Rendering trimmed NURBS with adaptive forward differencing. SIGGRAPH '88, ACM Comput. Graphics *22*, 189–198 (1988).
[29] Sheng, X., Hirsch, B. E.: Triangulation of trimmed surfaces in parametric space. Comput. Aided Des. *24*, 437–444 (1992).
[30] Vries-Baayens, A. E.: CAD product data exchange: conversions for curves and surfaces. Dissertation, Delft, University Press 1991.
[31] Vries-Baayens, A. E., Seebregts, C. H.: Exact conversion of a trimmed nonrational Bézier surface into composite or basic nonrational Bézier surface. In: Topics in surface modeling (Hagen, H., ed.), pp. 115–143. Philadelphia: SIAM 1992. (short version of [30]).

D. Lasser, G.-P. Bonneau
Computer Science
University of Kaiserslautern
D-67653 Kaiserslautern
Federal Republic of Germany

Computing Suppl. 10, 243–251 (1995)

Computing

© Springer-Verlag 1995

Control Point Representations of Trigonometrically Specified Curves and Surfaces

H. Müller, Dortmund, and **B. Gnadl**, Freiburg

Abstract. A representation of trigonometric curves and surfaces is introduced that allows intuitive control of the shape. The representation is obtained by transferring the Bernstein-Bézier representation of polynomials in geometric modeling to trigonometrically represented shapes.

Key words: Trigonometric curves and surfaces, control point representation, computer aided geometric design.

1. Trigonometric Polynomials, Curves and Surfaces

A real-valued trigonometric polynomial is defined as

$$f(x) = a_0/2 + \sum_{j=1}^{m} (a_j \cos(2\pi j x) + b_j \sin(2\pi j x)), \quad a_j, b_j \in \mathbb{R},$$

a trigonometric curve in parametric representation as

$$\mathbf{k}(t) = \mathbf{a}_0/2 + \sum_{j=1}^{m} (\mathbf{a}_j \cos(2\pi j t) + \mathbf{b}_j \sin(2\pi j t)), \quad \mathbf{a}_j, \mathbf{b}_j \in \mathbb{R}^d, t \in [0, 2\pi),$$

and a trigonometric surface in parameter representation as

$$\mathbf{s}(u, v) = \mathbf{a}_{0,0}/2 + \sum_{p=0}^{m} \sum_{q=0}^{n} \mathbf{a}_{p,q} \cdot \cos(2\pi(p \cdot u + q \cdot v)) + \mathbf{b}_{p,q} \cdot \sin(2\pi(p \cdot u + q \cdot v))$$

$$= \mathbf{a}_{0,0}/2 + \sum_{p=0}^{m} \sum_{q=0}^{n} \mathbf{a}_{p,q} \cdot (\cos(2\pi p u)\cos(2\pi q v) - \sin(2\pi p u)\sin(2\pi q v))$$

$$+ \mathbf{b}_{p,q} \cdot (\sin(2\pi p u)\cos(2\pi q v) + \cos(2\pi p u)\sin(2\pi q v))$$

for $(u, v) \in [0, 2\pi)^2$. For an introduction into trigonometric numerical analysis cf. [3]. Although the influence of the coefficients of these representations can be explained geometrically, they are not quite intuitive for controlling the geometric shape, for instance for the purpose of interactive geometric modeling. In particular, it is difficult to avoid the typical waviness induced by the trigonometric functions. The analogous experience is known from modeling with polynomials. Here, the Bernstein-Bézier representation has proven to be a suitable alternative, cf. [1]. In this contribution we show that this representation is useful for trigonometric modeling, too.

A trigonometric curve can also be represented as

$$\mathbf{k}(t) = \sum_{j=0}^{m} \sum_{l=0}^{j} \mathbf{c}_{j-l,l} \cdot \sin^l 2\pi t \cdot \cos^{j-l} 2\pi t, \quad \mathbf{c}_{i,k} \in \mathbb{R}^d.$$

This is obtained by replacing $\cos 2\pi kt$ and $\sin 2\pi kt$ by the polynomials in $\cos 2\pi t$ and $\sin 2\pi t$ obtained from the real and imaginary term, respectively, of the identity

$$\cos 2\pi kt + i \sin 2\pi kt = \exp(2\pi ikt) = (\cos 2\pi t + i \sin 2\pi t)^k.$$

For $d = 2$, setting

$$a = \cos 2\pi t, \quad b = \sin 2\pi t, \quad \mathbf{k}(t) = (k_1(t), k_2(t))$$

and eliminating a and b from the algebraic system of equations

$$x = k_1(t(a, b)), \quad y = k_2(t(a, b)), a^2 + b^2 = 1$$

yields an implicit representation

$$g(x, y) = 0$$

for two-dimensional trigonometric curves. $g(x, y)$ is a polynomial in x and y. By transforming $g(x, y)$ into Bernstein-Bézier representation, the shape of the curve can be influenced by the resulting control values.

The set $\{(a, b) | a = \cos t, b = \sin t, t \in [0, 2\pi]\}$ describes a circle. The same holds for the representation

$$a = \frac{-(1 - t)^2 + t^2}{(1 - t)^2 + t^2}, \quad b = \frac{\pm 2t(1 - t)}{(1 - t)^2 + t^2}, \quad t \in [0, 1].$$

Replacing $\cos t$ and $\sin t$ in $\mathbf{k}(t)$ by these expressions yields a rational parametric representation consisting of two parts. This representation can be transferred into a rational Bernstein-Bézier representation, allowing to influence the shape of the curve by control points.

The disadvantage of both approaches is that the degree of the polynomial representations increases with transformation. More annoying, however, is that arbitrary variation of the control values may lead to curves that cannot longer be represented by a trigonometric polynomial. The semi-implicit representation used in the following sections avoids these disadvantages.

2. Semi-Implicit Trigonometric Modeling

Semi-implicit trigonometric modeling is a special case of a combination of implicit representation and parametric representation. A trigonometric curve $\mathbf{k}$ as introduced above can also be represented by

$$\mathbf{k} = \{\mathbf{f}(a, b) | a^2 + b^2 = 1\},$$

$$\mathbf{f}(a, b) = \sum_{j=0}^{m} \sum_{l=0}^{j} \mathbf{c}_{j-l,l} \cdot b^l \cdot a^{j-l}, \quad \mathbf{c}_{i,k} \in \mathbb{R}^d.$$

The curve can be considered as a curve on the surface $\mathbf{f}(a, b)$ defined by an implicit curve, the circle $a^2 + b^2 = 1$, in the parametric domain of the surface. For that reason we call this representation *semi-implicit trigonometric representation*.

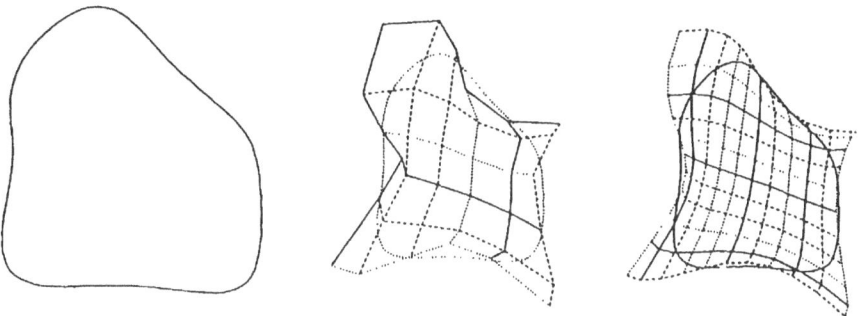

Figure 1. A trigonometric curve with control nets of size 6 × 6 and 11 × 11

A semi-implicit Bernstein-Bézier-representation is obtained by representing $\mathbf{f}(a, b)$ over $[-1, 1]^2$ in Bernstein-Bézier-representation,

$$\mathbf{f}(a, b) = \sum_{i=0}^{m} \sum_{j=0}^{n} \mathbf{k}_{i,j} B_i^m \left(\frac{1+a}{2}\right) B_j^n \left(\frac{1+b}{2}\right),$$

with the Bernstein polynomials $B_i^n(x) := \binom{n}{i} x^i (1-x)^{n-i}$. By modifying the control points $\mathbf{k}_{i,j}$ the shape of the surface changes and thus the curve. The resulting curves are still trigonometric. This can be seen by re-substituting $a = \cos t$ and $b = \sin t$. The curves lie in the convex hull of the control points. This results from the same well known property of Bernstein-Bézier surfaces. Figure 1 shows a curve with control nets of size 6 × 6 and 11 × 11.

A semi-implicit representation is not unique. Replacing square terms, e.g. b^2 by $1 - a^2$, yields new expressions describing the same shapes. In particular it is possible to achieve expressions containing one of the variables a and b with degree 1 at most,

$$\mathbf{f}(a, b) = \sum_{j=0}^{m} \sum_{l=0}^{j} \mathbf{c}_{j-l,l} \cdot b^l \cdot a^{j-l}$$

$$= \sum_{j=0}^{m} \left(\sum_{k=0}^{\lfloor j/2 \rfloor} \mathbf{c}_{j-2k,2k} b^{2k} a^{j-2k} + \sum_{k=0}^{\lceil j/2 \rceil - 1} \mathbf{c}_{j-2k-1,2k+1} b^{2k+1} a^{j-2k-1} \right)$$

$$= \sum_{j=0}^{m} \left(\sum_{k=0}^{\lfloor j/2 \rfloor} \mathbf{c}_{j-2k,2k} (1-a^2)^k a^{j-2k} + \sum_{k=0}^{\lceil j/2 \rceil - 1} \mathbf{c}_{j-2k-1,2k+1} (1-a^2)^k a^{j-2k-1} b \right)$$

$$= \mathbf{g}(a) + b \cdot \mathbf{h}(a).$$

This so-called *reduced semi-implicit representation* has an intuitive geometric interpretation. Let be

$$\mathbf{f}_1(a) := \mathbf{g}(a) - \mathbf{h}(a),$$

$$\mathbf{f}_2(a) := \mathbf{g}(a) + \mathbf{h}(a).$$

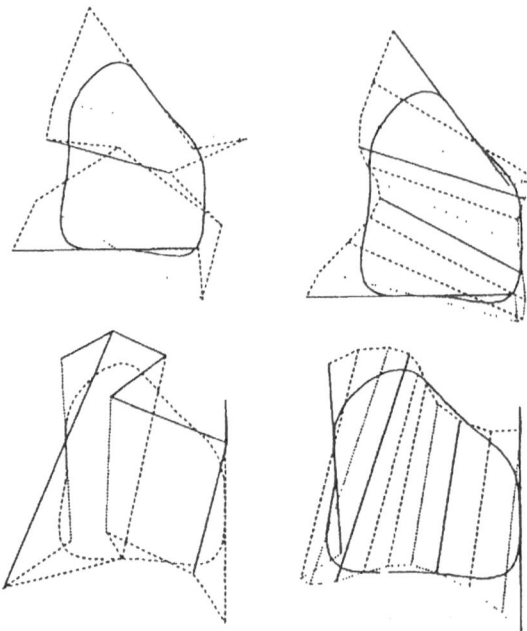

Figure 2. A trigonometric curve with control polygons of size 2 × 6 and 2 × 11, with maximum degree reduced to 1 for a respectively b

Then

$$\mathbf{f}(a, b) = \frac{1 - b}{2} \cdot \mathbf{f}_1(a) + \frac{1 + b}{2} \cdot \mathbf{f}_2(a).$$

Thus the trigonometric curve results from two polynomial curves $\mathbf{f}_1$ and $\mathbf{f}_2$ by a weighted average of points located "oppositely" on the two curves. A trigonometric representation of such a curve is obtained by replacing a and b by the trigonometric functions.

Especially for $a \in \{-1, 1\}$, we have $b = 0$, so that the respective curve points lie just in the middle, for $a = 0$ we have $b \in \{-1, 1\}$. In this case the respective points of the trigonometric curve lie on $\mathbf{f}_1$ and $\mathbf{f}_2$, respectively. The Bézier control net results from control polygons of $\mathbf{f}_1$ and $\mathbf{f}_2$ of equal degree, by connecting oppositely located vertices with line segments. Figure 2 shows the control polygon net of the example with two times 6 respectively 11 control points. In one case, the variable a, and in the other case, the variable b is reduced to maximum degree 1.

Surfaces can be treated analogously to curves. The semi-implicit representation of a trigonometric surface $\mathbf{s}$ is

$$\mathbf{s} = \{\mathbf{f}(a, b, c, d) | a^2 + b^2 = 1, c^2 + d^2 = 1\}.$$

$\mathbf{f} : \mathbb{R}^4 \to \mathbb{R}^d$ is a polynomial in four variables. By a suitable transformation an expression can be derived containing b and d of degree 1 at most,

$$\mathbf{f}(a, b, c, d) = \mathbf{g}(a, c) + b \cdot \mathbf{h}(a, c) + d \cdot \mathbf{p}(a, c) + b \cdot d \cdot \mathbf{q}(a, c).$$

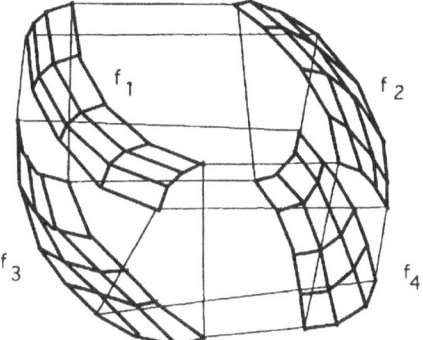

Figure 3. Control net for trigonometic surfaces in reduced semi-implicit representation

Let be

$$\mathbf{f}_1(a, c) := \mathbf{g}(a, c) - \mathbf{h}(a, c) - \mathbf{p}(a, c) + \mathbf{q}(a, c),$$

$$\mathbf{f}_2(a, c) := \mathbf{g}(a, c) + \mathbf{h}(a, c) - \mathbf{p}(a, c) - \mathbf{q}(a, c),$$

$$\mathbf{f}_3(a, c) := \mathbf{g}(a, c) - \mathbf{h}(a, c) + \mathbf{p}(a, c) - \mathbf{q}(a, c),$$

$$\mathbf{f}_4(a, c) := \mathbf{g}(a, c) + \mathbf{h}(a, c) + \mathbf{p}(a, c) + \mathbf{q}(a, c).$$

Then

$$\mathbf{f}(a, b, c, d) = \frac{1-d}{2} \cdot \left(\frac{1-b}{2} \cdot \mathbf{f}_1(a, c) + \frac{1+b}{2} \cdot \mathbf{f}_2(a, c) \right)$$

$$+ \frac{1+d}{2} \cdot \left(\frac{1-b}{2} \cdot \mathbf{f}_3(a, c) + \frac{1+b}{2} \cdot \mathbf{f}_4(a, c) \right).$$

Thus a trigonometric surface can be described by four polynomial surfaces $\mathbf{f}_1, \mathbf{f}_2, \mathbf{f}_3$ and $\mathbf{f}_4$. From the Bernstein-Bézier representation of these surfaces a control net is obtained that can also be used for the manipulation of the trigonometric surface. The control net is composed of the control polygons of the four surfaces. Oppositely located points are connected with line segments (Fig. 3).

As an example,

$$\{(x, y, z) | x = (2 + \cos \psi) \cdot \cos \phi, y = (2 + \cos \psi) \cdot \sin \phi,$$

$$z = \sin \psi, \phi, \psi \in [0, 2\pi)\}$$

is a torus centered at the origin. With $a = \cos \phi, b = \sin \phi, c = \cos \psi, d = \sin \psi$, we get

$$\mathbf{f}_1(a, c) = \begin{bmatrix} a \cdot (2 + c) \\ -(2 + c) \\ -1 \end{bmatrix}, \quad \mathbf{f}_2(a, c) = \begin{bmatrix} a \cdot (2 + c) \\ 2 + c \\ -1 \end{bmatrix},$$

$$\mathbf{f}_3(a, c) = \begin{bmatrix} a \cdot (2 + c) \\ -(2 + c) \\ 1 \end{bmatrix}, \quad \mathbf{f}_4(a, c) = \begin{bmatrix} a \cdot (2 + c) \\ 2 + c \\ 1 \end{bmatrix},$$

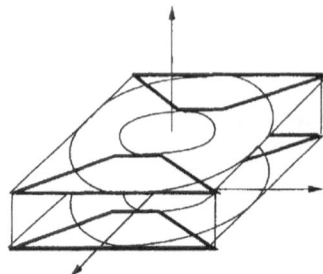

Figure 4. Sketch of a control net of a torus centered at the origin. f_1 and f_2 as well as f_3 and f_4 yield rings which are combined to form the torus

The control points of f_1 are

$$\begin{bmatrix} -1 \\ -1 \\ -1 \end{bmatrix}, \begin{bmatrix} -3 \\ -3 \\ -1 \end{bmatrix}, \begin{bmatrix} 1 \\ -1 \\ -1 \end{bmatrix}, \begin{bmatrix} 3 \\ -3 \\ -1 \end{bmatrix},$$

those of f_2

$$\begin{bmatrix} -1 \\ 1 \\ -1 \end{bmatrix}, \begin{bmatrix} -3 \\ 3 \\ -1 \end{bmatrix}, \begin{bmatrix} 1 \\ 1 \\ -1 \end{bmatrix}, \begin{bmatrix} 3 \\ 3 \\ -1 \end{bmatrix}.$$

The control points of f_3 and f_4 differ from those of f_1 and f_2, respectively, only in the third coordinate which is 1. Figure 4 shows a sketch of this control net. The combination of f_1 and f_2 as well as that of f_3 and f_4 each yields a planar ring. The combination of the two rings induces the torus.

Intermediate forms of surfaces that are topologically equivalent to a cylinder or a sphere can be obtained analogously. We show that in both cases two Bézier surfaces are sufficient.

Cylindric surfaces can be described by functions $f(a,b,c)$, $f: \mathbb{R}^3 \to \mathbb{R}^d$ over the unit cylinder $\{(a,b,c)\,|\,a^2 + b^2 = 1,\ -1 \leq c \leq 1\}$. A trigonometric representation is obtained from the substitution

$$a = \cos 2\pi s, \quad b = \sin 2\pi s$$

in $f(a,b,c)$. For a polynomial $f(a,b,c)$ it is again possible that the variable b appears with degree 1 at most,

$$f(a,b,c) = g(a,c) + b \cdot h(a,c).$$

Let be

$$f_1(a,c) := g(a,c) - h(a,c),$$

$$f_2(a,c) := g(a,c) + h(a,c).$$

Then

$$f(a,b,c) = \frac{1-b}{2} \cdot f_1(a,c) + \frac{1+b}{2} \cdot f_2(a,c).$$

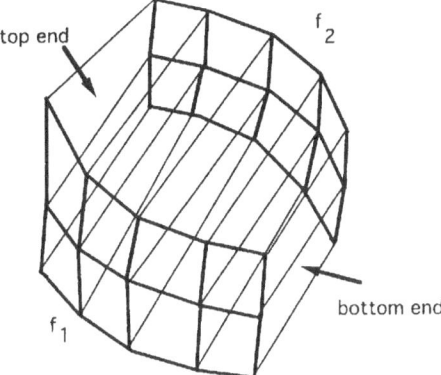

Figure 5. Control net of trigonometric cylindric surfaces in reduced semi-implicit representation

Thus the cylindric surface results from two polynomial surfaces $\mathbf{f}_1$ and $\mathbf{f}_2$ by taking weighted averages of oppositely located points. The Bézier control net results from the control polygons of $\mathbf{f}_1$ and $\mathbf{f}_2$ of equal degree, by connecting oppositely located points by line segments (Fig. 5).

Sphere-like surfaces can be considered as functions $\mathbf{f}(a, b, c)$, $\mathbf{f}: \mathbb{R}^3 \to \mathbb{R}^d$, from the unit sphere $a^2 + b^2 + c^2 = 1$ into the three-dimensional space. These functions can be represented trigonometrically and parametrically with the substitution

$$a = \cos 2\pi u \cos 2\pi v, \quad b = \cos 2\pi u \sin 2\pi v, \quad c = \sin 2\pi u, \quad (u, v) \in [0, 2\pi]^2.$$

A control net for a polynomial $\mathbf{f}$ is obtained through an intermediate step over cylindric surfaces. Let be

$$\mathbf{c}(a, b, c) := (\mathbf{f}(-ra, -rb, c) + \mathbf{f}(ra, rb, c))/2 + (\mathbf{f}(-ru, -rb, c) - \mathbf{f}(ra, rb, c))/(2r),$$

with $c^2 + r^2 = 1$. For polynomial $\mathbf{f}$, $\mathbf{c}$ is a polynomial in a, b, c, since r occurs only with even exponent in the first term, and only with odd exponent in the second term. The division by r of the second term results in even exponents also for this term. Thus r can be replaced by c in the whole expression by substituting $r^2 = 1 - c^2$.

Now

$$\mathbf{f}(ra, rb, c) = \frac{1 - r}{2} \cdot \mathbf{c}(a, b, c) + \frac{1 + r}{2} \cdot \mathbf{c}(-a, -b, c).$$

By understanding $\mathbf{c}$ as a function over the unit cylinder $a^2 + b^2 = 1$, we get an intuitive geometric interpretation (Fig. 6). The function $\mathbf{c}$ can be represented in the form

$$\mathbf{c}(a, b, c) = \frac{1 - b}{2} \cdot \mathbf{c}_1(a, c) + \frac{1 + b}{2} \cdot \mathbf{c}_2(a, c),$$

as described for cylinders. The control nets of $\mathbf{c}_1$ and $\mathbf{c}_2$ finally represent the spherical surface given by $\mathbf{f}$.

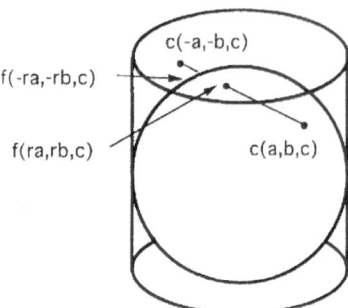

Figure 6. Mapping between a spherical surface and a cylindrical surface.

For the unit sphere, we have

$$
\mathbf{c}_1(a,c) = \begin{bmatrix} -a \\ 1 \\ c \end{bmatrix}, \quad \mathbf{c}_2(a,c) = \begin{bmatrix} -a \\ -1 \\ c \end{bmatrix}.
$$

The control points are the vertices of the axes-parallel cube with edge length 2 centered at the origin.

3. An Application of Semi-implicit Trigonometric Modeling

One advantage of the semi-implicit representation lies in the fact that a favorable low variation can be achieved because the variation diminishing property is inherited from the Bernstein-Bézier representation.

Figure 7. Trigonometric approximations of a closed polygonal chain of degrees 1, 2, 3, 5 and 10

For instance, a desired curve can be modeled by starting with a similar polygonal chain. Kuhl and Giardina [2] have derived a Fourier expansion

$$\mathbf{p}(t) = \mathbf{a}_0/2 + \sum_{j=1}^{\infty} (\mathbf{a}_j \cos(2\pi jt) + \mathbf{b}_j \sin(2\pi jt))$$

of closed polygonal chains. Figure 7 shows approximations of closed polygonal chains of degrees 1, 2, 3, 5 and 10. The curve used in the example above is that of degree 5. By cutting off its Fourier expansion, the polygonal chain is approximated by a trigonometric curve. This may result in a quite wavy curve. By manipulating the control net of this curve it can be tried now to reduce the waviness. Of course there is the alternative to develop the curve directly from an arbitrary initial control net.

References

[1] Farin, G.: Curves and surfaces for geometric design. San Diego: Academic Press 1993.
[2] Kuhl, F. P., Giardina, C. R.: Elliptic Fourier features of a closed contour. Comput. Graphics Image Proc. *18*, 236–258 (1982).
[3] Stoer, J., Bulirsch, J.: Introduction to numerical analysis. Berlin Heidelberg New York: Springer 1983.

H. Müller
Informatik VII
Universität Dortmund
D-44221 Dortmund
Federal Republic of Germany

B. Gnadl
Institut für Informatik
Universität Freiburg
D-79104 Freiburg
Federal Republic of Germany

Computing Suppl. 10, 253–274 (1995)

Towards Optimality in Automated Feature Recognition

M. J. Pratt, Troy

Abstract. Various approaches to the automated recognition of form features from geometric models are reviewed, and their relative advantages and disadvantages analysed. A synthesis is then made of the best aspects of the methods surveyed, and suggestions put forward for the development of improved methods for the future.

Key words: Form features, feature recognition.

1. Introduction

It is now well known that a form feature is a local geometric configuration on the surface of a manufactured object. Such a configuration has some semantic significance; thus, for the designer who originally creates it the feature has a functional purpose, while for the manufacturing engineer it may represent a requirement to remove certain volumes of material from an initial blank shape. Other activities in the design and manufacturing cycle may also be viewed in terms of form features, e.g. finite element mesh generation [36], assembly design [42], dimensioning and tolerancing [38], selection of joining methods [28], design of machining fixtures [11] and inspection [27], to mention but a few. The designer's functional features will not in general correspond to the features appropriate for subsequent activities [34], and hence some means is desirable for the automatic recognition of features for the downstream activities.

2. Automated Feature Recognition

This paper is concerned mainly with automated feature recognition from 'pure' geometric models, the subject of much research during the past ten years or more. During that time experimental systems have achieved a certain limited success, but all workers in the field would agree that much remains to be done. In the future, the use of feature-based design systems is likely to become commonplace. These generate product models containing information regarding the designer's functional features in addition to the geometric shape description. Some remarks are made in Section 5 on the advantage afforded by the presence of design feature data in the recognition of features for manufacturing and other downstream applications.

A 'pure' geometric model, then, is a geometric model containing no feature information. It may take one of several forms, most notably

- a set-theoretic or constructive solid geometry (CSG) model,

- a 2D drawing, or
- a boundary representation (B-rep) solid model.

Feature recognition from set-theoretic models is subject to various difficulties, some of them related to the fact that a particular solid does not have a unique representation in set-theoretic terms. The theory of one possible approach was suggested by Woo [49], though several years elapsed before it was further developed into a practical method [44, 19]. Lee and Fu [22], in another theoretical paper, study the problems of recognising just one type of form feature, a rounded edge, from a CSG data structure, and of reorganising that data structure so that the primitives involved in defining the feature can be grouped together as a substructure giving it a unified representation within the overall CSG tree. Woodwark [50] has suggested a further approach, whose implementation by Bowyer and Parry-Barwick [2] involves searches in a space of ten dimensions for the identification of features on 3D objects. The problems associated with automated feature recognition in the set-theoretic context has led several investigators using CSG modelling systems to generate B-rep models from their initial CSG representation and to use techniques based on relationships not between volumes but between faces and edges [6, 7, 20].

Another aspect of feature recognition from set-theoretic models is that it initially generates pure volumes, with no associated feature semantics. Further processing is therefore required in order to recognise those volumes as specific feature types and to augment them appropriately with other data pertinent to their intended application area. As an example, the semantic data for machining features will concern cutting access directions and related matters. The pure set-theoretic methods have much in common in the above respect with a class of cell decomposition methods which have been popular for several years. Early examples are provided by the work of Grayer [13] and Armstrong et al. [1]. The former automatically generated NC machining instructions for 2.5D parts by decomposing the removed material into a set of laminae. The latter arrived at a machining strategy by extending the faces of the part model to decompose the material lying between it and the boundaries of the stock material into a collection of cells which were then machined away individually. Perng et al. [30] and Sakurai and Chin [39] provide more recent examples of similar types of approach. The first of these references illustrates a technique sometimes known as DSG (destructive solid geometry) in which the original CSG tree structure is recast using de Morgan's laws so that the only Boolean operations in the revised part description are subtractions. Each such operation is then regarded as implying the removal of a volume of material by machining. In the opinion of the present writer these methods are at their best when dealing with parts having few features and simple geometry; combinatorial problems multiply very rapidly in more complex situations.

A final comment on CSG approaches is that a part or feature representation expressed purely in terms of volumetric primitives does not permit the attachment of attribute information to individual faces or edges of the model. Indications are that the use of such attributes in feature-based systems for design and manufacture is likely to be of great importance, and in a CSG context their use will require the generation of a B-rep model from the primary set-theoretic model.

For reasons given in the last three paragraphs, and since commercial developers of CAD systems have in recent years consistently rejected the CSG approach in favour of boundary representation, the remainder of the paper will be concerned solely with feature recognition from non-CSG models.

There has been little published work on feature recognition from 2D drawings, though this is a potentially important area, at least in the short and medium term, since the great majority of the design data residing in computers currently represents drawings. The use of this data directly for the automated recognition of manufacturing features will therefore open the way to automated process planning for the large number of companies who current restrict their CAD activities to 2D draughting. Contributors in this area include Brun [3] and Meeran [26]. More will be said about the work of Meeran later, in Section 3.6.

At the time of writing the most successful automated feature recognition systems are those which derive their data from boundary representation solid models. The paper will therefore concentrate on approaches based on boundary representation modelling, with the object of suggesting fruitful directions for further improvement.

3. Some Feature Recognition Techniques

This section contains a comparison of some feature recognition methods using B-rep modellers. It is not a complete survey, covering only a relatively small number of feature recognition experiments which in the author's opinion represent significant conceptual advances in the subject.

3.1. Kyprianou

Kyprianou [21] developed the first practical system for automated feature recognition, using the boundary representation modeller BUILD developed at Cambridge University in England. His method is composed of four phases, whose essentials are as follows:

1. The edges of the model are classified and labelled as being *convex, smooth* or *concave*, depending on whether the faces meeting at those edges, viewed from outside the object, meet at an angle of greater than, equal to, or less than 180° respectively. Smooth edges must be further examined with regard to the curvature of the surfaces where they meet and then reclassified appropriately. The necessary geometric calculations are easy in cases where the object faces lie on planes or natural quadrics such as cones and cylinders.
2. The faces of the model are now classified as either *primary* or *secondary*. A primary face is one which exhibits concavity, either by possessing concave edges or by lying on a concave surface, or one which contains interior boundary loops of edges. The primary faces are ordered into a list, firstly according to the number of inner loops they contain and secondly according to the number of concave edges they own; both are measures of 'morphological content' in that

they indicate the presence of form features in which the faces concerned play some role.

3. Thirdly comes the partitioning phase, creating a secondary data structure which makes reference to topological entities in the primary B-rep structure. The faces of the object are assigned to *facesets*, represented in the secondary data structure by nodes of a graph whose links indicate faceset connectivities. The initial faceset is built up starting from the face with highest morphological content. Faces connected to this around its outer periphery are collected into the faceset, and further new faces connected across their peripheral edges are added until no more are available. Further facesets are then constructed starting from inner edge loops of faces with a high morphological content; at this stage the facesets generated will generally correspond to features or combinations of features. The process runs recursively until every face of the model is assigned to a faceset. Three types of faceset are defined, namely *root*, *node* and *leaf*; the root may be regarded as the origin of the faceset graph, while a node has two or more connections in the graph and a leaf only one. The resulting graph is in general a network containing closed cycles; a cycle will occur, for example, for any through hole faceset since it will be linked back to the root by two different paths in the graph.

4. The final phase is that of the actual feature recognition. The facesets found will correspond in some cases to individual features, but in other cases they will represent combinations of interacting features which have to be disentangle from each other. The implemented system is oriented mainly towards the finding of machinable (depression) features. A search is started from that face having the highest morphological content in each faceset, and the faces connected to it by concave relationships are collected recursively until the process terminates, which implies that the collection of faces is now bounded by a loop of edges which are all convex. This collection is regarded as defining a depression feature; Kyprianou's work did not go much further towards a more detailed subclassification of feature types, which demands interrogation also of the geometry of the faces involved.

Although it does not generate detailed feature descriptions, Kyprianou's system has further capabilities in the recognition of gross exterior shapes of objects as being either prismatic or rotational. It also automatically generates a Pitney-Bowes group technology code for rotational parts. The essential observations underlying the actual feature recognition process are that a depression feature is (usually) bounded by a closed loop of convex edges, and that, conversely, a protrusion is bounded by a loop of concave edges. Through holes, as mentioned above, are identified from the presence of cycles in the faceset graph.

The conceptual advances made by Kyprianou towards automated feature recognition are as follows:

- For the first time, features are automatically recognised in a solid modelling system and explicitly represented in terms of sets of object faces.
- The work represents the first use of edge convexity and concavity in the recognition and characterisation of features.

- Features are hierarchically ordered into a graph structure indicating adjacency relationships. These relationships are important for many applications, including process planning where they help to determine the order of machining operations.
- Some internal feature information is generated and stored, notably a distinction between 'base' and 'wall' faces of a feature, which is likely to be of use in the transmutation of feature models into other forms for different applications purposes [35].

The major limitation of Kyprianou's pioneering work is that it generates no detailed classification of features, though it certainly lays the necessary groundwork for that goal to be achieved.

3.2. Henderson

Henderson, in his research at Purdue University [14, 15], took a different approach and made further significant advances in feature recognition. He used the commercial boundary representation modeller ROMULUS, and confined his attention to prismatic parts exhibiting depression and through hole features. Henderson's aim was the recognition of machining features for process planning. The features are assumed to be defined upon a part which has the basic shape of a rectangular block, so that its outer faces coincide with the surfaces of the stock material from which it is to be made. The first step is a Boolean subtraction of the part from the enclosing block in order to determine the *cavity volume*, or volume of material to be machined away. The object is to decompose this volume into a set of subvolumes each corresponding to a machining feature.

The representation of the cavity volume in terms of the topological elements generated by ROMULUS is written out automatically as a set of facts in the PROLOG language. Sets of rules defining various feature classes, also written in PROLOG, form the rule base of the system. The PROLOG inferencing mechanism is used to scan the knowledge database and find sets of observed topological elements matching the prescribed feature rules, thereby recognising features as sets of topological entities. Once a feature has been recognised it is reconstructed as a ROMULUS boundary representation model which is then subtracted from the current cavity volume to give a *reduced cavity volume*. The process then repeats until a null cavity volume results.

Henderson notes that the recognition process he uses is akin to the 'syntactic pattern recognition' procedure known in robotics and other fields, though this term is generally used in connection with two-dimensional rather than three-dimensional problems. Some slightly earlier work at Purdue University, described by Staley et al. [43] explains the use of syntactic pattern recognition in the classification of axially symmetric holes in terms of 2D profiles composed of straight line segments.

Like Kyprianou, Henderson implements a graph structure for representing adjacency relationships between features; it is now widely accepted that this is a necessary capability of any feature-based modelling system. One major difference is that Henderson models features as volumes. Such a volume is bounded by a combination of part faces and *transparent faces* which do not belong to the part. An example is provided by a blind cylindrical hole normal to a planar face; the faces of the feature volume are the cylindrical wall face, the bottom face (perhaps conical) and a planar top face which closes off the feature as a self-contained volume. The first two are part faces, but the third is a transparent face generated by ROMULUS in its reconstruction of the feature as a volume. From the machining point of view these faces provide access for the cutting tool to enter the feature volume. Feature adjacencies are represented in the feature graph by links between matching pairs of transparent faces.

The method of feature volume reconstruction is significant. Henderson has limited his feature types to those which can be represented in terms of linear or rotational sweeps of 2D profiles. When the existence of a feature has been recognised by the system then the underlying profile is extracted. The facilities of ROMULUS, which include both types of sweep operations, are then used to generate a boundary representation of the feature volume. It is noteworthy that this approach can reconstruct the full volume of a feature which has an interaction with a neighbouring feature. As an example, consider a blind cylindrical hole drilled obliquely in the bottom of a rectangular pocket, and assume that the pocket volume has already been reconstructed. If the cylindrical hole is then recognised its reconstructed volume will be that generated by the rotation of a straight line profile, or alternatively a linear sweep of a circle. In either case a complete cylinder will be generated, and due to the obliquity of the hole this volume will actually protrude into the pocket volume so that the two have a non-null intersection. Thus the interaction of the two features is readily detected, and the normal or non-intersecting versions of the features also determined.

Once the cavity volume has been completely decomposed into feature volumes, the resulting feature graph has much in common with a CSG representation of the part, in which all the Boolean operations are subtractions. The main difference is that the volumes subtracted are all fully evaluated B-rep models in their own right.

The major advances made by Henderson may be summarised as follows:

- For the first time a declarative rule-based approach is used for feature recognition. This permits the straightforward setting up of a 'library' of generic definitions of recognisable feature classes.
- Features are generated in a volumetric representation which has a number of practical advantages, including the possibility of investigating feature interactions in volumetric terms by using the standard Boolean operations of solid modelling [32, 33].
- The volumetric approach also permits the generation of a sequence of 'in-process' models, representing the part at various stages during its manufacture.
- Henderson goes further than Kyprianou in assigning attributes to the component faces of his features, to help in the subsequent process planning operation.

The main limitations of Henderson's work are the use of a stock volume having surfaces coincident with part faces and the restriction to 'swept' feature volumes (though by the nature of their manufacturing process most machined features are of this type).

3.3. Comparison of the Methods of Kyprianou and Henderson

Kyprianou and Henderson provide an interesting contrast in implementational methods. Kyprianou's feature recogniser is written in the procedural language ALGOL 68, and uses 'hard-coded' logic. Henderson, on the other hand, uses the declarative language PROLOG and a rule-based approach. This has the advantage that it is easy to extend or modify the rule base by adding new feature descriptions or editing existing ones.

There is also a contrast in recognition strategy. Kyprianou starts by identifying general depressions and protrusions, with the intention of subsequently refining the feature classification in a top-down manner. Henderson, on the other hand, uses sets of rules defining very specific types of features. For example, his rules for recognising a cylindrical blind hole have the form

> **if** *a circular transparent face exists*
>   **and** *a concave cylindrical wall face is attached to it*
>   **and** *a valid hole bottom face is attached to the wall face*
> **then** *the faces constitute a valid cylindrical hole feature.*

In this context, a valid hole bottom face is planar, hemispherical or conical; in either of the latter two cases its axis must coincide with that of the cylinder. Depending on the modelling system used, the 'cylindrical wall face' may actually be composed of two or more faces lying on the same surface; ideally, any feature recognition system should be able to take differences of this type into account.

There is an obvious assumption in the rules quoted above that the hole is defined in a planar face. Furthermore, the hole must be perpendicular to that face, otherwise the first condition will fail because the transparent face is elliptical. It appears, then, that for the recognition of an oblique hole another set of rules will be needed, differing only slightly from those given above. If holes are to be permitted in surfaces more general than a plane a further modified set of rules will be needed. Hence there is a danger with an approach of this nature that a large number of rule sets will be required to handle all the numerous special cases which can arise. Many of these rule sets will contain identical rules, as in the cases given above, where all three feature variants will share a common requirement for a cylindrical wall face and a valid hole bottom face.

One way of avoiding the disadvantage cited above is by deferring examination of the shape of the transparent face until the existence of some form of cylindrical hole has been established; it may then be used in a final step to distinguish between the three cases mentioned. This implies the application of rules in a hierarchical manner, initially to identify broad classes of features and later to refine their detailed classifications. Such a modification would go some way towards reconciling the differences between Kyprianou's top-down approach and Henderson's bottom-up approach. The point made in this paragraph seems to be a fundamental one, and it will recur later in the paper.

Neither method provides all the information which is required concerning material removal for process planning. The features recognised are those which occur on the surface of the part in its final machined state (assuming that method of manufacture). In practice, a part will be machined from a stock shape which initially encloses it, so that some of the material volumes to be removed are determined at least partially by the faces of the initial blank [35]. Henderson's approach comes slightly closer than Kyprianou's to accommodating this fact, but in practice his initial rectangular block is likely to have been produced from a larger block by facing operations generating smooth machined surfaces. However, even now few feature recognition publications can be cited which take considerations of this kind into account. The fact is that machining is a time-dependent process; features may be created and destroyed during that process for several reasons, including the simple requirement to have some convenient geometric configuration enabling the part to be clamped or otherwise fixtured at any stage of the overall process [11]. There is ample scope here for further research.

At this stage a few more recent approaches to feature recognition will be briefly reviewed; they have been chosen since in this writer's opinion they illustrate significant (and in some cases essentially new) principles.

3.4. Dong

Dong's work at Rensselaer Polytechnic Institute [6, 7] has several innovative aspects. Firstly, a rule-based hierarchical description of features classes is employed, suitable for a top-down recognition process and avoiding the repeated invocation of identical rule sets in the recognition of closely related feature types. This overcomes a disadvantage of Henderson's method noted in Section 3.3. Secondly, this work employs both procedural and declarative techniques, the first in the early stages for the identification of general depressions and the second for refining the classification of detected features in later stages. Thirdly, a Feature Description Language devised by Dong is implemented to facilitate the specification of new classes of feature within the system.

The recognition of features for process planning in Dong's system starts with a developed version of Kyprianou's algorithm. The general depressions identified are then decomposed if necessary into smaller volumes corresponding to individual features—this stage provides a limited capability for identifying interacting features using a profile-based approach and volume reconstruction as in Henderson's method. Fully refined feature classifications are now built up through the successive application of increasingly more detailed rule sets, whose rules mainly concern topological and geometrical interrelations between the faces composing the features. The system is based on the use of a hierarchical structure of feature frames, one for each generic feature class to be recognised and one for each instance of a recognised feature. The hierarchy is based on a feature taxonomy schema, and has inheritance of properties from the broader to the narrower feature classes defined. Frames are also implemented for the representation of topological entities from the original solid model.

The Feature Description Language provides a user-friendly method for defining new generic feature frames to act as templates for the recognition process. This avoids the necessity for using general programming languages or database utilities in configuring the system to be suitable for any particular operational context. The need for a facility of this kind is now becoming widely acknowledged, and Dong's system was amongst the first to implement one.

The order in which the system searches for feature types can be specified by the user, but no guidance is given for the choice of a 'good' strategy. This matter will be discussed later in the paper. Vandenbrande [46] points out some deficiencies in the earlier, graph-based, stages of Dong's method.

The underlying solid modeller is PADL-2, a CSG system which generates an associated boundary representation model; the latter is used by the feature recogniser. A translator, written in FORTRAN 77, translates entities from the B-rep model into the appropriate frame-based representation, and the remainder of the system is written in Common LISP.

3.5. Vandenbrande

Vandenbrande's research was carried out at Rochester University, NY, and at the University of Southern California [46, 47]. The fundamental underlying principle is that the presence of features in the model must be evidenced by what are called *hints*. For example, the presence of a (concave) cylindrical face is a hint that a cylindrical hole may exist in the model, while two parallel planar faces with oppositely directed normals may indicate the presence of a slot or groove. On the basis of such low-level initial hypotheses, further evidence is sought to confirm or disprove the existence of the postulated features. A particular feature, once its presence is confirmed, is reconstructed as a volume to allow investigation of its interactions with other features. The system is intended to recognise machinable features, and particular attention is paid to cutting tool access to the feature volumes identified.

While Vandenbrande restricts his attention primarily to pure geometric feature hints he notes that in a design-by-features system the designer's features may be taken as hints for the presence of corresponding machining features. Attribute information could also provide hints; for example, the presence of a 'thread' attribute in the model is likely to indicate the existence of a threaded hole.

The idea of the feature hint is a powerful one, transcending the effects of interactions between features. Each distinguishable feature type is characterised by certain fundamental geometrical properties used as hints; if the interaction of a particular feature with other features is such that those hints are no longer present in the model, then for practical purposes the feature no longer exists. Conversely, if the hint is present the feature may exist, albeit possibly in some highly degraded form. The use of hints also overcomes certain problems arising from the existence of features in variant forms, as discussed in Section 4 below.

Vandenbrande's system embodies a hierarchy of simple and composite features, with associated rules for each type. Following the generation of each feature hint a probability for the occurrence of the indicated feature is assigned by a *hint classifier*. This performs an initial evaluation by determining whether the faces locally adjacent to the entity giving rise to the hint are compatible with the rules for validity of the feature it suggests. Hints are categorised as *promising, unpromising* or *rejected*. The system incorporates a blackboard; unpromising hints are posted on the blackboard for possible future consideration, whilst promising hints are first followed up with a view to verification. This strategy is the primary influence on the order in which the features are recognised. The final verification of feature presence is achieved by matching against the full set of rules for each postulated feature.

In the author's opinion the notion of the feature hint is a very significant contribution to feature recognition technology. It gives rise naturally to a generate-and-test strategy having advantages over other approaches in that the system knows what it is testing for, and is not blindly seeking to match some undetermined subset of a large number of entities against one of possibly many rule sets. Feature interactions do not pose a major problem using this approach; if a feature is present, the model will provide a hint as to its existence, and Vandenbrand's verification techniques are not greatly affected by the presence of interactions. A further novel aspect of Vandenbrande's approach lies in the flexible opportunistically determined order of feature recognition.

As in Dong's work the PADL-2 CSG modeller is used to provide part models, and the primary information used in feature recognition is derived from the boundary representation it generates. In this implementation PADL-2 is tightly coupled with Knowledge Craft, an artificial intelligence environment built on Common LISP.

3.6. Some Further Methods

A few other methods will be briefly reviewed in this section, primarily to indicate the main lines of development since the early days of feature recognition.

Henderson's was the first of numerous feature recognition projects making use of rule bases and declarative programming languages. Kyprianou's work has also had a very strong influence on subsequent research, and there have been many variations on his basic ideas and on the data structures created to aid in the recognition process. A group of Italian researchers has described related methods based on a 'face adjacency hypergraph' [8, 10]; their work has in turn influenced ongoing research at Heriot-Watt University in Scotland [4, 5]. A modified version of the same idea has been used for the identification of form features in sheet metal components [23]. Joshi [17, 18] uses the similar concept of an 'attributed adjacency graph' in which the nodes represent faces and the links edges, the latter being marked with the attributes of convexity or concavity. His method works with a two-level hierarchy of feature definitions, the first level being concerned with matching graph patterns for families of feature types with local regions of the overall graph and the second with refinement of the feature description. Thus at the first level a 'strip' of three faces connected by two concave edges has the pattern of a slot; at the second level the angle between the walls and the floor is examined to determine whether it has vertical walls or is a dovetail slot. A more comprehensive graph-matching technique has been investigated by Pinilla et al. [31]. They represent each feature class by a standard face-edge subgraph, and also provide a means of generating all possible modifications of this subgraph which could arise due to interactions with other features. Feature recognition then requires the matching of subgraphs in any of the standard or modified forms with local regions of the part model graph. It appears that this method is subject to problems of combinatorial complexity. Regli and Nau [37] suggest a formalisation of the problem of recognising a class of machinable features possessing faces having planar, cylindrical, conical, spherical or toroidal geometry, using a hint-based technique similar to Vandenbrande's. Sakurai and Gossard [40] adopt a novel approach in which the user first identifies a feature on the part model interactively from a graphical display. The system then searches for similar patterns of entities in the overall data structure to identify further features of the same type. The above is just a small selection from amongst the many graph-based methods which have been proposed for automated feature recognition. Further relevant references may be found in a recent paper by Shah [41].

This partial survey will be concluded with a brief description of Meeran's work [25, 26], since it was the author's involvement with this project which stimulated the writing of the present paper. Meeran's method uses neutral file encodings of 2D drawings for its input. Patterns characteristic of various feature types are automatically recognised in the three orthogonal views of the drawing, and verification of the presence of a feature requires the appropriate spatial correlation between the patterns in those views. For simple prismatic parts most features have a 'key' pattern in one view, the patterns in the other views being rectangles. For example, the top view of a vertical cylindrical hole in a horizontal face is a circle, while that of a rectangular pocket is a rectangle with rounded corners. Both features generate pure rectangles in the front and side views, however, and so the key view provides the distinguishing information.

Meeran's system is written in PROLOG, and the edges composing the drawings are initially written out as a sequence of PROLOG facts. Rule sets are provided for the recognition of a range of commonly occurring machining features. As features are recognised the edges occurring in their patterns are deleted from the database. This results in a smaller search space for the recognition of subsequent features. Isolated features are recognised first, and when no more of them can be found the system attempts to identify and reconstruct partial patterns belonging to interacting features. There is a further final stage in which entities not belonging to any features for which rules have been implemented are grouped together into general protrusion or depression features. During the first two phases the order in which the types of features are recognised is determined simply by the order of the corresponding rule sets in the rule base, which does not conform to any particular logical scheme.

This work raises several questions:

1. The process starts, like Henderson's, with a bottom-up search for very specific feature types. This allows simplification of the database by rejection of the entities they contain, thereby aiding the search for the more difficult interacting cases. Such a strategy offends those who are more attracted to a top-down approach with progressive refinement. But what is the real virtue of the top-down approach other than intellectual elegance? And is it outweighed by the advantages of database simplification offered by Meeran's approach?
2. Assuming that only simple non-interacting features are present, is it possible to define an optimal order in which feature types should be recognised? This of course raises the further question as to what is meant by optimality in the feature recognition context.
3. How easy is it to generalise Meeran's approach to more general parts where features may be defined on sloping faces and have axes not aligned with the coordinate axes? For example, a cylindrical through hole with general orientation, emerging at both ends from non-planar faces, will now give rise to a more complex set of patterns. The circular 'key' pattern of the simple vertical hole translates into a loop which may have some indeterminate geometry, and such loops will now in general occur in more than one view. The only universal character in these patterns will be the presence in at least two views of pairs of parallel lines, connected at their ends by other lines, curves or loops. However, such parallel pairs are also characteristic of other feature types, and intensive geometric reasoning will be needed to distinguish between the various possibilities. The important thing is the occurrence of a clue to the existence of some kind of feature, which at least provides a starting-point for further investigation. Such a clue corresponds to the notion of the feature hint in Vandenbrande's 3D recognition system—it is a kind of lowest common denominator.

These considerations, and some of those raised earlier in the paper, set the scene for the discussion in the next section.

4. Feature Recognition Strategy

In this section some of the questions raised in the earlier parts of the paper will be addressed. The first topic is multiple invocation of feature recognition rules.

4.1. Hierarchical Rule Sets

In the discussion of Henderson's method (Section 3.2) it was noted that a cylindrical hole perpendicular to a planar face is characterised by a certain set of rules which differs in minor respects from the sets of rules pertaining to cylindrical holes in other situations. One rule common to *all* rule sets for cylindrical holes is the requirement for a concave cylindrical wall face; this may seem obvious, but the observation is worth making for two reasons. Firstly, to put the matter the other way round, if there is no cylindrical wall face there is no cylindrical hole. Secondly, if there is a cylindrical wall face then some form of cylindrical hole is undoubtedly present, albeit possibly in a degenerate form due to interactions. This example therefore illustrates the essence of Vandenbrande's feature hint idea.

The same example also leads on to the notion of a hierarchy of feature classes, in which 'cylindrical hole' is a class at one level, while various more specialised forms of cylindrical hole (perpendicular to a planar face, oblique to a planar face, defined on a non-planar face, ...) exist at lower levels in the hierarchy. This has the virtue that the 'concave cylindrical wall face' rule need be invoked once only in the recognition of any one of these various types of cylindrical hole. If the hierarchy were not implemented then the system would have to run through multiple rule sets all containing that same rule, with obvious implications on efficiency.

Feature hierarchies have been proposed by various researchers (examples include [12, 24, 29, 48]). In some cases the intention has been to provide some help in the automatic generation of group technology codes or in the selection of manufacturing processes, in others to impose a structure on the feature classes available in the design process. Possibly the first suggestion for a hierarchical approach to automated feature recognition was that of Dong [6], already mentioned in Section 3.4. The considerations of the last paragraph show that this is potentially a valuable means for achieving enhanced efficiency.

4.2. More on Feature Hints

The virtues of the feature hint concept in the recognition of interacting features have already been outlined. Examination of the effects of minor configurational variations on the recognition of form features provides further evidence for the power of this approach.

Consider the case of the rectangular pocket. The great majority of feature recognisers treat this as an assembly of five faces, a rectangular floor and four rectangular walls, giving rise to a characteristic topological pattern in the context of an edge-based or face-based data structure (see Fig. 1a). In such systems the walls occur in parallel pairs and meet at right angles in the corners of the pocket; similarly, the walls are required to be perpendicular to the floor, so that the volume associated with the pocket is that of a simple rectangular block.

Unfortunately, from the manufacturing point of view this is a very unrealistic pocket. If it is to be manufactured by machining, the corners must be rounded since the cutting tool has a finite radius. In practice there is often a cylindrical fillet between the wall faces and the floor face; this is easy to generate in the machined part by the use of a radiused end-mill, a cutting tool with a toroidal region blending a cylindrical side face with a planar end face. If the blending faces are included in the geometric model the corresponding topological pattern is now as shown in Fig. 1b. The rules for recognition of a realistic pocket from this data structure are considerably more complicated than for the unblended case if they are based upon face connectivities and perpendicularity relations. The walls are no longer connected

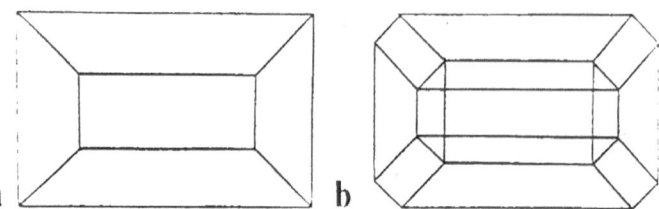

Figure 1. Edge graphs of a rectangular pocket with square (**a**) and rounded (**b**) internal corners

directly to the floor, for example, and all perpendicularity relations are defined not between immediately adjacent planar faces but between planar faces separated by blending faces.

It may be possible, with some ingenuity, to devise a single set of rules which will recognise the topological patterns of Fig. 1 as both being characteristic of four-sided pockets. Further classification of the pocket as being rectangular will require the perpendicularity relations to be established. But in fact the essential character of a rectangular pocket lies in its possession of two mutually perpendicular pairs of parallel planar faces with oppositely (inwardly) directed normals. If the existence of these two 'hints' is detected first, then other aspects of the pocket classification become easy. If the wall faces are directly connected then the pocket is truly rectangular; if not, the presence of blending faces must be confirmed. The floor face may or may not be planar—this is immaterial initially, but will emerge as a detail at a later stage. The floor may be directly connected to the walls in the simplest case; if not, the presence of further blending faces must be sought. The recognition process based on the use of feature hints therefore seems more straightforward than that based on the use of a topological graph with possible variant forms, and it also allows the subclassification of different forms of the feature in a natural way. The feature recognition procedure suggested by Regli and Nau [37] is one of the few which takes variant forms into account.

The points made above are illustrated by the variant forms of a rectangular protrusion shown in Fig. 2. From the graph-based point of view this feature in its various manifestations is bounded by (a) a loop of four concave edges, (b) an incomplete loop of three concave edges, (c) two unconnected concave edges, (d) a closed loop of two concave and two convex edges and (e) a closed loop of two smooth and two convex edges. From the feature hint point of view, however, the protrusion is uniformly characterised by the possession of two orthogonal pairs of outward-facing parallel faces. This simple example amply illustrates the problems to be faced in using graph-matching techniques for automated feature recognition in any comprehensive sense. The precise topological details of edge and face connectivity are clearly much less important than the *geometrical* relations between faces which are not necessarily connected in the model.

As a further point, it may be noted that some modellers allow corner blends and fillets to be represented implicitly, i.e. by simply *labelling* the edges concerned as

(a)   (b)   (c)

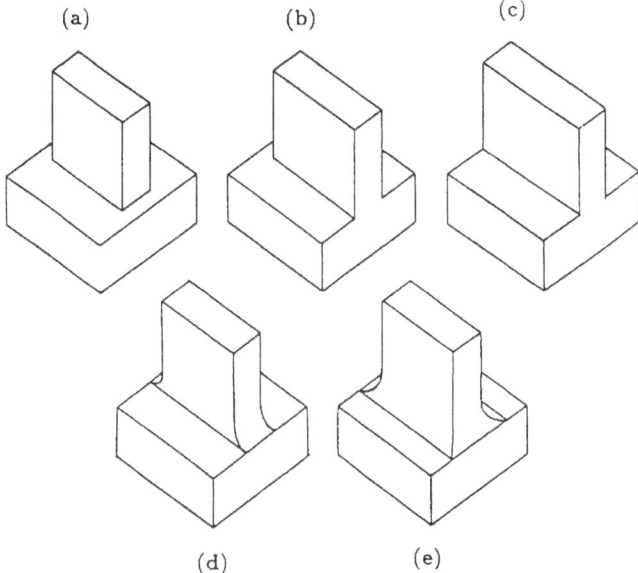

(d)   (e)

Figure 2. Five variant forms of a rectangular protrusion feature

being rounded even though they are represented geometrically as being sharp. This is sufficient for some applications purposes, such as the automatic choice of appropriate cutter geometries so that the specified roundings occur as a natural result of the machining process. It is less suitable for other applications, including the generation of realistic graphical renderings or the computation of mass properties. Feature recognition based on the use of hints will clearly not be affected by the use of this alternative means of representing variant forms of features.

The use of edge labelling is not available in CSG part models, which must therefore represent rounded edges explicitly. This can be done [22], but the resulting models may contain large numbers of cylindrical, spherical and toroidal primitives used in constructing the blends. Most CSG-based feature recognition work has avoided this complexity by restricting attention to unrealistic sharp-edged features.

4.3. Feature Interactions

The automatic recognition of form features with mutual interactions is currently still seen as a major research issue, although contributions towards this goal have been made in most of the references cited earlier. Van Houten [45] gives a partial list of the effects of feature interactions, the most important one being that an interaction may render a feature incapable of fulfilling the function originally intended for it. That function will usually require the feature to have faces with a certain geometry and a certain geometric relation to each other – it is unlikely to depend primarily upon the way edges or faces are connected together. Thus once again topological patterns appear to be of secondary importance, while the geometry required for the provision of functionality appears the logical basis for the provision of feature hints. If the functionality is destroyed by an interaction, then one or more of the crucial hints indicating the possible presence of the feature providing the functionality will be lacking. Feature hints therefore exist independently of feature interactions.

The work of Pinilla et al. [31] in the use of modified topological graphs for the recognition of interacting features has already been mentioned. This method will clearly be subject to the same combinatorial problems when applied to the recognition of features having variant forms as discussed in the last section.

4.4. Feature Definition Languages

Implementations of feature definition languages have been reported in several publications, including [6, 16, 45]. Such languages provide a convenient means for defining or editing feature class specifications without the requirement for writing code or using general database utilities. Once a particular class of design features has been defined it should be possible to invoke instances in the model of features belonging to that class. Feature classes relevant to other applications downstream of design may also be definable by such means; in this case the class descriptions should play the part of templates for automated feature recognition.

This paper will not dwell on the topic of feature definition languages beyond making two points concerning them:

1. It is generally agreed that feature-based systems should be configurable by the use of such languages, since different organisations will wish to work in terms of feature definitions best suited to their particular modes of operation, and
2. Such a language should take into acount the requirement identified earlier in this paper for a hierarchical structuring of feature types in the interests of efficiency in the process of automated recognition. An object-oriented approach is one obvious possibility, with inheritance of feature properties from broader be narrower classes.

Another possible form of configurability is that of flexible feature recognition logic. This could also be defined by the user to suit the requirements of his particular organisation, since companies vary widely in the range of parts they manufacture, the mix of feature types they exhibit and the prevalence of feature interactions. However, if a hierarchical classification of features is used as suggested above then the classification itself guides the recognition strategy to some extent, and flexible logic therefore results in a natural way provided the feature definition language is capable of defining appropriate hierarchies.

5. An Approach to Optimality?

The question mark in the heading of this section is intended to indicate that criteria for optimality in the present context are not completely clear-cut. Intuitively, we are seeking some combination of the following:

1. Speed: this requires amongst other things the organisation of the rule-base to permit efficient and non-repetitive invocation of the rules. It also implies the avoidance of techniques giving rise to combinatorial explosions when applied to complex parts.
2. Reliability, in the sense that all occurrences of feature types defined in the system will be correctly recognised.
3. Configurability, permitting user specification of feature classes and their hierarchical organisation, the latter in turn influencing the recognition logic as previously mentioned.

These are all matters which have been addressed in previous sections, where in particular the virtues of the feature hint approach to reliability and the hierarchical organisation of feature definitions for efficient access to the rule-base have been stressed.

5.1. Principal Orientations

Many mechanical engineering parts are prismatic and have most of their planar faces lying in orthogonal planes. It therefore seems worthwhile to reduce the overall recognition problem initially to a set of sub-problems, in effect by viewing the model normal to these planes. To this end, *principal orientations* may be defined for each specific component. Then for each principal orientation the system can determine such things as

- All horizontal faces and their planes.
- All faces lying on cylinders with vertical axes, with their radii and the positions of their axes.

The horizontal planes can be ordered in the vertical direction, so that it makes sense to talk of *successive planes*. Coaxial cylinders can be similarly ordered according to increasing radius. Hints or clues (the present writer prefers the latter term) concerning the existence of features can then be derived from the filtered information.

There are various possibilities for actually determining principal orientations, but for prismatic parts the simplest is to count the numbers of planar faces with various normal directions. The direction which occurs most frequently in the model may then be chosen as 'vertical' for the initial principal orientation. Note that in this context oppositely directed normals are regarded as equivalent for the purpose of finding principal orientations, though their senses become important later in the generation of feature hints.

Subsequent principal orientations may similarly be determined. For many simple parts the second and third will prove to be orthogonal to the first, though other significant orientations may be found in more complex cases.

For purposes of the scheme proposed here it will best if the underlying geometric modeller represents each plane in terms of a point and a normal vector, these in turn being represented as separate entities in the database. The point should be the point on the plane lying closest to the origin. All parallel planes will then have pointers to the same normal vector entity (the system should enforce this), so that no computation will be needed to determine sets of parallel planes. Further, distances from the origin will immediately give the spatial ordering of the planes. Cylinder axes should be represented in terms of points and axial vectors. Vertical axes in the principal orientation will then all share a common vector with the horizontal planes. The virtues of such geometric definitions in reducing the need for numerical computation in applications of geometric modellers were spelled out by Faux [9].

5.2. Nature of Clues Generated by the System

The major clues will be as follows:

- Cylindrical faces – these may be classed as convex or concave, and also according to their maximal angular extent. For example, a full cylinder implies a cylindrical hole or boss, a half cylinder may imply a slot end, a quarter cylinder may represent the rounded corner of a rectangular pocket, and so on.
- Concentric pairs of cylindrical faces (these could hint at a counterbore or a circular slot, amongst other possibilities),
- Conical face (countersink? bottom of blind hole? tapered boss?....)
- Pair of (successive) parallel planes—

  same normal direction: a depression feature (for example a step, slot, open or closed pocket,....),

  oppositely directed, inwards: could imply wall faces of a pocket or slot whose floor is vertical in this orientation,

  opposite directions, outwards: could indicate wall faces of a rectangular projection.

Some other aspects of the proposed method are summarised below:

- The method is top-down in the sense that a single clue may suggest a multiplicity of features; the possibilities are narrowed down by considering such matters as the angular extent of cylindrical faces, numbers of faces having identical geometry and the occurrence of hints in regular patterns (as will occur, for example, with the four cylindrical faces at the corners of a realistic rectangular pocket). Clues from different principal orientations will also conspire to support or refute possible interpretations.
- Once an entity has been used in the construction of a confirmed feature it may be marked to prevent its further consideration; this has the effect of reducing the size of the database used for the establishment of further feature identities.
- Hints or clues are assigned probability ratings with respect to the features they suggest (as in Vandenbrande's work). Reinforcement or contradiction of suggestions from consideration of subsequent principal orientations will modify these ratings.
- Clues based on cylindrical faces are considered to provide the most reliable suggestions since there is a greater likelihood that successive planes may relate to two different features which are non-local in the model. For validity of a parallel-plane feature clue there must be some overlapping area between the two faces concerned when they are viewed in a perpendicular direction.

- Postulated features will be confirmed or rejected using detailed rule sets, which will be organised in a hierarchical manner to allow the economical representation of families of related feature types and their variant forms.

5.3. Potential Use of Non-Geometric Data

The method as described is suitable for use with a 'pure' boundary representation modeller containing no design feature or technical attribute data. In a design-by-features system the designer's features will provide the most important initial information; every such feature should first be examined to see whether it is also a feature for the application of current interest. This will significantly reduce the scale of the overall problem. As Vandenbrande points out, attribute data (such as a 'thread' label attached to a cylindrical face) is also potentially very useful in providing hints or clues. Tolerance attributes such as 'cylindricity', 'parallelism' or 'perpendicularity', if present in the model, will also be helpful in reinforcing the geometric clues.

6. Summary and Conclusions

A survey of methods for automated feature recognition from pure geometric models has indicated that graph-based methods face very significant difficulties in the automated recognition of variant forms of features and of features subject to interaction. Hierarchical techniques based on this approach frequently operate in a top-down manner in which general protrusions and depressions are first identified and then decomposed into individual features of standard types. This requires the analysis of many special cases. By contrast, the approach based on the use of hints or clues recognises the canonical forms of features from the outset and allows the details of variant forms and interactions to be established subsequently in a relatively straightforward manner. The third class of feature recognition methods briefly discussed has been that based on cellular decomposition; this, like the graph-based approach, will be subject to combinatorial problems when applied to realistic complex parts.

Hint or clue-based methods therefore seem to give the best hope for the future. This is hardly surprising. Clues are concerned primarily with the faces of a part, and it is faces which provide functionality, faces which must be produced by manufacturing processes and faces which are toleranced and inspected. Feature recognition methods based on identifying patterns of edges are therefore not making use of the most fundamental information available in the part model.

Vandenbrande's hint-based method therefore seems a good way forward, being based on faces (or their underlying surfaces) and relationships between them. The proposed method is an extension of this approach, using the concept of principal orientations to reduce the overall problem initially to a set of sub-problems. Results from successive principal orientations may reinforce or contradict each other. As in Vandenbrande's system a generate-and-test, strategy will be used, with priority for testing based on reliability estimates. Confirmed features will be reconstructed volumetrically to enable analysis of feature interactions.

A case has been made for the use of a hierarchical method to avoid duplication and multiple invocation of the same rules. But the hierarchy does not have to start with general protrusions or depressions and finish with fully detailed features. The suggestion made here is that the hierarchy should be based on relationships between features classes having partial characterisation by the same clues.

A feature-based system should be configurable by its user organisations to operate in terms of whatever feature definitions best meet their needs. The definitions will of course be in terms of the hints or clues used in recognising features, which also give rise to the primary elements needed for feature creation in a feature-based design system. A user-friendly feature definition language is therefore required, which might eventually become standard for all systems. It should enable (and preferably enforce) the flexible specification of feature hierarchies.

References

[1] Armstrong, G. T., Carey, G. C., de Pennington, A.: Numerical code generation from a geometric modelling system. In: Solid modelling by computers (Picket, M. S., Boyse, J. W., eds.), pp. 139–157. New York: Plenum Press 1984. (Proc. General Motors Solid Modeling Symposium, Warren, MI, Sept. 1983).

[2] Bowyer, A. D., Parry-Barwick, S.: Multi-dimensional set-theoretic feature recognition. Comput. Aided Des. (in press).

[3] Brun, J.-M.: B-Rep to CSG conversion as a feature recognition process. In: Advanced geometric modelling for engineering applications (Krause, F.-L., Jansen, H., eds.). Amsterdam: North-Holland 1990. (Proc. GI/IFIP International Symposium, Berlin, Germany, Nov. 1989).

[4] Corney, J., Clark, D. E. R.: A feature recognition algorithm for multiply connected depressions and protrusions in $2\frac{1}{2}D$ objects. In: Solid modeling foundations and CAD/CAM applications. Association for Computing Machinery, New York, NY, 1991 (Proc. ACM Symposium, Austin, TX, June 1991).

[5] Corney, J., Clark, D. E. R.: Face-based feature recognition: generalizing special cases. Int. J. Comput. Int. Manuf. 6, 39–50 (1993).

[6] Dong, X.: Geometric feature extraction for computer aided process planning. PhD thesis, Rensselaer Design Research Center, Rensselaer Polytechnic Institute (published as Technical Report No. 88029), 1988.

[7] Dong, X., Wozny, M. J.: Feature volume creation for computer aided process planning. In: Geometric modeling for product engineering (Wozny, M. J., Turner, J. U., Preiss, K., eds.), pp. 385–403. Amsterdam: North-Holland 1990 (Proc. IFIP WG5.2/NSF Working Conf., Rensselaerville, NY, Sept. 1988).

[8] Falcidieno, B., Giannini, F.: Automatic recognition and representation of shape-based features in a geometric modelling system. Comput. Vision Graphics Image Proc. 48, 93–123 (1989).

[9] Faux, I. D.: Modelling of components and assemblies in terms of shape primitives based on standard dimensioning and tolerancing surface features. In: Geometric modeling for product engineering (Wozny, M. J., Turner, J. U., Preiss, K., eds.), pp. 259–275. Amsterdam: North-Holland 1990. (Proc. IFIP WG5.2/NSF Working Conf., Rensselaerville, NY, Sept. 1988).

[10] De Floriani, L.: Feature extraction from boundary models of three-dimensional objects. IEEE Trans. Pattern Anal. Machine Intell. 11, 785–798 (1988).

[11] Gilman, C. R., Dong, X., Wozny, M. J.: Feature-based fixture design and set-up planning. In: Artificial intelligence in optimal design and manufacturing (Dong, Z., ed). Englewood Cliffs: Prentice-Hall 1993.

[12] Gindy, N. N. Z.: A hierarchical structure for form features. Int. J. Prod. Res. 27, 2089–2103 (1989).

[13] Grayer, A. R.: The automatic production of machined components starting from a stored geometric description. In: Advances in computer-aided manufacturing (McPherson, D., ed.), pp.

137–152. Amsterdam: North-Holland 1977 (Proc. PROLAMAT 76 Conf., Stirling, Scotland, June 1976).

[14] Henderson, M. R.: Extraction of feature information from three dimensional CAD data. PhD thesis, Purdue University, West Lafayette, IN, 1984.

[15] Henderson, M. R., Anderson, D. C.: Computer recognition and extraction of form features: a CAD/CAM link. Comput. Ind. 5, 315–325 (1984).

[16] Hummel, K. E.: Coupling rule-based and object-oriented programming for the classification of machined features. In: Proc. ASME Computers in Engineering Conf., Anaheim, CA, July/Aug 1989, Vol. 1, pp. 409–418. American Society of Mechanical Engineers, New York, NY, 1989.

[17] Joshi, S.: CAD Interface for automated process planning. PhD thesis, Purdue University, West Lafayette, IN, 1987.

[18] Joshi, S., Chang, T. C.: Graph-based heuristics for recognition of machined features from a 3D solid model. Comput. Aided Des. 20, 58–66 (1988).

[19] Kim, Y. S.: Recognition of form features using convex decomposition. Comput. Aided Des. 24, 461–476 (1992).

[20] Kung, H.-K.: An investigation into the development of process plans from solid geometric modeling representation. PhD thesis, Oklahoma State University, 1984.

[21] Kyprianou, L. K.: Shape classification in computer aided design. PhD thesis, Computer Science Laboratory, Cambridge University, 1980.

[22] Lee, Y. C., Fu, K. S.: Machine understanding of CSG: Extraction and unification of manufacturing features. IEEE Comput. Graphics Appl. 7, 20–32 (1987).

[23] Lenz, D. H., Sowerby R.: Feature extraction of concave and convex regions and their intersections. Comput. Aided Des. 25, 421–437 (1993).

[24] Mäntylä, M., Opas, J., Puhakka, J.: Generative process planning of prismatic parts by feature relaxation. In: Advances in design automation (Ravani, B., ed.), pp. 49–60. American Society of Mechanical Engineers, New York, 1989. (Proc. 15th ASME Design Automation Conf., Montreal, Canada, Sept. 1989).

[25] Meeran, S.: Automated feature recognition from 2D CAD models. PhD thesis, School of Industrial and Manufacturing Science, Cranfield Institute of Technology, England, 1991.

[26] Meeran S., Pratt, M. J.: Automated feature recognition from 2D drawings. Comput. Aided Des. 25, 7–17 (1993).

[27] Merat, F. L., Radack, G. M.: Automatic inspection planning within a feature-based CAD system. Rob. Comput. Int. Manuf. 9, 61–69 (1992).

[28] Nieminen, J., Kanerva, J., Mäntylä, M.: Feature-based design of joints. In: Advanced geometric modelling for engineering applications (Krause, L., Jansen, H., eds.). Amsterdam: North-Holland 1990 (Proc. GI/IFIP International Symposium, Berlin, Germany, Nov. 1989).

[29] Ovtacharova, J, Pahl, G., Rix, J.: A proposal for feature classification in feature-based design. Comput. Graphics 16, 187–195 (1992).

[30] Perng, D.-B., Chen, Z., Li, R.-K.: Automatic 3D machining feature extraction from 3D CSG solid input. Comput. Aided Des. 22, 285–295 (1990).

[31] Pinilla, J. M., Finger, S., Prinz, F. B.: Shape feature description and recognition using an augmented topology graph grammar. Proc. NSF Engineering Design Research Conf., University of Massachusetts, Amherst, MA, June 1989.

[32] Pratt, M. J.: Synthesis of an optimal approach to form feature modelling. In: Proc. ASME Computers in Engineering Conf., San Francisco, July/Aug. 1988. American Society of Mechanical Engineers, 1988.

[33] Pratt, M. J.: A hybrid feature-based modelling system. In: Advanced geometric modelling for engineering applications (Krause, F. L., Jansen, H., eds.), pp. 177–189. Amsterdam: North-Holland 1990 (Proc. GI/IFIP International Symposium, Berlin, Germany, Nov. 1989).

[34] Pratt, M. J.: Applications of feature recognition in the product life-cycle. Int. J. Comput. Int. Manuf. 6, 13–19 (1993).

[35] Pratt, M. J.: Automated feature recognition and its role in product modelling. Computing [Suppl.] 8, 241–250 (1993).

[36] Razdan, A., Henderson, M. R., Chavez, P., Erickson, P. W.: Feature-based object decomposition for finite element meshing. Visual Comp. 5, 291–303 (1989).

[37] Regli, W. C., Nau, D. S.: Building a general approach to feature recognition of material removal shape element volumes (MRSEVs). In: Second Symposium on Solid Modeling and Application. (Rossignac, J., Turner, J. U., Allen, G., eds.). Association for Computing Machinery, New York, NY, 1993. (Proc. ACM Symposium, Montreal, Canada, May 1993).

[38] Roy, U., Liu, C. R.: Feature based representational scheme of a solid modeler for providing dimensioning and tolerancing information. Robotics Comput. Int. Manuf. 4, 335–345 (1988).

[39] Sakurai, H., Chin, C.-W.: Form feature recognition by spatial decomposition and composition. In: Geometric modeling for product realization (Wilson, P. R., Wozny, M. J., Pratt, M. J., eds.), pp. 189- 203. Amsterdam: North-Holland 1993. (Proc. IFIP WG5.2 Workshop, Rensselaerville, NY, 27 Sep–1 Oct 1992).

[40] Sakurai H., Gossard, D. C.: Shape feature recognition from 3D solid models. Proc. ASME Computers in Engineering Conf., San Francisco, July/Aug. 1988; American Society of Mechanical Engineers, 1988.

[41] Shah, J. J.: Assessment of features technology. Comput. Aided Des. 23, 331–343 (1991).

[42] Sodhi, R., Turner, J. U.: Representing tolerance and assembly information in a feature-based design environment. In: Advances in design automation. American Society of Mechanical Engineers, New York, NY. 1991. (Proc. 1991 Design Automation Conf., Miami, FL: ASME Publication DE-Vol. 32-1, pp. 101–106. Also published as Technical Report 91005, Rensselaer Design Research Center, Rensselaer Polytechnic Institute, Troy, NY, Oct. 1991).

[43] Staley, S. M., Henderson, M. R., Anderson, D. C.: Using syntactic pattern recognition to extract feature information from a solid geometric data base. Comput. Mech. Eng. 61–66 (Sept. 1983).

[44] Tang, K., Woo, T.: Algorithmic aspects of alternating sum of volumes, Part 1: Data structure and difference operation. Comput. Aided Des. 23, 357–368 (1991).

[45] van Houten, F. J. A. M.: PART: A computer aided process planning system. PhD thesis, Laboratory of Production Engineering, University of Twente, Enschede, The Netherlands, 1991.

[46] Vandenbrande, J. H.: Automatic recognition of machinable features in solid models. PhD thesis, Electrical Engineering Dept., University of Rochester, Rochester, NY, 1990.

[47] Vandenbrande, J. H., Requicha, A. A. G.: Spatial reasoning for automatic recognition of interacting form features. In: Computers in engineering. (Kinzel, G. I. et al., eds.), pp. 251–256. American Society of Mechanical Engineers, New York, NY, 1990. (Proc. 1990 ASME Computers in Engineering Conf., Boston, MA).

[48] Wilson, P. R., Pratt, M. J.: A taxonomy of features for solid modeling. In: Geometric modeling for CAD applications, (Wozny, M. J., McLaughlin, H. W., Encarnacao, J., eds.), pp. 125–136. Amsterdam: North-Holland 1988 (Proc. IFIP WG5.2 Working Conf., Rensselaerville, NY, May 1986).

[49] Woo, T. C.: Feature extraction by volume decomposition. In: CAD/CAM Technology in mechanical engineering. Cambridge: MIT Press 1982 (Proc. Conf. at Massachusetts Institute of Technology, March 1982).

[50] Woodwark, J. R.: Some speculations on feature recognition. Comput. Aided Des. 20, 189–196 (1988).

M. J. Pratt
Center for Advanced Technology
Rensselaer Polytechnic Institute
CII 7015, Troy
NY 12180–3590, U.S.A.

Computing Suppl. 10, 275–284 (1995)

© Springer-Verlag 1995

Solid Modeling with Constrained Form Features

D. Roller, Stuttgart

Abstract. This paper presents a method for the interactive generation of parametric form features. Furthermore, an approach for the modeling of constraints between features is introduced. In the engineering world, form features typically occur with counterparts in designs. Using the method presented herein, the user is able to create libraries of specific form features. These features can then be used in designs. Constraints between features can be set by the user and are then automatically maintained throughout design modifications. Capturing the intended design concepts in an intelligent CAD data structure, this approach constitutes a major step forward in CAD technology. Due to this approach, many errors are presented that conventional systems are prone to during design changes.

Key words: CAD, parametrics, solid modeling, feature-based design, form features.

1. Introduction

In data structures of contemporary mechanical CAD systems only the result of a geometric description of a design is kept. Thus, essential information about design decisions and the design context in terms of semantics are not registered at all. As a consequence, no intelligent help is provided for the designer. Furthermore, the integration into a CIM environment is restricted by the lack of relevant information within the CAD data structure. Gaps in information are either closed at the production engineering stage or through automatic regeneration or simply remain open. In order to resolve this problem, various kinds of form feature concepts have been developed [1–4]. In this context a form feature is defined as a design-related geometric part that is significant to the manufacturing process chain. Examples of such form features are BOLTS, CHAMFERS, THROUGH HOLES and GEAR RIMS (cf. Fig. 1).

For the purpose of efficient integration of current CAD-Systems into a CIM environment, procedures were developed that automatically recognize specific form features from the CAD data upon completion of the design process [5]. A typical example of the application of this method is the automatic generation of manufacturing process plans from CAD data. It shall be noted that all current form feature identification methods suffer from major shortcomings in their ability to cover various form feature categories as well as in their performance for an interactive user environment.

The objective should therefore be the development of CAD systems that enable the user to create designs complete with integrated form features. These form features are then to be mapped in the data structures. Of the existing methods for three-dimensional modeling, solid modeling provides the most complete geometry description. We will therefore focus on solid modeling complemented by form features.

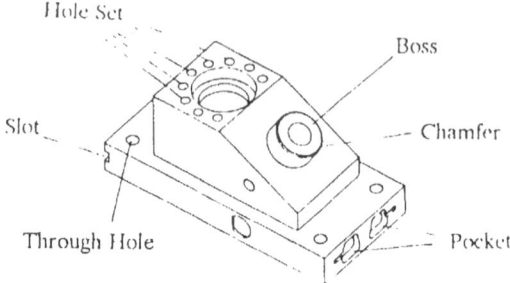

Figure 1. Example of form features

A representation of the structural and semantic knowledge supplied by form features in the data structure provides a foundation for a knowledge-based support during the design phase. The inherent potential of this method includes the ability to automatically assess such criteria as production costs, manufacturability and maintainability, to provide the user with associated feedback and even to prevent inefficient or faulty designs.

An expert system can be employed to evaluate the structural and semantic knowledge contained in the design. The purpose of the expert system is to link that knowledge to the appropriate field of application. An explicit representation of form features allows the creation of more efficient linkages to subsequent or parallel processing stages such as finite element analysis or process plan generation. However, since design engineers cannot realistically be expected to include in the design all form features that are relevant to the overall process, a sophisticated method for automatic form feature identification still is essential.

The support of form features in a CAD system has to satisfy the following prime requirements:

- The user should be able to interactively create form features.
- A library of standard form features should be provided for specific applications.
- Means for the efficient modification of any form features used in a design should be provided.
- Mutual constraints between form features should be modelable.

The latter two requirements are of particular significance because in real-life engineering a large number of designs are derived through modifications of existing designs. It is therefore desirable to automatically recognize specific relationships between form features in any design modifications. Dimensional proportions are typical examples of such relationships between form features; any change in diameter of a shaft should automatically cause the associated bearing diameter to be adjusted. Powerful modification techniques for form features bear the potential of at least partial automation or optimization of the design process.

On the following pages, a concept for computer aided design using form features is introduced that satisfies the prime requirements discussed above.

2. A System for Designing with Parametric Form Features

The first issue to be addressed is the representation of form features in a solid modeling system. An overview of relevant criteria is provided in [6]. The project presented in this paper was based on a solid modeling system using boundary representation (B-rep) [7], which implies that all surfaces of the model are explicitly included in the data structure.

In our approach, form features are represented by boundary surfaces plus associated generation specification. Any form feature can be selected either by picking a form feature component or by entering a name.

In this concept, the form feature representation is primarily based on the use of the method known as "workplane and machining commands" for 3D modeling [8]. A *workplane* is defined as a plane in model space that is always perpendicular to the viewing direction. Any design operations in the workplane are thus reduced to a two-dimensional problem. In a workplane, profiles are created through conventional 2D methods. *Machining commands* are used to build a three-dimensional model from these profiles. Typical machining commands include MILL, TURN and BORE. System-internally, such machining commands are implemented as a succession of elementary 3D commands.

In order to provide powerful methods for subsequent modifications to form features in a design, we will define parametric form features, namely with dimension and position parameters. The form features are then generated through the following steps:

- Interactive creation of a profile of variable dimensions.
- Use of the parameterized profile in parameterized machining commands.
- Optional use of the parameterized profile to create constrained form features.

Figure 2 shows the command structure of a software prototype of a design system based on form features as implemented for the study of the concept discussed in this paper. After selecting FORM FEATURES, the user can either call a predefined form feature from the library into the design or, alternatively, modify parameters of existing form features in the design. From the library the user can select either a single (i.e. unconstrained) form feature or a form feature together with a constrained form feature serving as a counterpiece.

The options DELETE, POSITION CHANGE and DIMENSION PARAMETERS are provided for modifications to form features included in the design. Also the option CREATE for generating specific new form features is supported.

2.1. Interactive Generation of Parametric Form Features

The following parameters for form features are introduced:

- Dimension parameters of profiles that are used to create form features.
- Dimension parameters for the third dimension that will be included in a machining command.

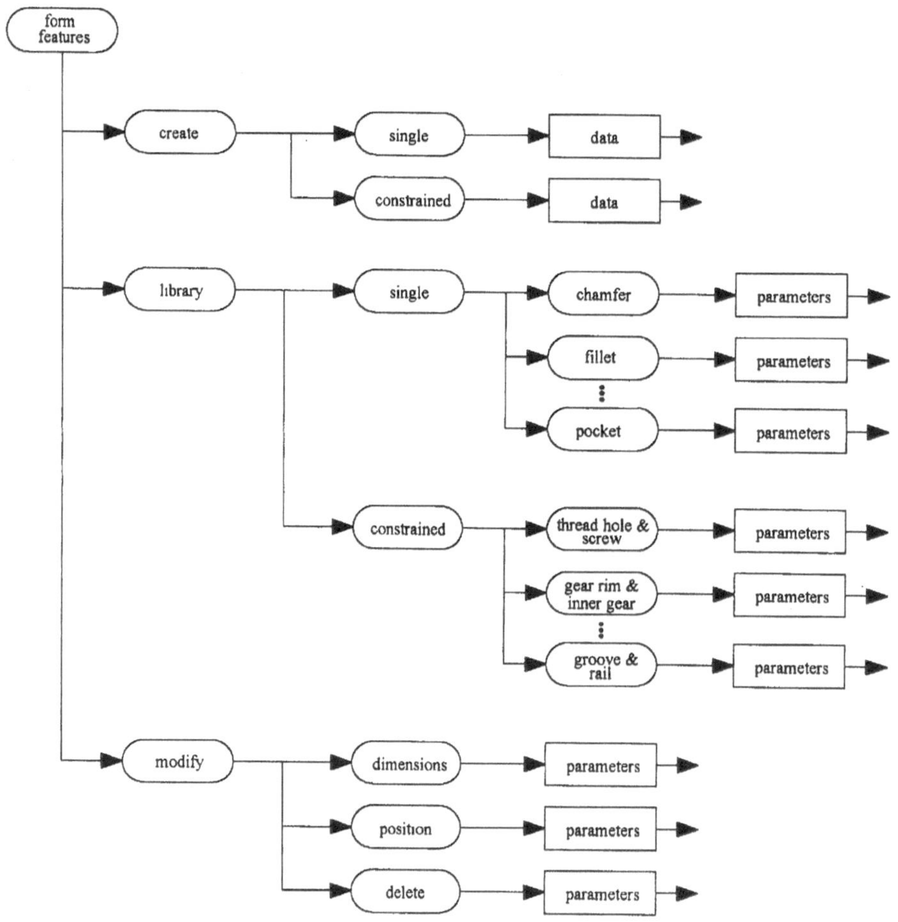

```
form
features
    ├──▶ create ─────▶ single ─────▶ data ▶
    │                 └──▶ constrained ──▶ data ▶
    │
    ├──▶ library ────▶ single ─────▶ chamfer ──▶ parameters ▶
    │                 │             fillet ──▶ parameters ▶
    │                 │               ⋮
    │                 │             pocket ──▶ parameters ▶
    │                 │
    │                 └──▶ constrained ──▶ thread hole & screw ──▶ parameters ▶
    │                                      gear rim & inner gear ──▶ parameters ▶
    │                                        ⋮
    │                                      groove & rail ──▶ parameters ▶
    │
    └──▶ modify ─────▶ dimensions ──▶ parameters ▶
                      position ──▶ parameters ▶
                      delete ──▶ parameters ▶
```

Figure 2. Command structure for feature based design

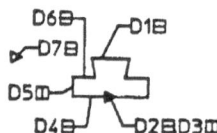

Figure 3. Profile parameters

■ Positional parameters for the relative location of a form feature following its addition to the design.

The parameterized profiles are generated by an interactive variation design method described in [9]. Figure 3 shows a parameterized guide rail section as an example. The parameters D1 through D7 are distance parameters, i.e. variable lengths. The symbols following the parameters designate the "horizontal" and "vertical" constraints relevant in this example.

The 3D dimension parameters specify the internal use of a profile in a machining command. An example of such a parameter is DEPTH for the MILL operation on a profile. A form feature is uniquely and fully described by a profile, the profile parameters and the specification of a machining operation. The selection of the machining operation and the sign (positive or negative) of the 3D parameter determine whether the form feature is positive or negative, i.e. whether the form feature 'adds or removes material'.

When a form feature is used in a design, any faces of the corresponding solid model that were created through that form feature are provided with a pointer to the form feature definition and to the associated position parameters.

2.2. Modeling Constraints between Features

Form features usually have a specific technical function. In a design, any form feature therefore tends to have an associated positive or negative counterpart. Such counterparts can be part of the same body, represent a self-contained body in their own right or be included in another body. A keyway plus its associated feather key is an example of this counterpiece concept. Another prime example is a gearing which in practice will always be provided with either a matching internal gear or a mated second gearwheel (cf. Fig. 4).

We will call such counterparts *constrained form features*. Similarly, the form featuresto which the constraints make reference are called *reference form features*. The simplest constraint is dimensional equality. Examples of general constraints are listed below:

- Relations between dimensions specified through a formula.
- Tolerance constraints such as the drive fit of a bearing bush.
- More than two dependent form features occurring in a number of bodies, e.g. a specific threaded bore that is used multiple times in the same machine.

gear rim   inner gearing

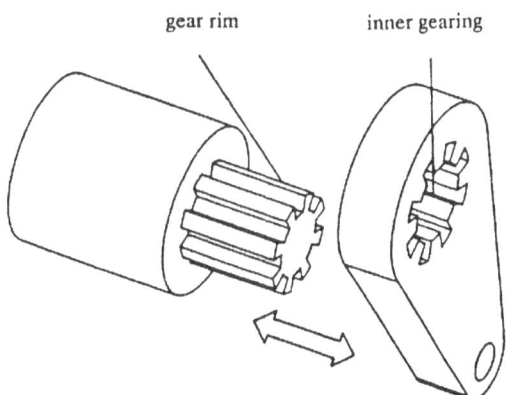

Figure 4. Example of constraints between features

form feature: SLOT

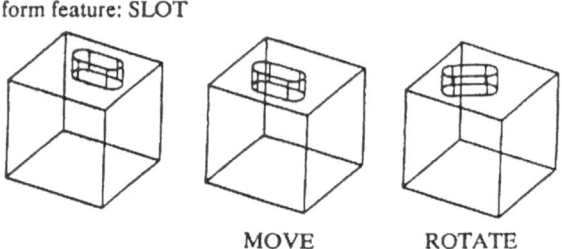

MOVE   ROTATE

Figure 5. Repositioning of form features

In the existing prototype implementation, the constraint used was dimensional equality. This constraint between form features is modeled through the following steps:

- Selecting a form feature with CONSTRAINED option from the library,
- Entering all parameters of the reference form feature,
- Entering the position parameters of the constrained form feature,
- Generating the model.

The model faces resulting from the constrained form features are marked as constrained and receive a pointer to the reference form feature. The definition of the constrained form feature is stored in the form feature library.

2.3. Design modification

Form features included in a design can be modified on two levels:

- Positional change within the reference face of the body
- Variation of arbitrary parameters

Any purely positional change is attained through *local operations*. With local operations the procedural definition of form features via machining commands remains unused. Since only the referenced faces of the solid model are concerned, these operations a very fast. Figure 5 illustrates the local operations MOVE and ROTATE applied to a form feature SLOT.

Figure 6 shows all parameters of the installed form feature SLOT for the same

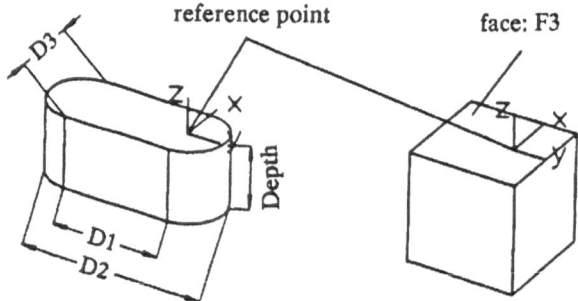

Figure 6. Dimension and position parameter of a form feature

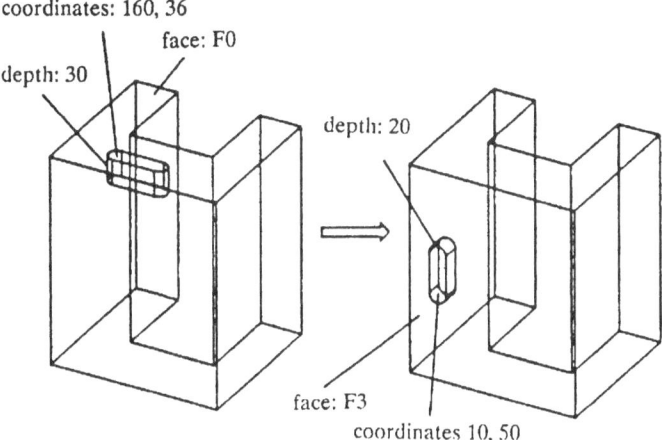

coordinates: 160, 36
face: F0
depth: 30
depth: 20
face: F3
coordinates 10, 50

Figure 7. Example for variation of dimension and position parameter

example. Any changes in dimension will first remove the form feature by executing a corresponding complementary form feature generation routine. Then the form feature displaying the changed dimension and position parameters is newly added.

Figure 7 illustrates the result of a modification through the DEPTH dimension parameter and all position parameters. The SLOT is reoriented, given different depth and then added to another reference face of the workpiece.

3. Example of a Design Using Constraints

The following description uses a simple example to illustrate the design procedure for a construction using constrained form features. Figure 8 shows a block and a channel section which serve as raw stock for an assembly. Such bodies can either be generated through conventional solid modeling techniques or called as ready-to-use form features from a library. The data structure for the conventional creation of such

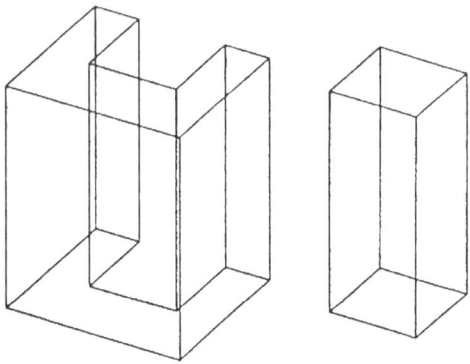

Figure 8. Starting bodies

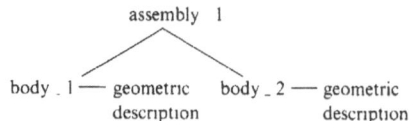

Figure 9. Schematic data structure for primitives

primitives is shown schematically in Fig. 9. If the user does not specify names for bodies or assemblies, the system assigns incremental names.

The SLIDING GROOVE form feature, option *constrained*, is called the form feature library. The user will then enter values for the dimension parameters shown in Fig. 4 and the length of the guide rail. The system supports these inputs by displaying the guide rail section complete with its associated parameters. Next, the user will select the body face to which the guide rail is to be added. The position will be chosen on the face on which the reference point of the selected form feature is to be located.

All parameters required for adding the GUIDE RAIL form feature are thus defined. The resulting model will be generated and displayed on the screen. The corresponding position parameters for the constrained form feature are then called. This constrained form feature representing a sliding groove is the counterpiece to the guide rail. In this example, equality of the section dimensions is the predefined constrained (idealized case). Figure 10 shows the workpiece after the guide rail and the sliding groove as its associated counterpart have been added. The data structure for the representation of the structural and semantic knowledge of the assembly through constrained form features is displayed in Fig. 11.

Both the dimension parameters of the guide rail and the position parameters of the sliding groove can be used for any modifications of the design. The guide rail dimension parameters are implicitly defined through the constrained relation and are therefore excluded from editing. Figure 12 shows the result of a change in sectional depth and sectional angle.

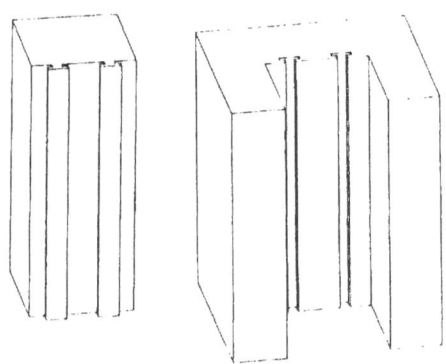

Figure 10. Adding constrained features GUIDE RAIL and SLIDING GROOVE

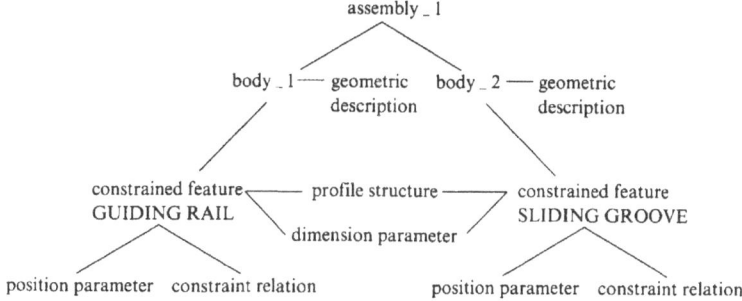

Figure 11. Schematic data structure for assembly representation

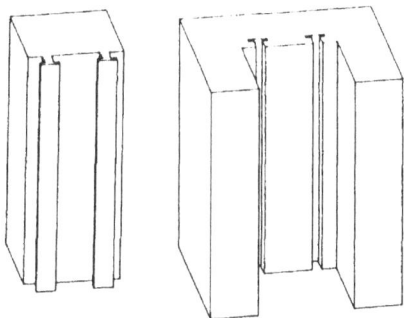

Figure 12. Modified guide rail with adapted counterpart

4. Conclusion

A system concept for designing with parameterized form features was introduced. This approach comprises the modeling of mutual constraints among form features are well as dimension and position parameters. This ensures a very powerful method for design changes through the modification of form features with automatic constraint recognition. The immediate advantages of this concept are increased efficiency in the design phase and a reduction of error sources in design changes.

An even greater potential of this approach is inherent in the utilization of semantic information derived from form features. It should therefore be the objective of future development to provide active design support in CAD systems through automatic evaluation of design knowledge.

References

[1] Shah, J. J., Rogers, M. T.: Expert form feature modelling shell. Comput. Aided Des., 20 (1988).
[2] Cunningham, J. J., Dixon, J. R.: Designing with features: the origin of features. Proceedings of the 1988 ASME International Computers in Engineering Conference and Exhibition. ASME, New York, pp. 237–243.

[3] Gossard, D., Zuffante, R., Sakurai, H.: Representing dimensions, tolerances and features in MCAE systems. IEEE Comput. Graphics Appl., 51–59 (1988).

[4] Hirschtick, J. K., Gossard, D. C.: Geometric reasoning and design advisory systems. Proceedings of the 1986 ASME International Computers in Engineering Conference and Exhibition, July 20–24, 1986, Chicago, pp. 263–270.

[5] Henderson, M. R., Anderson, D. C.: Computer recognition and extraction of form features: a CAD/CAM link. Comput. Indust. 6, 315–325 (1984).

[6] Pratt, M. J.: Aspects of form feature modelling. In: Geometric modelling, methods and applications (Hagen, H., Roller, D., eds.), pp. 227–250. Berlin Heidelberg New York Tokyo: Springer 1991.

[7] Roller, D.: Design by features: an approach to high level shape manipulations. Comput. Ind. *12*, 185–191 (1989).

[8] Roller, D., Gschwind, E.: A process oriented design method for three-dimensional CAD systems. In: Mathematical methods of computer aided geometric design (Lyche, T., Schumaker, L. L., eds.), pp. 521–528. New York: Academic Press 1989.

[9] Roller, D.: A system for interactive variation design. In: Geometric modeling for product engineering, proceedings of the 1988 IFIP/NSF Workshop on geometric modeling (Wozny, M., Turner, J., Preiss, K., eds.). Amsterdam: North-Holland 1989.

Institut für Informatik
Universität Stuttgart
Breitwiesenstrasse 20–22
D-70565 Stuttgart
Federal Republic of Germany

Computing Suppl. 10, 285–301 (1995)

A Hybrid Method for Shape-Preserving Interpolation with Curvature-Continuous Quintic Splines

N. S. Sapidis and **P. D. Kaklis**, Athens

Abstract. Quintic splines lead to efficient algorithms for constructing a curvature-continuous parametric interpolant to a given set of planar points, provided that a robust procedure is available for choosing (estimating) tangent vectors and curvature values at the given points. In this paper it is proved that, if this data is extracted from a convexity-preserving Non-Uniform Degree Polynomial Spline (a new spline recently proposed by the authors), the resulting quintic spline also preserves the convexity properties of the data points. A fully automatic procedure, based on the above spline, is proposed for constructing convexity-preserving curvature-continuous quintic interpolants to arbitrary planar points. Three examples are presented demonstrating the efficiency of the method.

Key words: Quintic spline, shape-preserving interpolation, convexity-preserving interpolation, polynomial splines of non-uniform degree, curvature, Bézier curves, control points, fair curves.

1. Introduction

This paper deals with one of the fundamental problems in CAD/CAM, the construction of a curvature-continuous (or G^2-continuous) polynomial spline that interpolates a set of planar points and, in a certain sense, preserves the shape implied by these points. Practical considerations impose three additional requirements on the sought-for solution (algorithm):

($\mathscr{A}$) The degree of the spline should be low, at most five, and the number of polynomial segments in the spline should be as small as possible.

($\mathscr{B}$) The algorithm should require the lowest possible level of technical (mathematical) competence on the part of the user, and it should be able to operate in a fully automatic mode where the user provides only the input data, i.e., the points and the boundary conditions.

($\mathscr{C}$) The constructed curve should be "fair" or "visually-pleasing". Although no universal definition exists for a fair curve, most researchers and CAD-users agree that a fair curve should be free of flat splots and sharp corners, unless these are implied by the data points. This observation can be directly translated into conditions on the curvature distribution of a curve: in order for a curve to be characterised as fair, its curvature plot should be as close as possible to a piecewise *monotone* function (see [13], [11], and references therein). [12] further strengthens this criterion by requiring linearity instead of mere monotonicity.

Existing techniques for constructing G^2-continuous interpolants—[15] offers a good review of the subject—have the common disadvantage of requiring considerable user-involvement for creating an interpolant that is free of unwanted inflection points. More specifically, the techniques based on polynomial splines-under-tension (see, e.g., [3, 10] and references therein) require that the user interactively selects appropriate values for certain "tension parameters" that determine the shape of the spline. (Designing with tension parameters is an old idea, initially proposed in 1974 [10], and has since found hardly any acceptance in the user community.) On the other hand, techniques based on assembling Bézier segments (see [2, 6, 15] and references therein) require that the user provides appropriate tangent directions and curvature values at the interpolation points. Naturally, this presumes that the user has the necessary mathematical background, i.e., these methods fail to satisfy requirement ($\mathscr{B}$) stated above.

Recently, Shirman and Séquin [15] introduced a "hybrid" method for data interpolation, where, first, an intermediate interpolant $\tilde{C}$ is calculated for the sole purpose of generating certain additional data required by the procedure constructing, in the second stage, the final interpolating spline C, the only curve provided to the user. More specifically, [15] describes a technique that uses first-derivative and curvature values from a tangent-direction-(or G^1)-continuous spline, calculated as in [14], to construct a G^2-continuous piecewise cubic interpolant, with two cubics between every pair of data points. Although Shirman and Séquin claim that this procedure produces "a spline of good visual quality" no formal statement is made or proved.

The present paper introduces also a hybrid method for solving the point interpolation problem, that can be described as follows. In the *first stage*, the points are interpolated with a Non-Uniform Degree Polynomial Spline (NUDS). This is a new C^2-continuous spline [5, 7, 8], where the polynomial degree is allowed to vary from segment to segment. This spline has the property that sufficiently large degrees produce a curve preserving the local-convexity properties of the data points. Appropriate segment degrees are automatically calculated by the algorithm $CONVNUDS$ [8], thus producing the convexity-preserving (c-p) NUDS $\mathbf{Q}_{\mathrm{nuds}}^{\mathrm{conv}}$. This spline satisfies requirement ($\mathscr{B}$) and, according to our experimentations, requirement ($\mathscr{C}$), but definitely not ($\mathscr{A}$), since segment degrees are often larger than five. In the *second stage* of the proposed technique, an augmented input data-set is created, by associating to each interpolation point a tangent direction and a curvature value both extracted from $\mathbf{Q}_{\mathrm{nuds}}^{\mathrm{conv}}$. This augmented data-set uniquely determines a G^2-continuous piecewise quintic interpolant $\mathbf{Q}_{\mathrm{q}}$, where each segment has a Bézier control polygon identical (up to collinear control points) with that of the corresponding segment of $\mathbf{Q}_{\mathrm{nuds}}^{\mathrm{conv}}$. The spline $\mathbf{Q}_{\mathrm{q}}$ *provably* preserves the convexity properties of $\mathbf{Q}_{\mathrm{nuds}}^{\mathrm{conv}}$ (see Theorem 3.2) and thus of the given input points. This hybrid methodology satisfies requirements ($\mathscr{A}$) and ($\mathscr{B}$) and it is claimed, on the basis of a large number of tests (two of which are included in this paper), that it also satisfies requirement ($\mathscr{C}$).

The paper concludes with some results giving a measure of the error between $\mathbf{Q}_{\mathrm{q}}$ and $\mathbf{Q}_{\mathrm{nuds}}^{\mathrm{conv}}$, which are employed by an interactive technique (variation of the second stage of the above interpolation method), that allows a user to minimize the error between $\mathbf{Q}_{\mathrm{q}}$ and $\mathbf{Q}_{\mathrm{nuds}}^{\mathrm{conv}}$. Note that this interactive technique should be used only when a user or an application require that the quintic interpolant approximates $\mathbf{Q}_{\mathrm{nuds}}^{\mathrm{conv}}$ as closely as possible (e.g., because Q_{nuds}^{conv} possesses a certain desirable characteristic that is lacking in the quintic Q_q automatically created as above).

2. Convexity-preserving NUDS

It is useful to begin by introducing some necessary notation. Let $\mathcal{D} = \{I_m = (x_m, y_m)^T, m = 1(1)N\}$ be a set of planar points with $I_m \neq I_{m+1}, m = 1(1)N - 1$, $\mathcal{L}_{\mathcal{D}}$ the polygonal arc connecting the points of $\mathcal{D}$, $L_m := I_{m+1} - I_m$, and $P_m := L_{m-1} \times L_m$, where $\times$ denotes the usual scalar cross product in $\mathbb{R}^2$. Finally, let $\mathcal{U} = \{u_1, u_2, \dots, u_N \in \mathbb{R}: u_1 < u_2 \cdots < u_N\}$ and $Q(u) = (Q_1(u), Q_2(u))^T$, $u \in [u_1, u_N]$, a G^2-continuous planar parametric spline, which interpolates $\mathcal{D}$ with parameterization $\mathcal{U}$ (i.e., $Q(u_m) = I_m$), and satisfies standard boundary conditions $\mathcal{B}$, namely type-I boundary conditions (given unit tangent vectors t_1, t_N at u_1, u_N, respectively), or type-II' boundary conditions ($\ddot{Q}_1 = \ddot{Q}_N = (0., 0.)^T$, where the dot accent denotes differentiation with respect to u), or, in the case of closed data ($I_1 = I_N$), periodic boundary conditions ($\dot{Q}_1 = \dot{Q}_N$, $\ddot{Q}_1 = \ddot{Q}_N$). We are now ready to precisely define the notion of convexity-preserving interpolation used in this paper.

Definition 2.1. Let $Q(u)$ be a G^2-continuous planar parametric spline, interpolating a data set $\mathcal{D}$ with parameterization $\mathcal{U}$ and satisfying boundary conditions $\mathcal{B}$. $Q(u)$ will be called convexity-preserving (c-p) provided that the curvature $\kappa(u)$ of $Q(u)$ satisfies the following condition: if $P_m P_{m+1} > 0$ then $\kappa(u) P_m > 0$ for $u \in [u_m, u_{m+1}]$.

[8] proposed using $\bar{N}on$-$\bar{U}niform$ $\bar{D}egree$ $Polynomial$ $\bar{S}plines$ (NUDS) for c-p interpolation. This family of curves, also denoted by $\Gamma(\mathcal{K}, \mathcal{U})$, $\mathcal{K} = \{k_1, k_2, \dots, k_{N-1}\}$, contains polynomial splines admitting of the following representation in $[u_m, u_{m+1}]$:

$$Q(u) = I_m(1 - t) + I_{m+1}t + h_m^2 \ddot{Q}_m F_m(1 - t) + h_m^2 \ddot{Q}_{m+1} F_m(t),$$
$$u \in [u_m, u_{m+1}], \quad t \in [0, 1], \qquad (2.1a)$$

where

$$F_m(t) = \frac{t^{k_m} - t}{k_m(k_m - 1)}, \quad t = \frac{u - u_m}{h_m}, \quad h_m = u_{m+1} - u_m, \qquad (2.1b)$$

and $\ddot{Q}_m = (\ddot{Q}_{1m}, \ddot{Q}_{2m})^T$, $\ddot{Q}_{im} = \ddot{Q}_i(u_m)$, $i = 1, 2$. The segment degrees k_m are in general different from segment to segment.

In order for $Q(u)$ to be a C^2-continuous curve interpolating $\mathcal{D}$, $\{\ddot{Q}_m, m = 1(1)N\}$ must be the unique solution of a linear system of N equations, corresponding to a symmetric and strongly diagonally dominant matrix, enforcing continuity of $\dot{Q}(u)$ at the interpolation points (for details, see [8]). Note that, if $k_m = 3, m = 1(1)N - 1$, then $Q(u)$ reduces to the classical C^2-continuous cubic interpolatory spline.

The following algorithm gives a simple iterative technique for constructing a c-p C^2-continuous interpolant in $\Gamma(\mathcal{K}, \mathcal{U})$.

The Algorithm *CONVNUDS*
Step 0. Let $\mathcal{I} = \{1, 2, \dots, N\}$. Determine the set $\mathcal{I}_1 = \{m \in \mathcal{I}: P_{m+n} P_{m+n+1} > 0$ for either $n = -1$ or $0\}$ and set $j = 0, k_m^{(j)} = 3, m = 1(1)N - 1$.

Step 1. Construct the spline $Q^{(j)} \in \Gamma^{(j)} = \Gamma(\mathcal{K}^{(j)}, \mathcal{U}) \cap C^2[u_1, u_N]$, $\mathcal{K}^{(j)} = \{k_1^{(j)}, k_2^{(j)}, \ldots, k_{N-1}^{(j)}\}$, which interpolates the data set $\mathcal{D}$ and satisfies the boundary conditions $\mathcal{B}$.

Step 2. Compute the quantities

$$\lambda_m^{(j)} = \dot{Q}_m^{(j)} \times \ddot{Q}_m^{(j)}, \quad m \in \mathcal{I}_1, \tag{2.2}$$

and

$$\mu_{m,m+1}^{(j)} = \ddot{Q}_{m+1}^{(j)} \times \ddot{Q}_m^{(j)}, \quad m, m+1 \in \mathcal{I}_1. \tag{2.3}$$

If these quantities satisfy the conditions:

$$\lambda_m^{(j)} P_m > 0, \quad \text{for } m \in \mathcal{I}_1, \tag{2.4}$$

and

$$|\mu_{m+n,m+n+1}^{(j)}| < h_{m-n}^{-1} 2^{k_{m+n}^{(j)}-1} (k_{m+n}^{(j)} - 1) \cdot \min\{|\lambda_{m+n}^{(j)}|, |\lambda_{m+n+1}^{(j)}|\},$$
$$\text{for } \mu_{m+n,m+n+1}^{(j)} P_m < 0, \quad m \in \mathcal{I}_1, \quad n = -1 \text{ and/or } 0, \tag{2.5}$$

then $Q^{(j)}$ is convexity-preserving in the sense of Definition 2.1 and stop.

Step 3.1. If, for some $m \in \mathcal{I}$, $\lambda_m^{(j)}$ does not fulfill condition (2.4), then $k_n^{(j+1)} = k_n^{(j)} + 1$, for $n = m - 1, m$.

Step 3.2. If, for some pair $m, m+1 \in \mathcal{I}_1$, $\mu_{m,m+1}^{(j)}$ does not fulfill condition (2.5), then $k_n^{(j+1)} = k_n^{(j)} + 1$, for $n = m - 1, m, m + 1$. Set $j = j + 1$ and go to step 1.

In [8] it is proved that the above algorithm converges after a finite number of iterations on the basis of the fact that, for sufficiently large segment degrees $k_n, n = 1(1)N - 1$, the curvature of $Q(u)$ satisfies the shape-preserving criterion introduced in Definition 2.1. This fact is a direct result of the fundamental property of a NUDS $Q(u) \in \Gamma(\mathcal{K}, \mathcal{U})$ that, as $k_n \to \infty$, then $Q(u), u \in [u_n, u_{n+1}]$, tends to the linear interpolant $I_n(1 - t) + I_{n+1}t, t \in [0, 1]$ [8].

3. Convexity-preserving G^2-continuous Quintic Interpolation Using NUDS

The present section aims to exploit the convexity-preserving capabilities of NUDS, which have been briefly summarized in the preceding section, for constructing convexity-preserving G^2-continuous *quintic* interpolating splines. Towards this aim, we first state and prove a general result, demonstrating a fundamental geometrical property of the Bézier control polygon of NUDS.[1]

Theorem 3.1. Let $b_n^m, n = 0(1)k_m (k_m \geq 4)$, be the Bézier control points of $Q(u) \in \Gamma(\mathcal{K}, \mathcal{U})$ in $[u_m, u_{m+1}]$. Then:

$$\Delta^l b_j^m = 0, \quad j = 1(1)k_m - 3, \quad l = 2(1)k_m - j - 1, \tag{3.1}$$

$$\Delta b_j^m = \frac{h_m}{k_m} \dot{Q}_m + \frac{h_m^2}{k_m(k_m - 1)} \ddot{Q}_m, \quad j = 1(1)k_m - 2, \tag{3.2}$$

---

[1] This property has been communicated to the authors by Costantini [1], who has been working on similar interpolation techniques. The formulation and proof of Theorem 3.1 are due to Karavelas [9].

where Δ denotes the so-called forward difference operator.

Proof: By straightforward differentiation on the right-hand side of the representation formula (2.1a), we arrive at the following expression for

$$Q_m^{(l)} = h_m^{2-l}(-1)^l \ddot{Q}_m \frac{(k_m - 2)!}{(k_m - l)!}, \quad l = 2(1)k_m - 1, \tag{3.3}$$

where the superscript (l) denotes the l-th derivative with respect to the global parameter u. Recalling standard Bézier-curve theory (see, e.g., [4]), we can write

$$Q_m^{(l)} = \frac{1}{h_m^l} \frac{k_m!}{(k_m - l)!} \Delta^l b_0^m. \tag{3.4}$$

Substituting (3.3) into (3.4), we have

$$\Delta^l b_0^m = \frac{h_m^2(-1)^l}{k_m(k_m - 1)} \ddot{Q}_m, \quad l = 2(1)k_m - 1. \tag{3.5}$$

The above formula will be now used for establishing (3.1) by induction. First, we have to prove that (3.1) holds true for $j = 1$. Rearranging the terms of the equality

$$\Delta^{l+1} b_0^m = \Delta^l b_1^m - \Delta^l b_0^m, \quad l = 2(1)k_m - 2, \tag{3.6a}$$

as below

$$\Delta^l b_1^m = \Delta^{l+1} b_0^m + \Delta^l b_0^m, \tag{3.6b}$$

and using (3.5), we find

$$\Delta^l b_1^m = 0, \quad l = 2(1)k_m - 2,$$

i.e., (3.1) is true for $j = 1$. Assuming that (3.1) is true for $j = n, 1 \le n \le k_m - 4$, we prove its validity for $j = n + 1$. Noting that

$$\Delta^l b_n^m = 0, \quad l = 2(1)k_m - n - 1, \tag{3.7a}$$

$$\Delta^{l+1} b_n^m = 0, \quad l = 2(1)k_m - n - 2, \tag{3.7b}$$

the formula

$$\Delta^{l+1} b_n^m = \Delta^l b_{n+1}^m - \Delta^l b_n^m \tag{3.8}$$

implies that

$$\Delta^l b_{n+1}^m = 0, \quad l = 2(1)k_m - n - 2, \tag{3.9}$$

i.e., (3.1) is valid also for $j = n + 1$. Thus, by virtue of the induction principle, (3.1) is indeed valid.

We now proceed to establish (3.2). Setting $l = 2$ in (3.1), the latter gives

$$\Delta^2 b_j^m = 0, \quad j = 1(1)k_m - 3, \tag{3.10}$$

from which we have

$$\Delta b_{j+1}^m = \Delta b_j^m, \quad j = 1(1)k_m - 3. \tag{3.11}$$

In view of (3.11), it suffices to evaluate $\Delta \mathbf{b}_1^m = \mathbf{b}_2^m - \mathbf{b}_1^m$. Setting $l = 1$ and then $l = 2$ in (3.4), and taking into account that $\mathbf{b}_0^m = \mathbf{I}_m$, we obtain the equalities

$$\mathbf{b}_1^m = \mathbf{I}_m + \frac{h_m}{k_m}\dot{\mathbf{Q}}_m, \tag{3.12}$$

$$\mathbf{b}_2^m = \mathbf{I}_m + 2\frac{h_m}{k_m}\dot{\mathbf{Q}}_m + \frac{h_m^2}{k_m(k_m - 1)}\ddot{\mathbf{Q}}_m, \tag{3.13}$$

respectively. Subtracting now (3.12) from (3.13) we obtain:

$$\Delta \mathbf{b}_j^m = \Delta \mathbf{b}_1^m = \frac{h_m}{k_m}\dot{\mathbf{Q}}_m + \frac{h_m^2}{k_m(k_m - 1)}\ddot{\mathbf{Q}}_m, \quad j = 2(1)k_m - 2, \tag{3.14}$$

which completes the proof ot Theorem 3.1.   ∎

For $l = 2$, the above theorem implies that the segment of a NUDS $\mathbf{Q}(u)$ corresponding to $[u_m, u_{m+1}]$ has interior control points $\mathbf{b}_j^m$, $j = 1(1)k_m - 1(k_m \geq 4)$, that are *collinear* (see Eq. (3.1) for $l = 2$), *equidistant* (Eq. (3.2)), and *ordered* on their supporting line according to the order of the index set $\{1, 2, \ldots, k_m - 1\}$ (since $\Delta \mathbf{b}_j^m \cdot \Delta \mathbf{b}_{j+1}^m > 0$, for $j = 1(1)k_m - 2$). As a consequence, the structure of the Bézier control polygon of a NUDS is extremely simple, consisting of at most four non-collinear control points (vertices) per parameter segment. In other words, the Bézier control polygon of a NUDS is as complex as that of the standard C^2-continuous cubic spline. The fact is illustrated in Fig. 1, which depicts a set of four points (large bullets) interpolated by the C^2-continuous cubic spline (Fig. 1a) and the associated convexity-preserving $\mathbf{Q}_{\text{nuds}}^{\text{conv}}(u)$, provided by the algorithm *CONV NUDS* (Fig. 1b), along with their "composite" Bézier control-polygons (control points are depicted as small bullets). Both curves share the same parameterization (chord-length) and the same boundary conditions (type-II'). Finally, $\mathscr{K}^{\text{conv}} = \{4, 6, 6\}$ is the segment-degree distribution for $\mathbf{Q}_{\text{nuds}}^{\text{conv}}(u)$. Note that, as a result of imposing type-II' boundary conditions, the restriction of the Bézier control polygon of this curve within $[u_1, u_2]$ and $[u_3, u_4]$ contains only three non-collinear control points.

We now possess all the necessary tools for constructing a G^2-continuous quintic spline $\mathbf{Q}_q(u)$, which shares the same $\mathscr{D}, \mathscr{U}$ and $\mathscr{B}$ with $\mathbf{Q}_{\text{nuds}}^{\text{conv}}(u)$ and preserves its convexity properties. Let

$$\mathbf{Q}_q(u) = \sum_{j=0}^{5} \mathbf{c}_j^m B_j^5(t), \quad \text{where } t = \frac{u - u_m}{u_{m+1} - u_m}, \quad u \in [u_m, u_{m+1}], \quad t \in [0, 1], \tag{3.15}$$

with $B_j^5(t), j = 0(1)5$, denoting the Bernstein polynomials of degree five [4]. If $k_m = 3$ or 4 for some m, $1 \leq m \leq N - 1$, then the restriction of the sought-for quintic $\mathbf{Q}_q(u)$ in $[u_m, u_{m+1}]$ can be readily obtained by applying degree elevation on $\mathbf{Q}_{\text{nuds}}^{\text{conv}}(u), u \in [u_m, u_{m+1}]$. In view of this remark, and without loss of generality, we shall henceforth assume that $k_m \geq 5$ for all $m = 1(1)N - 1$, unless otherwise noted.

Lemma 3.1. *Let $\mathscr{D}, \mathscr{U}, \mathscr{B}$ and the associated $\mathbf{Q}_{\text{nuds}}^{\text{conv}}(u)$ be given, with $k_m \geq 5$, $m = 1(1)N - 1$. Let $\mathbf{b}_n^m$, $n = 0(1)k_m$, be the Bézier control points of $\mathbf{Q}_{\text{nuds}}^{\text{conv}}(u)$ in*

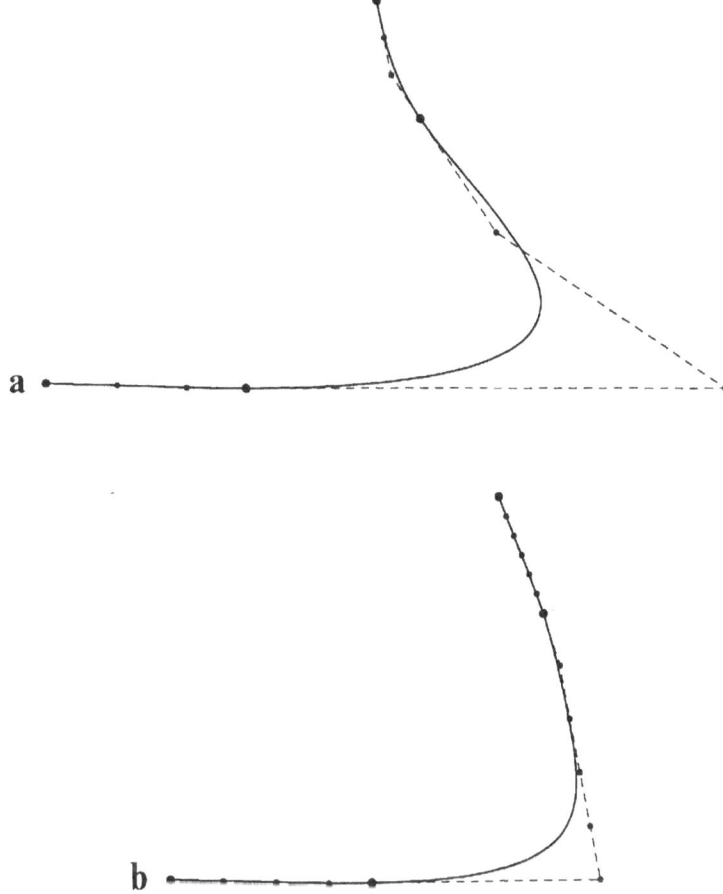

Figure 1. Cubic spline **(a)** and NUDS **(b)** interpolation: the curve and its composite Bézier control polygon

$[u_m, u_{m+1}]$, *and let* $\mathbf{Q}_q(u)$ *be the quintic spline defined by* (3.15). *Also, let* $\kappa_{\text{nuds}}^{\text{conv}}(u)$ *and* $\kappa_q(u)$ *be the curvature distribution of* $\mathbf{Q}_{\text{nuds}}^{\text{conv}}(u)$ *and* $\mathbf{Q}_q(u)$, *respectively. Then, the conditions:*

(i) $\mathbf{c}_0^m = \mathbf{I}_m$, $\mathbf{c}_5^m = \mathbf{I}_{m+1}$, $m = 1(1)N - 1$,

(ii) $\mathbf{c}_1^m = \mathbf{b}_1^m$, $\mathbf{c}_4^m = \mathbf{b}_{k_m - 1}^m$, $m = 1(1)N - 1$,

(iii) $\kappa_q(u_m +) = \kappa_{\text{nuds}}^{\text{conv}}(u_m)$, $m = 1(1)N - 1$, $\kappa_q(u_m -) = \kappa_{\text{nuds}}^{\text{conv}}(u_m)$, $m = 2(1)N$,

(iv) $\Delta\mathbf{c}_j^m \times \Delta\mathbf{c}_{j+1}^m = \mathbf{0}$, $\Delta\mathbf{c}_j^m \cdot \Delta\mathbf{c}_{j+1}^m > 0$, $j = 1, 2$, $m = 1(1)N - 1$,

ensure that $\mathbf{Q}_q(u)$ *possess the following properties:*

(a) $\mathbf{Q}_q(u)$ *is* G^2-*continuous and interpolates the point set* $\mathcal{D}$,

(b) $\mathbf{Q}_q(u)$ *satisfies the boundary conditions* $\mathcal{B}$,

(c) $\mathbf{Q}_q(u)$ *preserves the convexity properties of* $\mathbf{Q}_{\text{nuds}}^{\text{conv}}(u)$,

(d) *at each interpolation point* $\mathbf{I}_m$, $\mathbf{Q}_q(u)$ *has the same unit tangent vector and the same curvature value with* $\mathbf{Q}_{\text{nuds}}^{\text{conv}}(u)$.

Proof: First, conditions (i) imply that $\mathbf{Q}_q(u)$ interpolates $\mathscr{D}$ with parameterization $\mathscr{U}$. Second, conditions (ii) and (iii) directly establish that $\mathbf{Q}_q(u)$ is G^2-continuous and satisfies the boundary conditions $\mathscr{B}$. Conditions (iv), in conjunction with conditions (ii) and Theorem 3.1, imply that, if we ignore collinear control points, the Bézier control polygons of $\mathbf{Q}_q(u)$ and $\mathbf{Q}_{nuds}^{conv}(u)$ coincide. Then, recalling the *variation diminishing property* of Bézier curves, we readily arrive at (c). Finally, (d) is a direct result of conditions (i), (ii) and (iii). ∎

The question now arises whether one can specify the vertices $\mathbf{c}_j^m$, $j = 0(1)5$, $m = 1(1)N - 1$, so that conditions (i)–(iv) of the above lemma are simultaneously fulfilled. As far as (i) and (ii) are concerned, the answer is obviously in the affirmative. It remains to examine the solvability of conditions (iii) and (iv). For this purpose, we recall from standard Bézier-curve theory (see, e.g., [4]) the following formula for the nodal curvature $\kappa(u_m)$ of a Bézier curve $\mathbf{Q}(u) = \sum_{j=0}^{l} \mathbf{d}_j^m B_j^l(t)$, $u \in [u_m, u_{m+1}]$, of degree l:

$$\kappa(u_m) = \frac{l-1}{l} \frac{\mathbf{t}_0^m \times (\mathbf{d}_2^m - \mathbf{d}_1^m)}{|\mathbf{d}_0^m - \mathbf{d}_1^m|^2}, \tag{3.16}$$

where $\mathbf{t}_0^m$ is the unit tangent vector of $\mathbf{Q}(u)$ at $u = u_m$. Let us focus our attention on a parameter segment $[u_m, u_{m+1}]$ and assume that the therein control points $\mathbf{c}_j^m$, $j = 0$, 1, 4, 5, have been chosen in conformity with conditions (i) and (ii) of the above lemma, i.e.,

$$\mathbf{c}_0^m = \mathbf{b}_0^m, \quad \mathbf{c}_1^m = \mathbf{b}_1^m, \quad \mathbf{c}_4^m = \mathbf{b}_{k_m - 1}^m, \quad \mathbf{c}_5^m = \mathbf{b}_{k_m}^m, \tag{3.17}$$

(see Fig. 2). We shall now attempt to satisfy the remaining two conditions (iii) and (iv). Letting $\mathbf{c}_2^m$ and $\mathbf{c}_3^m$ lie on the straight line through $\mathbf{c}_1^m, \mathbf{c}_4^m$, the first part of conditions (iv), namely $\Delta \mathbf{c}_j^m \times \Delta \mathbf{c}_{j+1}^m = 0$, $j = 1, 2$, is automatically satisfied. By virtue of the above assumptions and with the aid of (3.16), the requirement $\kappa_q(u_m +) = \kappa_{nuds}^{conv}(u_m)$ takes the following final form:

$$\mathbf{t}_0^m \times \left[\frac{4}{5}(\mathbf{c}_2^m - \mathbf{b}_1^m) - \frac{k_m - 1}{k_m}(\mathbf{b}_2^m - \mathbf{b}_1^m) \right] = 0. \tag{3.18}$$

A sufficient condition for satisfying (3.18) is the following equation

$$\frac{4}{5}(\mathbf{c}_2^m - \mathbf{b}_1^m) - \frac{k_m - 1}{k_m}(\mathbf{b}_2^m - \mathbf{b}_1^m) = 0, \tag{3.19}$$

which in its turn gives

$$\mathbf{c}_2^m = \mathbf{b}_1^m + \frac{4}{5} \frac{k_m - 1}{k_m}(\mathbf{b}_2^m - \mathbf{b}_1^m). \tag{3.20}$$

Working analogously for $\mathbf{c}_3^m$ we find

$$\mathbf{c}_3^m = \mathbf{b}_{k_m - 1}^m + \frac{4}{5} \frac{k_m - 1}{k_m}(\mathbf{b}_{k_m - 2}^m - \mathbf{b}_{k_m - 1}^m). \tag{3.21}$$

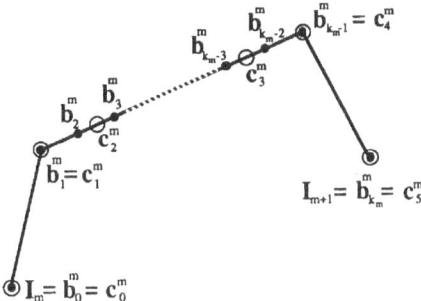

Figure 2. Constructing the vertices c_j^m, $j = 0(1)5$, of the Bezier control polygon of $Q_q(u)$ in the parameter segment $[u_m, u_{m+1}]$

Since

$$1 \leq \frac{5}{4} \frac{k_m - 1}{k_m} \leq 1.25 \quad \text{for } k_m \geq 5, \tag{3.22}$$

and the vertices $b_i^m, i = 1(1)k_m - 1$, are located on the line through b_1^m and b_{km-1}^m equidistantly, formulae (3.20) and (3.21) imply that c_2^m and c_3^m lie in between b_2^m, b_3^m and b_{km-3}^m, b_{km-2}^m, respectively. In any case, the so-obtained vertices c_j^m, $j = 1(1)4$, satisfy the second part of conditions (iv), namely $\Delta c_j^m \cdot \Delta c_{j+1}^m > 0, j = 1, 2$. We have thus succeeded in specifying the vertices $c_j^m, j = 0(1)5, m = 1(1)N - 1$, so that the sufficient conditions of Lemma 3.1 are satisfied. Conclusively, the construction process of $Q_q(u)$ can be described in the form of a theorem as below.

Theorem 3.2. *Let* $\mathcal{D}, \mathcal{U}, \mathcal{B}$ *and the associated* $Q_{nuds}^{conv}(u)$ *be given. Furthermore, let the Bézier control vertices* $c_j^m, j = 0(1)5, m = 1(1)N - 1$, *of* $Q_q(u)$ *(see Eq. (3.15)) be defined as follows:*

(i) *If* $k_m = 3$ *or* 4 *for some* $m, 1 \leq m \leq N$ 1, *then* $c_j^m, j = 0(1)5$, *should be the vertices of the Bézier control polygon of the curve resulting by applying degree elevation, from* k_m *to* 5, *to the restriction of* $Q_{nuds}^{conv}(u)$ *on* $[u_m, u_{m+1}]$.

(ii) *If* $k_m \geq 5$ *for some* m, *set* (Fig. 2):

$$\begin{cases} c_0^m = I_m, \\ c_1^m = b_1^m, \\ c_2^m = b_1^m + \frac{5}{4} \frac{k_m - 1}{k_m} (b_2^m - b_1^m), \\ c_3^m = b_{km-1}^m + \frac{5}{4} \frac{k_m - 1}{k_m} (b_{km-2}^m - b_{km-1}^m), \\ c_4^m = b_{km-1}^m, \\ c_5^m = I_{m+1}, \end{cases} \tag{3.23}$$

with $b_j^m, j = 0(1)k_m, m = 1(1)N - 1$, *denoting the vertices of the composite Bézier control-polygon of* $Q_{nuds}^{conv}(u)$. *The so-resulting* $Q_q(u)$ *is a* G^2-*continuous quintic spline, which interpolates the point set* $\mathcal{D}$ *with parameterization* $\mathcal{U}$, *satisfies the boundary conditions* $\mathcal{B}$ *and preserves the convexity properties of* $Q_{nuds}^{conv}(u)$. *Further-*

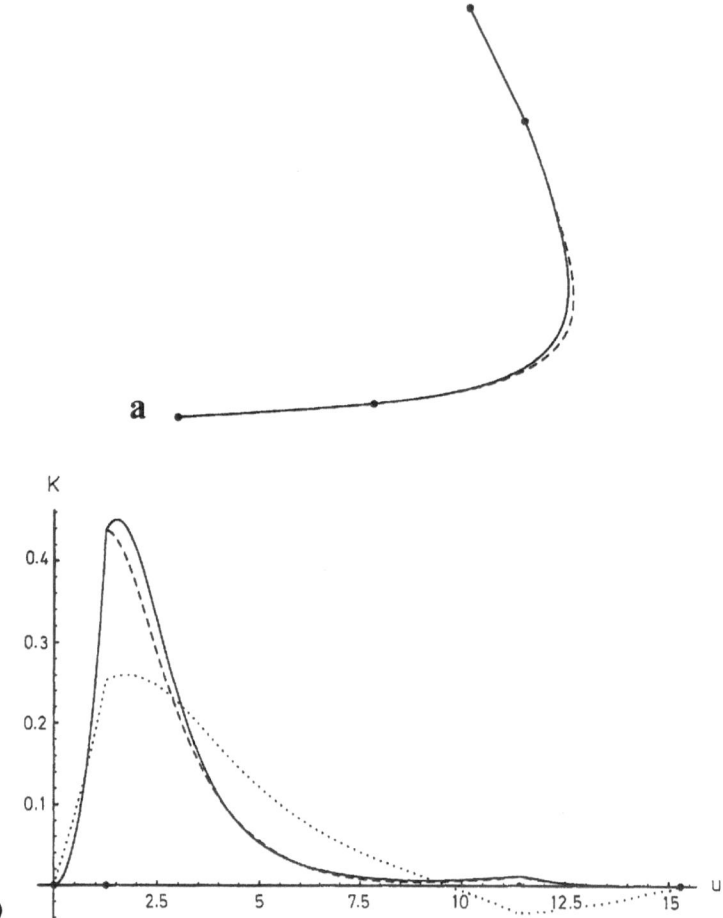

Figure 3. a Convexity-preserving NUDS (dashed curve) and quintic spline (solid curve) according to Theorem 3.2. **b** Curvature plots of the splines in **a** (dashed and solid curve) and of the spline in Fig. 1a (dotted curve)

more, at each interpolation point $I_m \in \mathcal{D}$, $Q_q(u)$ has the same unit tangent vector and the same curvature value with $Q_{nuds}^{conv}(u)$.

3.1. Examples

Example 3.1a. This example deals with the point set that was considered in Fig. 1. The dashed curve in Fig. 3a is the interpolant produced by the algorithm *CONVNUDS* (i.e., this curve is identical with the spline shown in Fig. 1b), while the solid curve is the G^2-continuous quintic spline produced by the procedure described in Theorem 3.2. The corresponding curvature plots (Fig. 3b) confirm that the quintic spline possesses the properties implied by Theorem 3.2. Figure 3b also includes the curvature plot (dotted curve) of the C^2-continuous cubic interpolant (see Fig. 1a),

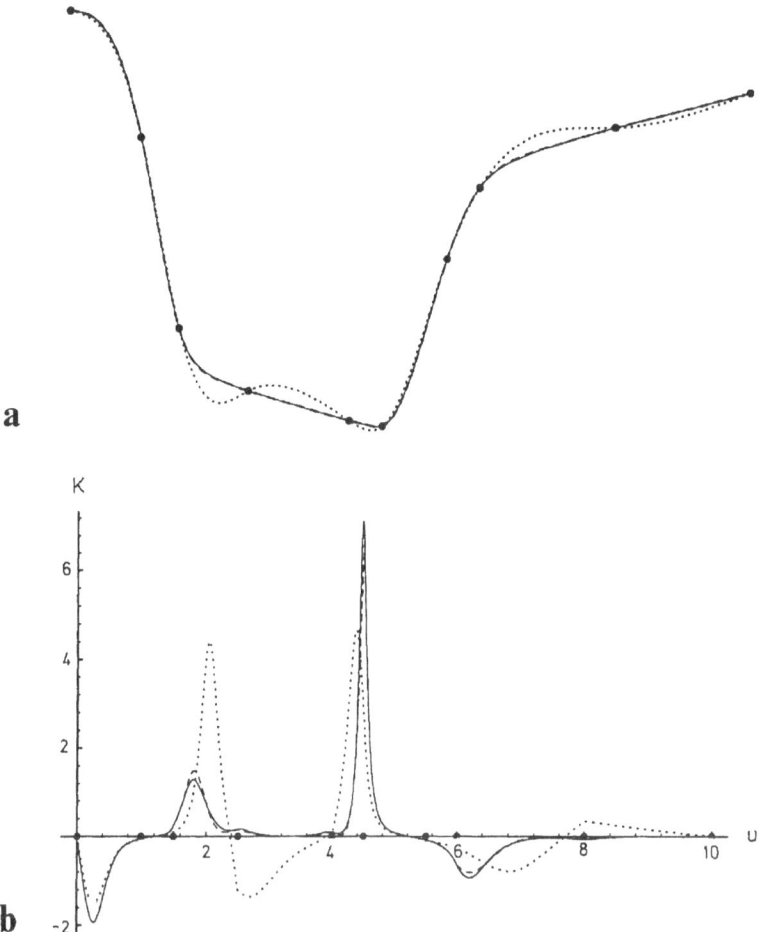

Figure 4. a Cubic spline (dotted curve), convexity-preserving NUDS (dashed curve), and quintic spline (solid curve) according to Theorem 3.2. **b** Curvature plots of the splines in Fig. 3.4a

demonstrating that this spline fails to preserve the convexity properties of the given data points.

Example 3.1b. We consider the set of ten points shown, with bullets, in Fig. 4a (this data set is used frequently in papers for testing spline-interpolation techniques). The dotted curve is the standard C^2-continuous cubic spline, the dashed curve is the NUDS produced by the algorithm *CONVNUDS* (with segment degrees $\{3,3,6,8,7,3,3,7,7\}$), and, finally, the solid curve is the G^2-continuous quintic interpolant according to Theorem 3.2. The curvature-plots of these curves are shown in Fig. 4b, and clearly demonstrate that the NUDS curve and the quintic spline have the same convexity properties and also the same curvature values at $u_1, u_2, \ldots, u_{10}$.

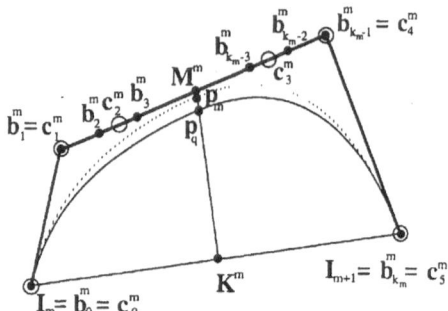

Figure 5. A NUDS (dotted curve) and $Q_q(u)$ (solid curve). The four points K^m, p_q, p_m, and M^m are collinear

4. Controlling the Error between Q_{nuds}^{conv} and Q_q

As Fig. 3a demonstrates, it is possible that the error between $Q_{nuds}^{conv}(u)$ and the quintic spline $Q_q(u)$ is quite large in a particular parameter segment (u_m, u_{m+1}). A user who is interested not only in creating a shape-preserving G^2-continuous interpolant but also in obtaining a curve that closely resembles (approximates) $Q_{nuds}^{conv}(u)$, e.g., because the latter satisfies certain aesthetic or function criteria and $Q_q(u)$ does not, will need to modify accordingly $Q_q(u)$. Below, some results are presented that provide a foundation for developing a variation of the technique described in Theorem 3.2 that, to some degree, controls the error between Q_{nuds}^{conv} and Q_q. More specifically, these results relate $Q_{nuds}^{conv}((u_m + u_{m+1})/2)$ and $Q_q((u_m + u_{m+1})/2)$ to the corresponding control points, thus leading to techniques that determine appropriate control points for Q_q so that the above "midpoints" are as close as possible to each other.

Simple algebraic calculations lead to:

Theorem 4.1. *Let* $\mathcal{D}, \mathcal{U}, \mathcal{B}, \mathcal{K}$, *and the associated* $Q(u) \in \Gamma(\mathcal{K}, \mathcal{U})$ *be given. Let* $b_n^m, n = 0(1)k_m, k_m \geq 5$, *denote the Bézier control points of* $Q(u)$ *in* $[u_m, u_{m+1}]$ *(naturally* $b_0^m = I_m, b_{k_m}^m = I_{m+1}$*) with* $K^m := (I_m + I_{m+1})/2$, *and* $M^m := (b_1^m + b_{k_m-1}^m)/2$ *(see Fig. 5). Also, let* $P_m := Q((u_m + u_{m+1})/2)$ *be the "midpoint" of the segment of* $Q(u)$ *corresponding to* $[u_m, u_{m+1}]$. *Then[2]:*

$$P_m = K^m \frac{1}{2^{k_m-1}} + M^m \left(1 - \frac{1}{2^{k_m-1}}\right) \quad \text{or} \quad K^m P_m = \left(1 - \frac{1}{2^{k_m-1}}\right) K^m M^m. \quad (4.1)$$

A similar result holds also for quintic polynomials. More specifically, the fact that, if the vectors **ab** and **cd** are equal then the midpoint of **ad** is identical with that of **bc**, and simple algebraic calculations establish:

Theorem 4.2. *Let* $d_n, n = 0(1)5$, *denote the Bézier control points of a quintic Bézier*

---

[2] In the remainder of the paper, for any two points **a** and **b** of the two-dimensional point space $\mathbb{E}^2$, **ab** will denote the unique vector that points from **a** to **b**. The x and y components of **ab** are calculated using: $ab = b - a$. A detailed discussion on points and vectors can be found in [4].

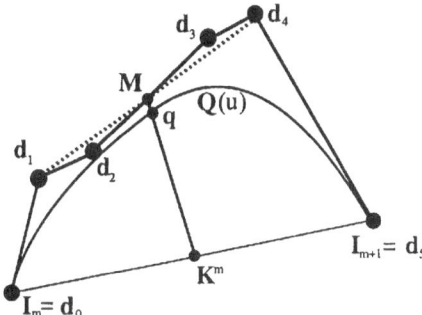

Figure 6. The points K^m, q, and M are collinear and in accordance with Eq. (4.2)

curve $Q(u)$, $u \in [u_m, u_{m+1}]$, having the properties: (a) the endpoints of the curve are $d_0 = I_m$, $d_5 = I_{m+1}$, and (b) the vectors $d_1 d_2$ and $d_3 d_4$ are equal (Fig. 6). Also, let K^m be the midpoint of $d_0 d_5$, and M that of $d_1 d_4$. Then, the "midpoint" $q := Q((u_m + u_{m+1})/2)$ is given by:

$$q = K^m \frac{1}{2^4} + M\left(1 - \frac{1}{2^4}\right) \quad \text{or} \quad K^m q = \left(1 - \frac{1}{2^4}\right)K^m M. \tag{4.2}$$

It is obvious that the quintic spline $Q_q(4)$ described in Theorem 3.2 satisfies the requirements of Theorem 4.2. Thus, we obtain:

Corollary 4.1. Let $\mathscr{D}, \mathscr{U}, \mathscr{B}$, and the associated $Q^{conv}_{nuds}(u)$ be given. Let $Q_q(u)$ be the corresponding quintic spline defined as in Theorem 3.2. Finally, let $k_m \geq 5$, and $p_m := Q^{conv}_{nuds}((u_m + u_{m+1})/2)$, $p_q := Q_q((u_m + u_{m+1})/2)$ the corresponding "midpoints" of the two spline segments. Then, p_m and p_q are collinear with the midpoint K^m of $I_m I_{m+1}$ and the midpoint M^m of $b_1^m b_{k_m-1}^m$ (Fig. 5). Also,

$$\frac{\|K^m p_q\|}{\|K^m p_m\|} = \frac{1 - 1/2^4}{1 - 1/2^{k_m - 1}}. \tag{4.3}$$

I.e., for $k_m > 5$, p_q lies closer to K^m than p_m does, and when $k_m = 5$, the two points are identical.

The above results suggest that, when in a particular parameter interval (u_m, u_{m+1}) the error between $Q^{conv}_{nuds}(u)$ and $Q_q(u)$, from Theorem 3.2, is large (one might say that this is the case in Example 3.1a), one may try to alleviate the problem by modifying $Q_q(u)$, in accordance with Theorem 4.2, so that the midpoint p_q of this curve becomes identical with or lies as close as possible to the midpoint p_m of $Q^{conv}_{nuds}(u)$. Since the point K^m (see Eq. (4.2)) cannot be modified, due to the interpolation condition $Q_q(u_j) = I_j$, one understands that the only way to "correct" p_q is to modify the control points $c_1^m, c_2^m, c_3^m, c_4^m$ of Q_q so that the midpoint M_q of $c_1^m c_4^m$ is "pulled" away from K^m making $\|K^m M_q\|$, and thus—see Eq. (4.2)—$\|K^m p_q\|$ larger, which brings p_q closer to the midpoint p_m of $Q^{conv}_{nuds}(u)$.

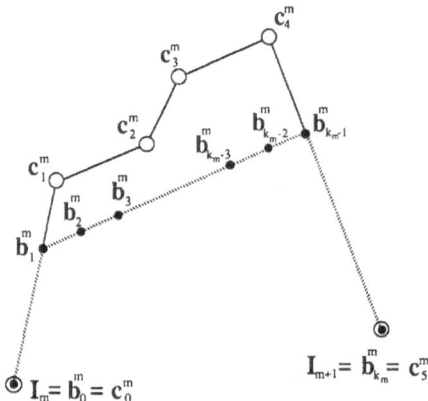

Figure 7. Determining the control points of $Q_q(u)$ according to Eq. (4.4)

4.1. A Modified Procedure for Producing a G^2-continuous Quintic Interpolant from a NUDS Curve

In this subsection, we formalise the methodology, highlighted above, for determining the control points c_j^m, $j = 0(1)5$, of $Q_q(u)$, $u \in [u_m, u_{m+1}]$. Naturally,

$$c_0^m = I_m \quad \text{and} \quad c_5^m = I_{m+1}. \tag{4.4.a}$$

A necessary and sufficient condition for $Q_{nuds}^{conv}(u)$ and $Q_q(u)$ to have the same unit tangents at $u = u_m$ and $u = u_{m+1}$ is (Fig. 7):

$$I_m c_1^m = \alpha_m I_m b_1^m \quad \text{and} \quad I_{m+1} c_4^m = \alpha_{m+1} I_{m+1} b_{km-1}^m, \quad \alpha_m, \alpha_{m+1} > 0. \tag{4.4.b}$$

Finally, iterating the analysis presented in Section 3, one can prove that a sufficient condition for $Q_{nuds}^{conv}(u)$ and $Q_q(u)$ to have the same curvature values for $u = u_m$ and $u = u_{m+1}$ is (Fig. 7):

$$c_1^m c_2^m = \frac{5}{4} \frac{k_m - 1}{k_m} \alpha_m^2 b_1^m b_2^m \quad \text{and} \quad c_4^m c_3^m = \frac{5}{4} \frac{k_m - 1}{k_m} \alpha_{m+1}^2 b_{km-1}^m b_{km-2}^m. \tag{4.4.c}$$

For given α_m and α_{m+1}, Eq. (4.4.b) determines uniquely c_1^m, c_4^m, and (4.4.c) determines the remaining points c_2^m, c_3^m. Note that setting $\alpha_m = \alpha_{m+1} = 1$ results in the procedure implied by Theorem 3.2. Also, in order for the quintic curve determined by (4.4) to satisfy the requirements of Theorem 4.2, one must set $\alpha_m = \alpha_{m+1} = \alpha$. Then, for any $\alpha > 0$, (4.4) defines a quintic curve $Q_q^\alpha(u)$, $u \in [u_m, u_{m+1}]$, whose midpoint $Q_q^\alpha((u_m + u_{m+1})/2)$ lies at the location specified by Theorem 4.2. As $\alpha (> 1)$ increases, c_1^m, c_2^m, c_3^m, and c_4^m are obviulsy pulled away from K^m, thus making $\| K^m p_q \|$ larger, which is the objective of the present procedure.

A program has been produced, including the method of Theorem 3.2 as well as the above technique, for constructing a G^2-continuous quintic interpolatory spline using the NUDS automatically calculated by the algorithm *CONVNUDS*. The first component of the program is also automatic and constructs Q_q according to Theorem 3.2 (the two examples discussed in Section 3 were produced using this code). The second component is invoked only when the user finds the error between Q_{nuds}^{conv} and Q_q unacceptable. This component is interactive and utilises the method of Eq. (4.4) with $\alpha_m = \alpha_{m+1} = \alpha$. The user identifies the segment of Q_q that must be modified, and then he/she selects a value for the parameter α. A modified curve Q_q^α is calculated, which is plotted along with its curvature distribution on the screen. The user accepts the curve and terminates the process or a new value for α is given and the process is iterated. It should be emphasized that the method of Eq. (4.4) does not guarantee that the constructed quintic spline preserves the convexity properties of Q_{nuds}^{conv} and thus of the given data. It is the responsibility of the user to compare the curvature plot of Q_q^α with that of Q_q or Q_{nuds}^{conv} to ensure compatibility of convexity properties.

Ensuring that the modified quintic spline Q_q^α is convexity-preserving would result in a completely automatic "error-control" procedure, which is the subject of our current research. We conclude this section with a demonstration of the method of

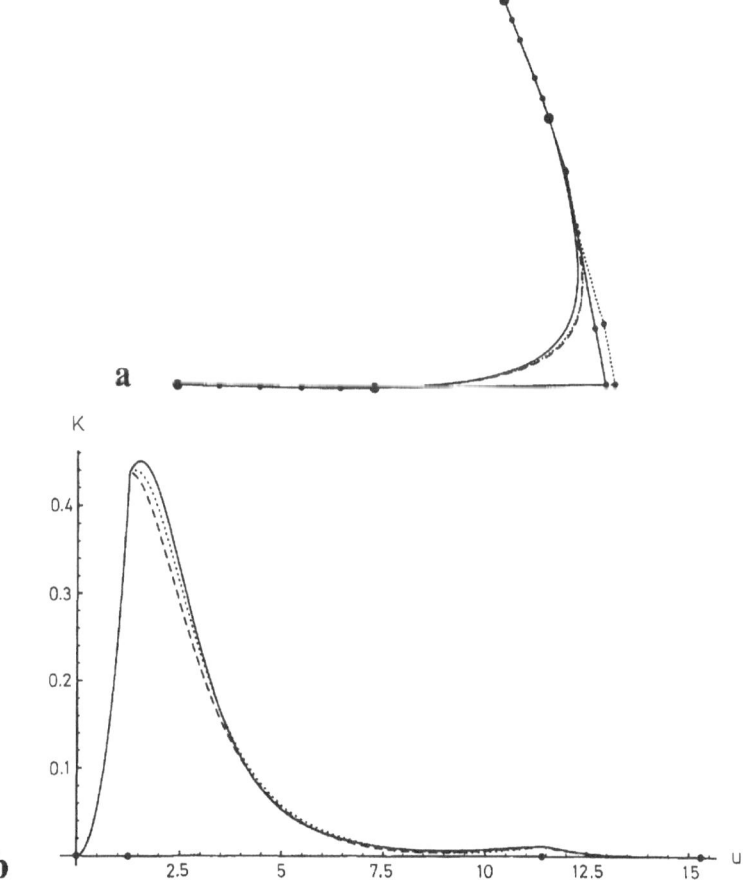

Figure 8. a NUDS (dashed curve), Q_q (solid curve), and Q_q^α (dotted curve). For the last two splines, the composite Bézier control-polygons are also shown. **b** Curvature plots of the splines in Fig. 4.4a

Eq. (4.4): Let us follow up on Example 3.1.a (Fig. 3a) assuming that one considers as unacceptable the error between $Q^{conv}_{nuds}(u)$ and $Q_q(u)$ in $[u_2, u_3]$. Figure 8a depicts the NUDS (dashed curve), the quintic Q_q according to Theorem 3.2 (solid curve), and the quintic Q^α_q calculated according to (4.4) for $\alpha = 1.04$ (dotted curve). This value of α has been determined according to the criterion of making the midpoint of the modified quintic interpolant identical with that of $Q^{conv}_{nuds}(u)$, $u \in [u_m, u_{m+1}]$, and clearly succeeds in making $Q^{conv}_{nuds}(u)$ and $Q^\alpha_q(u)$ practically indistinguishable. Figure 8a also depicts the control polygons corresponding to Q_q and Q^α_q with a solid and a dotted line, respectively. Notice that the second, third, fourth, and fifth control point of the modified quintic Q^α_q are no longer collinear. Finally, Fig. 8b shows the curvature plots of all three curves of Figure 4.4a demonstrating that application of (4.4) reduces also the error between the curvature distribution of Q^{conv}_{nuds} and that of the quintic interpolant.

5. Conclusions

A fully automatic procedure has been developed for interpolating a set of planar points with a shape-preserving G^2-continuous quintic spline. The method is hybrid, in the sense that it involves two spline-interpolation schemes: First, a non-uniform degree polynomial spline (NUDS) is constructed from which additional data (first-derivative vectors and curvature values) are obtained. Second, these data, along with the given points, uniquely determine a piecewise quintic curvature-continuous interpolant $Q_q(u)$ with a composite Bézier polygon identical (up to collinear Bézier points) with that of the NUDS curve; see Section 3. Using properties of the NUDS it is established that $Q_q(u)$ is free of unwanted inflection points and its curvature is everywhere compatible with the convexity information implied by the data points.

The paper concludes (Section 4) with a modified version of the above method that allows a user to control (reduce) the error between $Q_q(u)$ and the corresponding NUDS, when this is necessary.

Acknowledgements

This work was supported by the Ship-Design Laboratory of the Naval Architecture and Marine Engineering Department of NTUA. The authors thank P. Constantini, for pointing out that the Bézier control points of a NUDS are collinear, and M. Karavelas, for formulating and proving Theorem 3.1 as well as for developing the computer program used for producing the examples presented in the paper. All curve-plots were created using the *Mathematica* computer-algebra system.

References

[1] Costantini, P.: Personal communication 1993.
[2] de Boor, C., Hollig, K., Sabin, M.: High accuracy geometric Hermite interpolation. Comput. Aided Geom. Des. *4*, 269–278 (1987).
[3] De Rose, T., Barsky, B.: Geometric continuity, shape parameters, and geometric constructions for Catmull-Rom splines. ACM Trans. Graph. 7, 1–41 (1988).
[4] Farin, G.: Curves and surfaces for computer aided geometric design: A practical guide 3rd edn. New York: Academic Press 1993.

[5] Ginnis, A., Kaklis, P. D., Sapidis, N. S.: Polynomial splines of non-uniform degree: Controlling convexity and fairness. In: Designing fair curves and surfaces (Sapidis, N. S., ed.), pp. 253–274. Philadelphia: SIAM 1994.

[6] Goodman, T., Unsworth, K.: Shape preserving interpolation by curvature continuous parametric curves. Comput. Aided Geom. Des. 5, 323–340 (1988).

[7] Kaklis, P. D., Pandelis, D. G.: Convexity-preserving polynomial splines of non-uniform degree. IMA J. Numer. Anal. 10, 223–234 (1990).

[8] Kaklis, P. D., Sapidis, N. S.: Convexity-preserving interpolatory parametric splines of non-uniform polynomial degree. Comput. Aided Geom. Des. 12, 1–26 (1995).

[9] Karavelas, M. I.: Personal communication 1993.

[10] Nielson, G.: Some piecewise polynomial alternatives to splines under tension. In: Computer aided geometric design (Barnhill, R. E., Riesenfeld, R., eds.), pp. 209–235. New York: Academic Press 1974.

[11] Roulier, J., Rando, T., Piper, B.: Fairness and monotone curvature. In: Approximation theory and functional analysis (Chui, C. K., ed.), pp. 1–22. New York: Academic Press 1990.

[12] Sapidis, N. S.: Towards automatic shape improvement of curves and surfaces for computer graphics and CAD/CAM applications. In: Progress in computer graphics, vol. 1 (Zobrist, G. W., Sabharwal, C., eds.), pp. 216–253 (1992).

[13] Sapidis, N. S., Farin, G.: Automatic fairing algorithm for B-spline curves. CAD 22, 121–129 (1990).

[14] Shirman, L. A., Séquin, C. H.: Procedural interpolation with geometrically continuous cubic splines. CAD 24, 267–277 (1992)

[15] Shirman, L. A., Séquin, C. H.: Procedural interpolation with curvature-continuous cubic splines. CAD 24, 278–286 (1992).

Ass. Prof. P. D. Kaklis
Dr. N. S. Sapidis
National Technical University of Athens
Department of Naval Architecture and Marine Engineering
Ship-Design Laboratory
9 Heroon Polytechneiou
Zografou 157–73, Athens
Greece

Computing Suppl. 10, 303–321 (1995)

Scale-Invariant Functionals for Smooth Curves and Surfaces

C. H. Séquin, P.-Y. Chang, Berkeley, and **H. P. Moreton,** Mountain View

Abstract. Various functionals for optimizing the fairness of curves and surfaces are compared. Minimizing these functionals leads to the well-known Minimum Energy Curve (MEC) and Minimum Energy Surface (MES), to the more recently discussed Minimum Variation Curve and Surface (MVC and MVS), and to their scale-invariant (SI-) versions. The use of a functional that minimizes curvature variation rather than bending energy leads to shapes of superior fairness and, when compatible with any external interpolation constraints, forms important geometric modeling primitives: circles, helices, and cyclides (spheres, cylinders, cones, and tori). The addition of the scale-invariance property leads to stability and to the possibility of studying curves and surfaces that are determined only by their topological shape, free of any external geometrical constraints. The behavior of curves and surfaces optimized with the different functionals is demonstrated and discussed on simple representative examples. Optimal shapes for curves of various turning numbers and for some low-genus surfaces are presented.

Key words: Curves, surfaces, fairness, minimum variation, minimum energy, scale invariance, global optimization.

1. Introduction

Smooth, curved surfaces play an important role in data fitting, computer-aided geometric modeling, and manufacturing. Such surfaces can have simple well-known shapes such as cylinders or spheres as appear in shafts or ball bearings; alternatively, they can assume spline shapes as used in boat hulls and airplane wings; or they may assume aesthetically pleasing free-form surfaces as desired in the shape of vases or perfume bottles. Many methods have been developed to define interpolating surfaces with ease, precision, and speed. Most methods used today break a surface into several polynomial or rational patches. First the positions of the corners of the patches are determined, then the shape of the edges between them; finally the internal shapes of the patches are optimized individually. Much work has gone into guaranteeing that the seams between these patches remain visually smooth [3], typically providing tangent-plane continuity (G^1), sometimes even curvature continuity (G^2). Many of these approaches yield visually pleasing results, and several can achieve interactive speeds for shapes composed of tens or hundreds of patches.

However, if the resulting interpolating surfaces are analyzed carefully [2], e.g., by displaying lines of reflection or by pseudo-coloring the surfaces with its curvature value, it becomes evident that they are often less than perfect. While they may indeed maintain the desired degree of continuity (G^1 or G^2), these surfaces often show unnecessary wiggles and faint asymmetric bulges reflecting the network of geometry used to specify the surface's shape.

With the power of computers available to engineers and designers still growing at an exponential rate with no end in sight, it seems appropriate to try to use some of this power to form surfaces of maximum quality for a given set of constraints. In the most general case, this calls for a *global optimization*, using all available degrees of freedom of the chosen representation to first satisfy the given constraints and then to maximize the quality of the surface within these constraints. Global optimization relies on the choice of a *functional* to minimize. This may be total surface area, as in minimal surfaces, modeled in nature by soap films, or it may be the total bending energy as in thin metal plates.

What constitutes maximum quality cannot be defined absolutely, as it strongly depends on the application domain. However, there are a few general properties, that are almost always desirable. Foremost, in our context, there is geometrical smoothness, in particular, tangent and curvature continuity, and possibly continuity of higher order. But we also like surfaces to be "lean", avoiding unnecessary wiggles and curvature changes. Of course, they should be invariant under rigid body transformations and under uniform scaling, and should maintain all the symmetries of any specifying network of geometric constraints. Last but not least, we would like the solution shapes resulting from a set of geometric constraints to be *unique* and *consistent*. The latter means, that when additional external constraints are introduced that are compatible with the current optimal solution, the optimal shape does not change. This property cannot be achieved using local optimization and requires a global method.

The functionals used in the definition of these curves and surfaces possess other characteristics that affect their usefulness in a design environment. In particular this paper focuses on the *scaling properties* of functionals. Consider the classic minimum energy curve (MEC), resulting from minimizing bending energy:

$$\text{MEC} \Rightarrow \int \kappa^2 ds. \tag{1}$$

While the shape of the curve resulting from minimizing this functional is independent of scaling, the value of the functional is not. As a consequence, MEC are unstable in some configurations, and finite curves do not exist for all geometric specifications. The value of the functional of the minimum variation curve (MVC):

$$\text{MVC} \Rightarrow \int \left(\frac{d\kappa}{ds}\right)^2 ds \tag{2}$$

also varies with scale and possesses similar instabilities. To nullify the effect of scaling, we have developed scale-invariant versions of the MVC and MEC functionals [8, 9].

In this paper, we review our previous work on minimum variation curves (MVC) and surfaces (MVS), and extend our recent work on a scale-invariant version of the functional used for curves. We then present new work and results on a scale-invariant version of the minimum variation functional used for surface design. Finally, we propose some new functionals whose properties appear promising and warrant further investigation.

In our review we focus on properties and issues that are independent of the particular underlying representation. We discuss possible alternatives to achieving scale invariance and present some illustrative examples to demonstrate the different behaviors of the various functionals for curves, both the traditional MEC and the newer MVC, and their scale invariant counterparts. We then turn the discussion to surfaces and present examples to elucidate the characteristics of surface functionals and the effect of scale invariance. These examples involve modeling symmetric low-genus objects free of interpolatory constraints; as such they are well suited to demonstrate scale-invariant functionals and the properties of the associated surfaces.

While scale-invariant curves are reasonably well understood, work on scale-invariant minimum variation surfaces is still in its infancy. Some valuable insights have been gained by using the original program developed by Moreton [6, 8]. The lessons learned will guide further research in this area. A thorough review of previous work on minimum energy and minimum variation curves and surfaces, and on optimization for curved shape design may be found in [8].

2. Scale Invariance

Useful properties for functionals for curve and surface design include invariance under rigid body transformations and uniform scaling. The shape of MEC and MVC is independent of scaling, however, the value of these functionals is decreased when the defining geometric constraints are scaled up. As a result, both the MVC and MEC are unstable in certain configurations, expanding to infinity to reduce the functional value (Fig. 1).

In all these cases, the force that drives the curve towards unbounded size originates from the reduction in the value of the cost functional as the scale of the curve is increased. The arc length of the curve grows linearly with scale, and its curvature decreases reciprocally. The MEC functional integrates the square of curvature, and thus its value decreases in direct proportion to the scale factor. In the MVC functional, the derivative of curvature with respect to arc length falls as the square of the scale, and it appears squared in the functional. Thus, overall, the MVC functional decreases in proportion to the cube of the scale factor.

To obtain stability, the functional should be scale-invariant, so that the forces resulting from uniformly growing or shrinking an arc cancel out. Scale-invariance

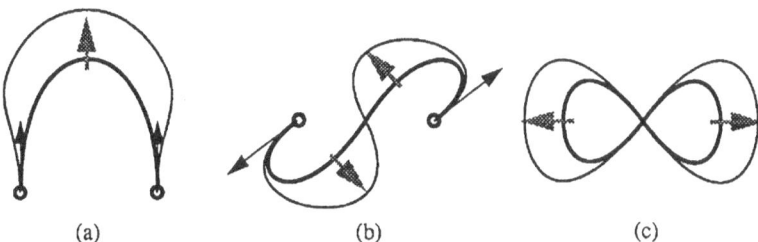

| (a) | (b) | (c) |

Figure 1. Conditions leading to unstable situations for minimum energy curves (a), and for minimum variation curves (b), (c)

then permits us to study closed curves that are totally unconstrained by any external geometric specifications other than their turning number, i.e., the total number of turns swept out by the tangent direction vector.

2.1. Scale-Invariant Functionals

Scale-invariance can be obtained by multiplying the integrated functional by a compensating power of the total arc length [8]. For the MEC the multiplication factor is proportional to arc length, whereas the MVC functional must be multiplied by the cube of the arc length to obtain a scale-invariant expression:

$$\text{SI-MEC} \Rightarrow \left(\int ds \right) \left(\int \kappa^2 ds \right) \tag{3}$$

$$\text{SI-MVC} \Rightarrow \left(\int ds \right)^3 \left(\int \left(\frac{d\kappa}{ds} \right)^2 ds \right). \tag{4}$$

This is equivalent to scaling each shape so that its total arc length becomes unity and evaluating the costs of those shapes.

In our original implementation, the curve is represented as a series of quintic Hermite segments. The program starts with an heuristic guess for the initial tangent directions and curvature values at all the joints between segments. The desired G^1- or G^2-continuity at the joints is maintained by construction. During the optimization process, the values of the selected functional and its partial derivatives with respect to all n free parameters are computed numerically. Starting from the heuristically established point in the n-dimensional solution space, Polak-Ribiere conjugate gradient descent is used to reduce the objective function until the gradient falls below a predetermined threshold. See Moreton's thesis [8] for details.

Experimentally we observe that the scale-invariant functionals converge more slowly than their scale-dependent originals. The cause seems to be a more global coupling of changes in one part of the curve to some other part via the arc length normalization. Thus one might consider an alternative, possibly faster, approach to obtaining scale-invariance. In particular, one might try to use a different power of curvature—or of the derivative of curvature—to compensate for the change in arc length under uniform scaling operations.

Curvature, indicated as the width of a "fin" on some arc segments in some of the following figures, decreases linearly with an increase in scale and could thus naturally balance a corresponding increase in arc length. However, if we simply use curvature as a cost functional, we find that any smooth deformation of a curve does not change the cost at all, since a local increase in positive curvature can be compensated for either by a decrease in curvature or by an increase in negative curvature somewhere else. Integrating the first power of curvature simply returns the tangent direction, and the integrated cost over a closed curve would just reflect the turning number of that curve, i.e., the total change in tangent direction, which is constant under smooth deformations.

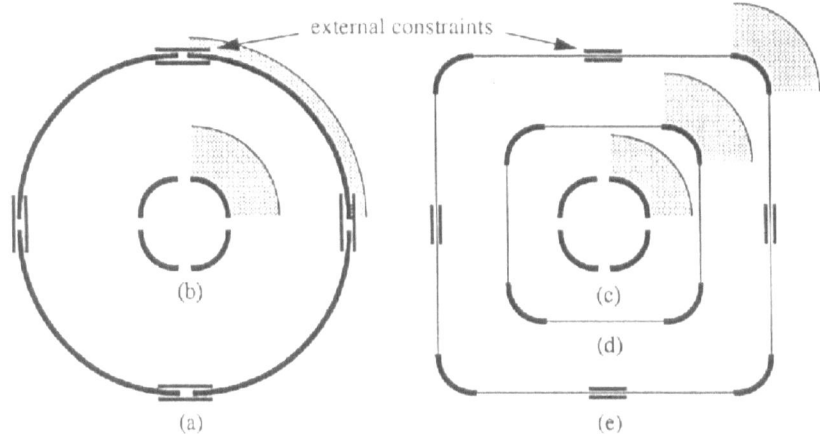

Figure 2. Cost comparisons for differently scaled figures

To obtain a cost functional more in the spirit of bending energy, we might try to use the absolute value of curvature, since the direction of the bend should not matter. But, as the example in Fig. 2 shows, this does not lead to the proper behavior either.

Using any properly scale-invariant cost functional based on the absolute value of curvature, a 90°-turn achieved with a certain curve shape has a fixed cost associated with it, which is independent of scale (Fig. 2a, b). Connecting these corner pieces to straight curve segments does not incur any additional costs if the connection is made with proper observance of tangent continuity. The straight pieces themselves do not have any cost associated with them when considered separately. Thus all five shapes in Fig. 2 have the same costs if we simply add the cost of the individual pieces. This result seems intuitively wrong—we certainly would not want the almost square shape (Fig. 2e) to have the same cost as the nearly circular shape of similar size (Fig. 2a); this would imply that Fig. 2a as well as Fig. 2e represent stable minimum cost solutions for the set of four tangential constraints shown with these two curves. This demonstrates that there may not exist any local cost functional based on *curvature* that will give us the proper behavior *and* provide the desired scale invariance. The reason is that simply considering the *linear* power of curvature with the overall constraint that the integral over curvature must remain constant (equal to the total change in tangent direction) does not discourage the formation of small regions of high curvature; we need to use a higher power to favor more uniform distribution of curvature which avoids excessive values.

For MVC, the above approach would do even worse, since we would have to use the *square root* of the curvature derivative to obtain the desired scaling behavior. This would also cause problems with the differentiation during the optimization process for values near zero.

These considerations force us to use an approach based on the normalization of the total arc length. If the total integrated curvature cost is multiplied with a suitable power of total arc length, the finite cost of the corner pieces is augmented if straight sections are inserted between them, thereby leading to increasing overall cost as the corners become "relatively sharper", i.e., as they take a smaller fraction of the overall curve length to make their turns (Fig. 2c, d, e). This exhibits the desired behavior and is equivalent to introducing an "optimal" arc-length constraint. It is worth pointing out that the use of the square of curvature also leads to curves that are infinitely differentiable between constraint points.

3. Behavior of Functionals

In this section we will demonstrate the behavior of the four cost functionals on representative curve geometries, and, in particular, compare the two new scale-invariant cost functions with their scale-dependent fore-runners. This will give us a good understanding of the typical shapes assumed by the different curve types and of their relative stabilities.

In all the experiments the curves are computed using standard optimization techniques. The starting point for the optimizations, the initial shapes of the curves, are specified with as few constraints as practical. After the initial shape is defined in this manner, unnecessary or undesirable constraints are removed, and the curve is left to evolve under the chosen functional and remaining constraints.

The classical arrangement to demonstrate the instability of the MEC is the arrangement by Horn [4]: two parallel tangential constraints; if the tangents are tilted slightly outward, or the curve is disturbed slightly outward from its metastable equilibrium point, the curve runs away. In [8] we reported that the other three functionals result in circular arcs. This seems rather obvious for the MVC and the SI-MVC since the circular shape reduces the functional to zero. However, for the SI-MEC it may at first appear surprising that the introduction of a finite cost for arc-length reduces Horn's arch to exactly a semi-circle. The reason has to do with the consistency of these globally optimized curves. For all functionals except the MEC, a circle of any size is a stable optimal configuration for a totally unconstrained curve of turning number 1. Thus for any set of compatible constraints, the circular shape will be assumed and will form an optimum. For the MEC, however, an unconstrained circle is not a stable optimum; it will want to expand indefinitely. Thus an MEC circle clamped between two opposing parallel tangential constraints will squeeze out in the perpendicular directions to form two joined Horn curves.

Here we present another illuminating example. It is a modified version of the Horn experiment discussed above. Two Horn segments of different sizes are joined together at one leg so as to form an inflection point (Fig. 3), then we let the curve evolve towards its nearest local minimum under the influence of the different functionals. The final curve shapes are shown with fins that indicate the value of curvature at any point along the curve; this makes any variations or discontinuities in curvature more visible.

In all six cases, the two curve segments are constrained by three points with vertical tangents. For the MEC we find—not surprisingly—two copies of the Horn curve

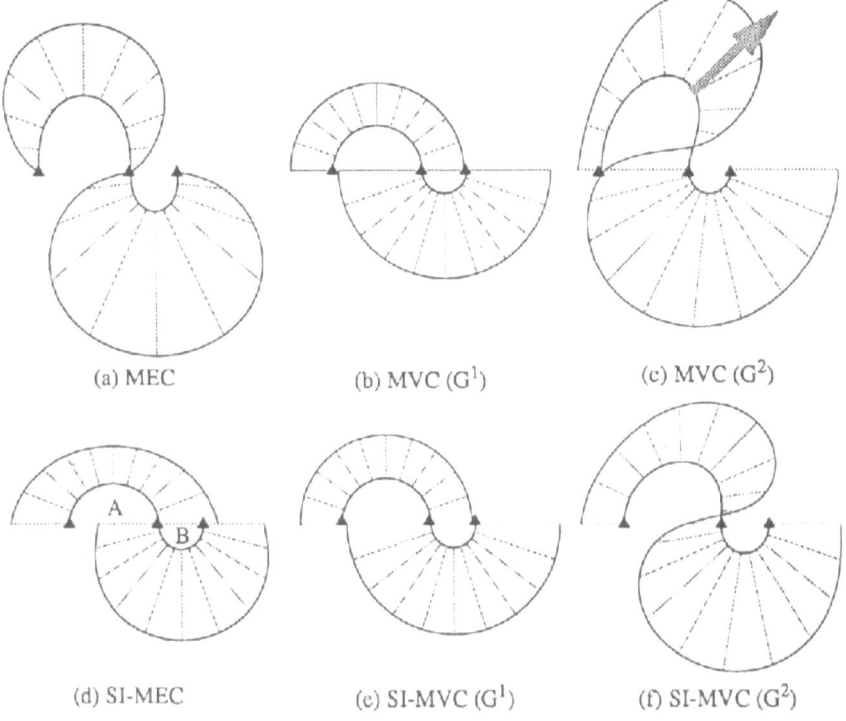

(a) MEC   (b) MVC (G^1)   (c) MVC (G^2)

(d) SI-MEC   (e) SI-MVC (G^1)   (f) SI-MVC (G^2)

Figure 3. Comparison of the behavior of four functionals (**a**) MEC, (**b, c**) MVC, (**d**) SI-MEC, and (**e, f**) SI-MVC on the asymmetric double Horn curve with three vertical tangential constraints and an inflection point. In (**a, b, d, e**) the intermediate joint was specified to be of G^1-continuity; in (**c, f**) it is a G^2-joint

properly scaled to fit the given constraints (Fig. 3a); the fixed tangent constraint at the intermediate joint effectively de-couples the two halves. For the two MVC segments joined by a G^1 vertex where the curvature can make a jump without penalty, we find two circular arcs (Fig. 3b). For the SI-MEC one might again expect two half-circles, but the multiplication with total arc-length creates a mutual influence between the two arcs. The larger arc gets shortened at the price of higher curvature, and the smaller arc grows slightly to achieve lower curvature cost and to balance the cross terms {curvature-cost(A)∗arc-length(B)} and {curvature-cost(B)∗arc-length(A)} (Fig. 3d). A similar balancing is absent in the SI-MVC (Fig. 3e) since the net cost of both pieces is truly zero, and one still obtains circular arcs. The MVC and SI-MVC segments coupled by a G^2-joint produce a smooth curve with continuously varying curvature (Fig. 3c, f). However, the MVC runs away, since the large arc is thrown out of the region of stability by the high curvature imposed by the small arc. The SI-MVC, of course, remains stable, but the larger arc becomes somewhat lopsided because of the constraints imposed by the small arc.

Table 1 summarizes what degree of continuity can be achieved by various curves at the constraint points. For completeness, we have also included MLC, the minimum-length curve, best exemplified by a rubber band.

Table 1. Achievable degree of continuity at constraint points

Curve\constraint	None	Position	Tangent	Curvature
MLC	G^x	G^0	not applicable	not applicable
MEC	G^{2v}	G^2	G^1	not applicable
MVC	G^∞	G^4	G^3	G^2

If the vertical tangent constraint at the intermediate shared vertex is removed after initialization, the shapes of the *scale-invariant* curves do not change materially. However, the other three curves become unstable and run away forming arcs of unbounded size on either the top or bottom half. The exact form of this process and the speed of running away depend on the details of the representation of the curve. The MEC typically runs away the fastest.

3.1. Loosely Constrained Closed Curves

The global minimum for curves of turning number zero is a figure-8 shape that looks like a lemniscate. However, the observed optimal shapes are slightly fatter (MEC, MVC, SI-MVC) or narrower (SI-MEC) than the true lemniscate, and the cost of the true lemniscate when measured with the various functionals is from a fraction of a percent (SI-MEC) to nine percent (MVC) larger than the global optimum. We refer to all these lemniscate-like shapes as "lemnoids".

As demonstrated in [8], a SI-MVC of turning number $T = 1$ can come to rest in the shape of a lopsided lemnoid with an inner osculating circle in one lobe (Fig. 4d) which does represent a true local minimum. Thus we study this configuration for all four functionals (Fig. 4). The circle is attached at the tip of the lemnoid lobe where the curvature is a maximum. This shape is initially defined with four points and vertical tangent constraints on the x-axis, but the constraint on the left of the circle is later removed. The constraints keep the size of the lemnoid fixed, but the attached loop can assume any shape it likes. For the MVC coupled by G^2-joints, it becomes an inner osculating circle on the minimum-cost lemnoid (Fig. 4c); attaching such a circle does not increase the cost of the shape. The SI-MVC produces a slightly smaller circle and also reduces the size of the lobe where it is attached (Fig. 4d), since the increase in arc-length is no longer free. The SI-MEC produces a considerably larger circle with a G^2-discontinuity where it is attached to the lobe of the lemnoid (Fig. 4b); the size of the final circle is such that the ratio of the arc lengths of the lemnoid and of the circle are the inverse of their respective curvature integrals; this balance minimizes overall cost. Finally, the MEC's circle grows without bound (Fig. 4a) because of the lack of scale-invariance or any other restraining force.

3.2. Evolution of Unconstrained Closed Curves

The SI-functionals permit us to study the optimization of closed curves that are defined by their original shape and are only constrained their turning number, T, but

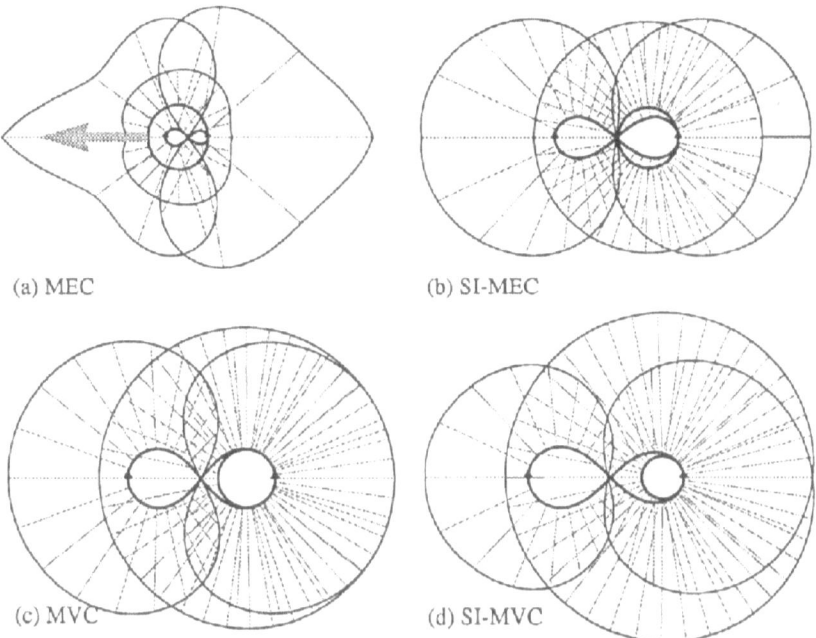

(a) MEC   (b) SI-MEC

(c) MVC   (d) SI-MVC

Figure 4a–d. A figure-8 with a loop is constrained by three position-tangent specifications, one on the left and two on the right (the scale of each plot may be derived from the fixed spacing of the constraints)

are otherwise completely free of any external constraints. The turning number is zero for a smooth figure-8, and ± 1 for a single-turn circle. All curves with the same turning number can be smoothly deformed into one another without ever assuming singularities such as kinks and cusps. Thus for any given turning number, there must be some shape that assumes the lowest possible cost of that class. The question is whether any curve of that class will automatically evolve to this global optimum under our optimization procedure or whether it will get hung up in a local minimum.

In general, curves without inflection points readily evolve to multiply traced circles under both scale-invariant functionals. The situation becomes more complicated, when there are inflections as in the curve shown in Fig. 5a which has turning number $T = 0$. In principle, the curve could evolve into a simple figure-8 shape as indicated in Fig. 5a through e. And indeed, this evolutionary path is followed by the SI-MEC. The MEC assumes a shape similar to the one in Fig. 5c while running off to infinity. The MVC functionals can also follow a different path indicated by Fig. 5f through i, ending up in a shape that is a lemnoid with an osculating circle in each lobe. Which of the two evolutionary paths is taken depends critically on the starting shape. Fig 5a is close to the decision saddle point; slightly flattening or bulging the two large arcs, will send the evolution one way or the other. The last state shown in Fig. 5i represents a true local minimum of the SI-MVC since the two osculating circles in the lemnoid lobes cannot be eliminated without subjecting the curve to some major, costly deformation. For the SI-MEC, however, this shape does not

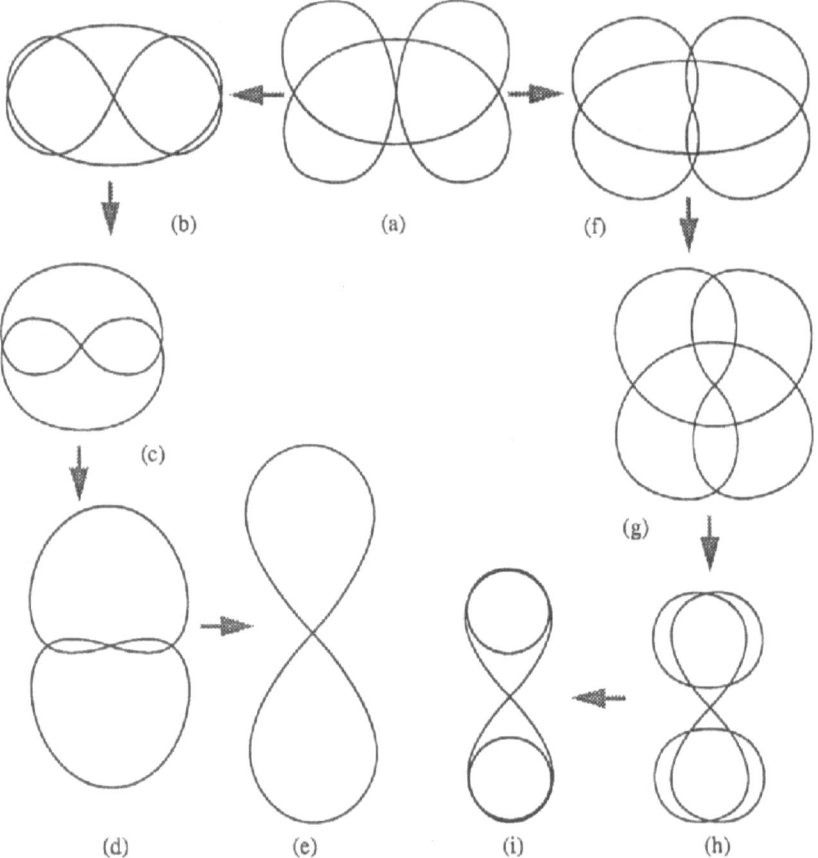

Figure 5. Evolution of a closed unconstrained curve under different functionals: (a→b→c→d→e) SI-MEC and SI-MVC; (a→f→g→h→i) SI-MVC under slightly different initial conditions

correspond to a local minimum; the two circles will expand, cross over each other, and unfold the figure into a simple figure-8 shape.

In our experiments we have not found any true local minimum for the SI-MEC that was not simply an artifact of the internal representation and/or the numerical techniques used in the optimization process. These local SI-MEC pseudo-minima disappear if the curve segments are suitably subdivided or combined or if the number of numerical integration points is changed.

4. Space Curves

The concepts developed for planar curves can readily be applied to curves embedded in R^3 or even R^n. In a first implementation, the scalar valued derivative of curvature was simply replaced with a vector valued derivative [8],

$$\int \left(\frac{d\kappa}{ds}\right)^2 ds \qquad (5)$$

where the curvature vector was defined by $\kappa = \kappa \mathbf{b}$ so that:

$$\left(\frac{d\mathbf{\kappa}}{ds}\right)^2 = \left(\frac{d}{ds}(\kappa\mathbf{b})\right)^2 = \left(\frac{d\kappa}{ds}\mathbf{b} + \kappa\frac{d\mathbf{b}}{ds}\right)^2 = \left(\frac{d\kappa}{ds}\mathbf{b} + \kappa\tau\mathbf{n}\right)^2 = \left(\frac{d\kappa}{ds}\right)^2 + \kappa^2\tau^2 \qquad (6)$$

which for the planar (torsion-free) case reduces to the expected $(d\kappa/ds)^2$.

Since functional (5) looks at the total change of the curvature vector, it also captures the effect of torsion to some degree. However, when we apply this to a sequence of points spaced in a regular helical sequence, e.g., $x = \cos t$, $y = \sin t$, $z = t$; for $t = 0, 1, 2 \ldots 8$, we find that we do not obtain a segment of a regular, infinitely long helix as we would like. The resulting space curve has torsion that vanishes towards the ends of the segment, i.e., at points 0 and 8.

The reason becomes obvious when we look at the expanded expression (6): torsion does not appear as a derivative. But since curvature and torsion are terms of the same order in describing the motion of the Frenet frame —they measure angular velocity around the binormal and the tangent vector, respectively— they should be treated equally. A more appropriate functional would thus be:

$$\text{Space} - \text{MVC} \Rightarrow \int\left(\left(\frac{d\kappa}{ds}\right)^2 + \left(\frac{d\tau}{ds}\right)^2\right)ds. \qquad (7)$$

This functional will lead to constant torsion for the above interpolatory example [8].

All the terms appearing in the above functionals have the same dependence on a uniform scale factor; they decrease with the 4th power of the scale factor. Thus multiplying the integrated cost with the cube of total arc length as in (4) will again achieve scale invariance.

Looking for generic minimum cost curve shapes in 3D is pointless. If the curve has the freedom to move freely through three-dimensional space, then it takes only a small asymmetry and any figure-8 curve will begin to warp, open up into the third dimension, and finally flatten out in the shape of a circle. Thus the study of "natural" unconstrained space curves is rather uninteresting; if the curves are allowed to cross themselves, they can all unfold into the basic circle.

5. Scale-Invariant Minimum Variation Surfaces

Analogous to the positive results obtained with MVC, we have developed techniques for modeling surfaces that also minimize the variation of curvature (MVS) and a corresponding scale-invariant version (SI-MVS). Here we review the former and introduce the latter functional.

5.1. Minimum Variation Surface

The simplest approach to introducing a cost functional based on curvature variation uses the in-line arc-length derivative of the normal curvature in the principal directions $\hat{e}_1$ and $\hat{e}_2$ [6]:

$$\text{MVS} \Rightarrow \int \left(\frac{d\kappa_1}{d\hat{e}_1}\right)^2 + \left(\frac{d\kappa_2}{d\hat{e}_2}\right)^2 dA. \tag{8}$$

This functional has the property that all cyclides, i.e., spheres, cones, cylinders, tori, and horned cyclides, have total cost zero because they all have circular lines of curvature. Thus, minimum variation surfaces (MVS) readily assume these circle-based shapes, commonly used in geometric modeling, whenever possible within the given external constraints. Furthermore, MVS exhibit enhanced fairness, because they distribute curvature uniformly, and they often lead to more pleasing shapes in situations where the surfaces are specified by only a few external constraints.

The first implementation uses an underlying representation based on a quilt of biquintic Bézier patches, where the given external constraints are normally coincident with a joint corner of several such patches. Again the system makes a heuristic guess to come up with an initial network of quintic curves which will later form the seams between the patches. These edges of the minimum variation network (MVN) [6, 8] are joined at the vertices of the network with shared geometric specifications for the normal direction, the principal directions and principal curvatures, so that G^2-continuity can be guaranteed at the shared vertices of the patches. G^1- and or G^2-continuity along the seams is enforced numerically during the optimization process with the use of penalty functions [6, 8].

While qualitatively the results look very good, the optimization procedure used to find these shapes is computationally very costly. Every biquintic patch has 36 control points and thus 108 degrees of freedom. Since the dimension of the solution space is typically very high, the optimization process, even for simple shapes with just a few patches, can run hours or days on a powerful workstation.

5.2. Introducing Scale-Invariance

MES are naturally scale invariant [5], since the square of curvature scales inversely to the total surface area over which the bending energy is integrated. The basic MVS functional (8), however, decreases with the square of the scale factor. It is thus desirable to construct a functional that is scale-invariant and which leads to stable surfaces even in the presence of sparse geometrical constraints.

The first approach to obtain scale invariance for MVS follows the construction of the scale-invariant functionals for MEC and MVC. The integrated functional is multiplied by a suitable power of the total surface area. The MVS functional requires a scale factor proportional to surface area:

$$\text{SI-MVS} \Rightarrow \left(\int dA\right)\left(\int \left(\frac{d\kappa_1}{d\hat{e}_1}\right)^2 + \left(\frac{d\kappa_2}{d\hat{e}_2}\right)^2 dA\right). \tag{9}$$

This approach leads to very smooth surfaces and provides stability even for surfaces with no constraints at all other than their topological connectivity. However, the additional implicit global coupling introduced by the multiplication with the integrated area makes convergence to the final shape even slower.

5.3. Results

Here we present some preliminary but representative results obtained with these functionals on unconstrained low-genus surfaces. For the evaluation of the simple low-genus surfaces of different symmetry classes, polyhedral models were first constructed. They were composed of all quadrilateral patches, even though the optimization program can handle triangles as well [7, 8]. These shapes were then converted to initial MVN and used as input to the optimization program. Symmetry was exploited to speed up the optimization process. The shapes presented were thus composed of just 2 to 4 different patches. Only the control points on all unique patches are optimized; the full shape then follows from replicating these patches with suitable transformations.

Figure 6 presents two interesting cases of genus-2 surfaces with different symmetries. Figure 6a shows a shape with three symmetrical handles exhibiting D_{3h} symmetry. Figure 6b depicts a disk with two holes with D_{2h} symmetry. It appears that the D_{3h} object has lower absolute cost, but still the D_{2h} object will not evolve into the first one. This does not imply, however, that this shape represents a true local minimum. Because our optimization program strongly exploits any existing symmetry, it also remains locked into the particular symmetry group implied by the initial constraints. More research is required to determine reliably the exact energy values of these shapes and to establish whether true local minima exist.

Among the genus-3 surfaces we compare a shape with four symmetrical handles exhibiting D_{4h} symmetry (Fig. 7a) and a tetrahedral frame with corresponding symmetry (Fig. 7b). These two shapes are even closer in their respective energy values for both the MVS and SI-MVS functionals.

We can obtain very similar shapes even with the MVS functional. Run-away can be prevented with a few judiciously specified interpolation points. For the shapes in

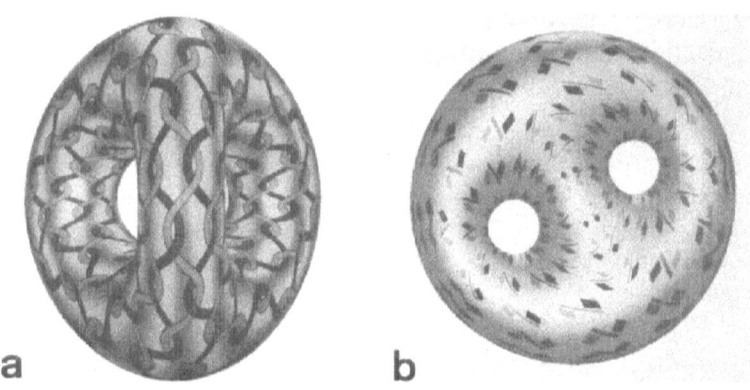

a   b

Figure 6a, b. Final states of some symmetrical genus-2 surfaces

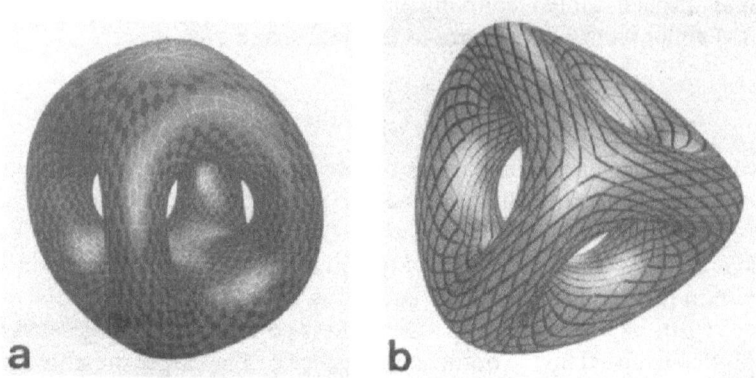

Figure 7a, b. Final states of some symmetrical genus-3 surfaces

Fig. 6a, b and Fig. 7a, two points are sufficient, one each on the north and south-pole of these shapes respectively; the tetrahedral shape (Fig. 7b) takes 4 fix-points. Qualitatively, the shapes obtained with the MVS and the SI-MVS functional are very similar. On closer inspection of the shape shown in Figure 6a, for instance, we observe that the MVS produces a shape that has three legs that are about 10% thicker, and 30% further from the center. Both types of legs are very close to circular at the major (horizontal) symmetry plane, with the minor half-axis of the SI-MVS leg deviating 4% form a true circle, and the MVS leg deviating 5%. The advantage of the SI-MVS functional is that one need not place any interpolation points, which, when improperly placed, could break the intended symmetry and thus distort the resulting shape.

6. Alternative Cost Functionals

The original MVS functional yields zero cost for all surfaces in the family of cyclides. This does not seem completely satisfactory. A cone is not as regular as a sphere, and the horned cyclid with its singularities is certainly less fair than a perfectly circular torus. We have thus started to look at other functionals as alternatives to the original minimum variation surfaces described above.

6.1. Other Curvature Variation Terms

The original MVS functional (8) measures the in-line change in normal curvature in the direction of the lines of curvature only. We could try to generalize that expression and to consider in-line curvature change in *all* directions. But because of Euler's relation:

$$\kappa_\theta = \kappa_1 \cos^2 \theta + \kappa_2 \sin^2 \theta \tag{10}$$

the normal curvature κ_θ in any arbitrary direction θ is uniquely determined by the two principal curvatures. If we average the variation of curvature over all angles, we obtain:

$$\frac{1}{\pi}\int_0^\pi \left(\frac{d\kappa_\theta}{ds_\theta}\right)^2 d\theta = \frac{3}{8}\left(\frac{d\kappa_1}{d\hat{e}_1}\right)^2 + \frac{3}{8}\left(\frac{d\kappa_2}{d\hat{e}_2}\right)^2 + \frac{1}{4}\left(\frac{d\kappa_1}{d\hat{e}_1}\right)\left(\frac{d\kappa_2}{d\hat{e}_2}\right) \tag{11}$$

This expression does not contain significant new information, it just combines the basic terms from (8) in a slightly different manner. In order to truly obtain some stronger discrimination against tori, we need to introduce some cross terms that measure the change along a direction φ of the normal curvature taken in a different direction θ,

such as the terms $\dfrac{d\kappa_1}{d\hat{e}_2}, \dfrac{d\kappa_2}{d\hat{e}_1}$ or even the more general $\dfrac{d\kappa_\theta}{ds_\varphi}$.

An even stronger discriminator against anything that is not an umbilic point is given by the *surface torsion* term. This term can also be averaged over all θ using (10):

$$\frac{1}{\pi}\int_0^\pi \left(\frac{d\kappa_\theta}{d\theta}\right)^2 d\theta = \tfrac{1}{2}(\kappa_1 - \kappa_2)^2 \tag{12}$$

Thus there seem to be a variety of "reasonable" choices for a curvature-variation functional for surfaces. A considerable amount of work lies ahead to evaluate and compare some of these functionals.

6.2. Other Means of Obtaining Scale Invariance

A generic way to obtain scale-invariance for all these possible functionals is to determine their cost change under uniform scaling and then compensate it by multiplication with a suitable power of the total surface area. However, there is another possibility.

In the case of curves, we have not yet found any useful functionals in which the power of the curvature derivative itself was adjusted to directly obtain the desired scale-invariance. However, for 2-dimensional manifolds the arguments are more involved and do not directly rule out the use of the first degree of curvature variation. Let us consider what would happen in a Gedanken-experiment similar to the one described in Section 2.1 applied to a minimum energy surface in the shape of a cube with rounded edges and corners. Let us assume that the "suitcase corners" are octants of a sphere. Because of the scale invariance of the MEC functional, they have a fixed cost associated with them regardless of scale. Similarly, we can take a piece of an edge in the form of a quarter cylinder and scale it uniformly without changing its cost. The six flat faces, of course, have cost zero. When we expand this "suitcase cube" in the style of Fig. 2c, d, e where all radii remain the same, the total cost still increases because the *length* of the cylinders has to be increased-an operation that is not free. Thus the existence of any flat pieces in the surface carries an implicit cost because of the associated edges surrounding it. The minimum energy "suitcase" is obtained when the cylinder length is zero, the shape consists of "corner pieces" only, and thus becomes a sphere. The existence of the scale-invariant MEC functional strongly suggests that we need not introduce an explicit cost for total surface area.

Because of this difference compared to curves, we may devise a functional that is measuring change in curvature and which is still scale-invariant by itself without multiplication by overall surface area. For the MVS functionals discussed above, this can be achieved by taking a *first* degree power of all the curvature derivative terms except surface torsion which becomes scale-invariant when taken to the second power.

Here is the generalized directional functional (11) that measures in-line curvature in all directions; it is made scale-invariant by using the root of the sum of the squared terms:

$$\int_A \sqrt{\frac{1}{\pi}\int_0^\pi \left(\frac{d\kappa_\theta}{ds_\theta}\right)^2 d\theta} dA = \int_A \sqrt{\frac{3}{8}\left(\frac{d\kappa_1}{d\hat{e}_1}\right)^2 + \frac{3}{8}\left(\frac{d\kappa_2}{d\hat{e}_2}\right)^2 + \frac{1}{4}\left(\frac{d\kappa_1}{d\hat{e}_1}\right)\left(\frac{d\kappa_2}{d\hat{e}_2}\right)} dA \quad (13)$$

Either (9) or (13) can be optionally enhanced with a surface torsion term which is itself scale-invariant:

$$\int_A \frac{1}{\pi}\int_0^\pi \left(\frac{d\kappa_\theta}{d\theta}\right)^2 d\theta dA = \int_A \frac{(\kappa_1 - \kappa_2)^2}{2} dA = \int_A (2H^2 - 2K) dA \quad (14)$$

In this expression, H is the mean curvature and K is the Gaussian curvature. The integration of the Gaussian curvature over a closed surface is a topological constant that depends only on the genus of that surface but not on its geometry. Adding the surface torsion term is thus essentially equivalent to also considering the classical MES functional.

6.3. Comparisons on Torus

As we have seen in the previous two sections, there is a choice of possible cost functionals. To help us decide which one of them might be the most appropriate one, we can study some of the minimum cost shapes produced with the different functionals. A particularly good test case is the regular torus, because its description is simple enough so that the functionals can be calculated rather than found via an optimization process.

For the MES functional there is an optimal torus that minimizes the total bending energy, known as the Clifford torus [5]; the minimum is $W = 2*\pi^2 = 19.72$ and occurs for a ratio of the two radii of $\sqrt{2}$. The simple MVS does not give any discrimination. For the MVS and the SI-MVS, all tori have cost zero regardless of their radii. When we use some of the more complete functionals discussed above, we again find that some intermediate tori have lower cost than extremely skinny tori which have a lot of curvature variation going around the circles of radius r, or than fat tori with very small holes which contribute much to the curvature variation integral.

For our calculations we use a normalized parameterized expression for the torus in which the ratio of the two radii appears as an explicit parameter: $\rho = r/R$:

$$X(u, v) = ((R + r \cos u) \cos v, (R + r \cos u) \sin v, r \sin u); (0 \leq u < 2\pi); (0 \leq v < 2\pi)$$

First we study a generalization of the variation of the in-line curvature according to functional (11). This is the square of the norm of the complete gradient in (u, v) of the normal curvature $\kappa(u, v, \theta)$. We average this functional over θ and integrate it over the whole area of the torus. In the expression below, the integrations over θ and over v have already been carried out:

$$\frac{3\pi}{4R^2} \int_0^{2\pi} \frac{(\sin u)^2}{\rho(1 + \rho \cos u)^3} du = \frac{1}{R^2} \frac{3}{4} \pi^2 \frac{1}{\rho} (1 - \rho^2)^{-3/2} \tag{15}$$

This term yields a minimum cost of $4\pi^2/\sqrt{3} = 22.79$ for $\rho = 1/2$.

If this is multiplied by the area of the torus, we obtain a scale invariant expression:

$$(4\pi^2 R^2 \rho) \frac{3\pi}{4R^2} \int_0^{2\pi} \frac{(\sin u)^2}{\rho(1 + \rho \cos u)^3} du = 3\pi^4 (1 - \rho^2)^{-3/2} \tag{16}$$

This term yields a minimum cost of $3\pi^4 = 292.2$ for $\rho = 0$.

We can obtain directly a scale-invariant expression based on this total derivative of the normal curvature if we take the first power of the norm:

$$\pi \int_0^{2\pi} \frac{|\sin u|}{1 + \rho \cos u} du = 2\pi \frac{1}{\rho} \log \frac{1 + \rho}{1 - \rho} \tag{17}$$

This term yields a minimum cost of $4\pi = 12.57$ for $\rho = 0$.

And finally, if we integrate surface torsion over the whole area of the torus,

$$\int_A \frac{1}{\pi} \int_0^\pi \left(\frac{d\kappa_\theta}{d\theta}\right)^2 d\theta dA = \int_0^{2\pi} \frac{\pi}{\rho + \rho^2 \cos u} du = \frac{2\pi^2}{\sqrt{\rho^2 - \rho^4}} \tag{18}$$

we find a minimum cost of $4\pi^2 = 39.48$ for $\rho = 1/\sqrt{2}$. This is the well known Clifford torus that minimizes the MES functional.

This concludes our little survey of possible alternative cost functionals which may prove useful in certain applications and which would give a stronger and more specific discrimination among the various cyclides. The four alternate cost functionals can be combined in various ways. For instance, if we combine the first power of the gradient norm (17) and add to it the surface torsion term (18), we obtain a new functional that assumes a minimum cost of 54.9 for $\rho = 0.6689$. Alternatively, if we add the surface torsion term (18) to expression (16), we obtain a minimum cost torus of $W = 403.2$ for $\rho = 0.2624$. By choosing appropriate weighting coefficients for the various expressions presented above, one can construct scale-invariant functionals that favor different features, e.g., more skinny or more fleshed out torus segments. The above list of the fundamental building blocks of such functionals should help other researchers with their experiments. We intend to introduce some of them into

Brakke's Surface Evolver [1], a public-domain program to study the minimum cost states of discretized curves and surfaces.

7. Conclusions

In this paper we present scale-invariant functionals for curves and surfaces for computer-aided geometric design. In particular we discuss ways to obtain scale-invariant versions of the minimum-energy curve (SI-MEC), of the minimum-variation curve (SI-MVC), and of the minimum-variation surfaces (SI-MVS).

The most practical approach to obtain scale-invariant curves is derived from the traditional functionals of the MEC and of the MVC by multiplying the respective functionals by suitable powers of the curve's arc length. Both scale-invariant functionals seem to offer practical design solutions of superior fairness and stability. The SI-MVC retains the attractive features of the MVC: its smoothness characterized by G3-continuity even in the presence of tangent constraints, its lean look that avoids unnecessary bumps and wiggles, and its tendency to assume circular arcs wherever they are compatible with the given constraints. To that it adds the unconditional stability of a scale-invariant curve which permits it to assume a stable minimum even in situations where no absolute constraints confine the size of the solution curve. The SI-MEC guarantees a lower degree of geometric continuity but also leads to stable and very pleasing shapes. It seems to have the attractive property, at least for an ideal implementation, that closed curves constrained only by their turning number, T, do not have any stable local minima, but always bulge out into the true global minimum: a lemnoid shape for $T = 0$, and circles traced T times in all other cases.

For surfaces, there is no unique winner either. The MES has the advantage of being naturally scale-invariant. On the other hand, the MVS functional or some related alternative seems preferable for many modeling applications, since it naturally leads to cylinders and spheres where these shapes are compatible with the given constraints.

The same approach taken for curves can again be used to obtain scale-invariant functionals: multiplying with a suitable power of the total area of the manifold. Some open research issues remain to determine which terms concerning curvature variation should be considered. It appears that taking into account the complete gradient of normal curvature offers a useful discrimination among various elliptical and toroidal surfaces. This form has some terms that are naturally scale-invariant and others that decrease proportional to the square of any scale factor. The latter terms can be made scale-invariant by using a first power of the norm rather than by multiplication with the total area. The answer as to which is the best approach awaits further investigation and may very well be application dependent. In either case, our current implementation based on bi-quintic Bézier patches, with penalty functions to force continuity at the seams, is painfully slow. A more direct approach where geometric continuity is achieved by construction even at the price of giving up some of the available degrees of freedom needs to be developed.

Bernoulli's "elastica" are an idealization of a behavior that is only approximated by nature as well as by the actual spline functions used in today's CAD tools. Similarly, the minimum-variation curves and surfaces discussed in this paper are idealizations of a behavior that may be difficult or expensive to implement exactly in a CAD system, but they serve as potential goals and benchmarks against which more efficient practical solutions can be compared.

References

[1] Brakke, K. A.: The surface evolver. Exp. Math. *1*, 141–165 (1992).
[2] Mann, S., Loop, C., Lounsbery, M., Meyers, D., Painter, J., DeRose, T., Sloan, K.: A survey of parametric scattered data fitting using triangular interpolants. In: Curve and surface design (Hagen, H., ed.) pp. 145–172. Philadelphia: SIAM 1992.
[3] Farin, G. E.: Curves and surfaces for computer aided geometric design, a practical guide. San Diego: Academic Press: 1990.
[4] Horn, B. K. P.: The curve of least energy. ACM Trans. Math. Software 9, 441–460 (1983).
[5] Hsu, L., Kusner, R., Sullivan, J.: Minimizing the squared mean curvature integral for surface in space forms. Exp. Math. *1*, 191–207 (1992).
[6] Moreton, H. P., Séquin, C. H.: Surface design with minimum energy networks. In: Proceedings symposium on solid modeling foundations and CAD/CAM applications (Rossignac, J., Turner, J., eds.), pp. 291–301. New York: ACM Press 1991.
[7] Moreton, H. P., Séquin, C. H.: Functional optimization for fair surface design. Comput. Graphics 26, 167–176 (1992).
[8] Moreton, H. P.: Minimum curvature variation curves, networks, and surfaces for fair free-forms shape design. Ph.D. Thesis, UC Berkeley, 1992.
[9] Moreton, H. P., Séquin, C. H.: Scale-invariant minimum-cost curves: fair and robust design implements. Proc. Eurographics' 93, Barcelona, 1993.

C. H. Séquin
P.-Y. Chang
Computer Science Division
EECS Department
University of California
Berkeley, CA 94720
U.S.A.

H. P. Moreton
Silicon Graphics Inc.
Mountain View, CA 94039 7311
U.S.A.

Computing Suppl. 10, 323–340 (1995)

© Springer-Verlag 1995

The C-Tree: A Dynamically Balanced Spatial Index

K. Verbarg, Würzburg

Abstract. We introduce an efficient and robust spatial index to support a set of different queries, which is developed from Günther's *Celltree* [6] and the *Monotonous Bisector* Tree* [12, 17]. In practice, huge scenes are manageable by using the *paging–concept*. For convex polyhedrons in N-dimensional real space and any L_p-metric ($1 \le p \le \infty$) we are able to show that the C-tree can be constructed in time $\mathcal{O}(n \log n)$, linear space and logarithmic height, where n denotes the number of objects. Dynamic insertion of objects is performed in $\mathcal{O}(\log^2 n)$ amortized worst-case time. Objects can be deleted in amortized $\mathcal{O}(n)$ or in amortized $\mathcal{O}(\log^2 n)$ if the actualization of cluster radii can be delayed. In all cases logarithmic height and linear space requirements are preserved.

Key words: Spatial index, geometric databases, clustering, computational geometry.

1. Introduction

In a wide range of applications the crucial problem is to represent the neighborhood structures in huge scenes of spatial objects in an adequate and efficient manner. These applications cover e.g. motion-planning in robotics, geographic information systems, computer graphics, managing the spatial structure of huge proteins (up to 100.000 amino acids per molecule) and clustering sequence spaces in biology.

Particular to these applications is first, that data represents geometric objects embedded in the topological real space $\mathbb{R}^N$. Second, objects are accessed through their position in space. This is done by different queries, like region queries, nearest-neighbor queries, point queries and exact-match. Third, spatial objects usually have a more complex structure that can not be represented in a static record. Furthermore the efficient management of huge data sets, as in DBMS, is a major issue.

In the past, numerous spatial indexes coping with these requirements have been developed. Overviews can be found in [6, 11, 13].

One example is the *Celltree* introduced by Günther. The Celltree is a dynamic rooted tree. It is designed for secondary memory, i.e. the nodes can be paged. Its basic shape is that of the B-tree [2], one of the classical tree structures for secondary memory. Each node represents a cell of the space of objects. It is partitioned completely into disjoint cells, each of which is the cell of one of its successors. The subsets are obtained by *binary space partitioning* (*BSP*) according to [3], i.e. the cell decomposition of each node is described by a BSP-tree. The assignment of objects to a cell may follow one of the well-known conflict strategies for extended objects, cf. [16].

The Celltree is affected with many open parameters and problems. In the following, we replace the BSP-tree by an improved alternative: the *Monotonous Bisector* Tree* [12, 17]. This results in a new structure, the C-tree, that combines the advantages of the Celltree and the Monotonous Bisector* Tree.

To speed up searching, Günther proposes containers, to give a tighter approximation of the objects of a subtree, than the given partitions do. Through using a Monotonous Bisector* Tree as BSP such a container is automatically realized as a sphere of the cluster center with cluster radius.

With the knowledge of the Monotonous Bisector* Tree we are able to use the balancing step instead of heuristically splitting overflowing pages with a plane sweep in l directions. This guarantees a partition into equal halfs and is even applicable in $\mathbb{R}^N$. The Günther-Celltree demands at least m sons for each internal node. But without a rebalancing strategy this rigid shape can not be guaranteed, since splitting may fail and cause overflow pages.

The remainder is organized as follows. Section 2 contains the definitions of the C-tree structures. In Section 3, an algorithm is presented creating C-trees within $O(n \log n)$ time, n the number of objects inserted. The queries efficiently supported by the C-tree are compiled in section 4. In Section 5 we show that objects can be inserted in $\mathcal{O}(\log^2 n)$ amortized worst-case time. Deletion of objects can be performed in $\mathcal{O}(n)$ amortized time, and in $\mathcal{O}(\log^2 n)$ amortized time if the actualization of the so-called cluster radii can be delayed.

2. The C-Tree

A Monotonous Bisector* Tree is a binary leaf-oriented rooted tree to represent a set S of arbitrary objects (not necessarily of the $\mathbb{R}^N$). The objects are only clustered according to their neighborship relation, modeled by a distance function. The partitioning of the object space is thereby described by a set E of simple auxiliary objects. Therefore this concept is superior to common spatial indices, because no embedding in the $\mathbb{R}^N$ is necessary. Only the topology of the objects space must be modeled by a distance function. The allows to handle more abstract (not geometric) or only partially geometrically described problems.

E is now a set of meaningful auxiliary points to create the directory. The only restriction on E is that an efficiently computable distance function $d: E \times S \mapsto \mathbb{R}_{\geq 0}$ must be available. The idea is to keep E simple, such that the computation of d is cheaper than computing the distance between two complex objects of S. On the other hand E must be chosen powerful enough to allow a flexible clustering.

A N-dimensional sphere $K(v)$ (induced by the metric d) is assigned to each node v in the tree. $K(v)$ is described by a *cluster center* (split value) and a *cluster radius*. The cluster of a node is the finite intersection

$$\text{Cluster}(v) := \bigcap \{K(\tilde{v}) | \tilde{v} \text{ lies on the path from the root to } v\}.$$

The partitioning is not disjoint and not complete. The conflicts are resolved with the strategy overlapping clusters in the following way: An objects is assigned to the cluster with the nearest cluster center. If the decision is not unique, any of the nearest clusters is chosen.

Thus, objects are partitioned with *bisectors*. The bisector of $e_1, e_2 \in E$ is defined as the set of all $s \in S$ with $d(e_1, s) = d(e_2, s)$. In the case of $E = \mathbb{R}^2$, $S \subset \mathscr{P}(\mathbb{R}^2)$ and the euclidean distance d the bisector is the mid-perpendicular of e_1 and e_2.

If S is a set of extended objects in $\mathbb{R}^N$, simple examples show that it is sometimes impossible to create a balanced tree with $S = E$ [8]. So the extension with artificial split values is sensitive, if it is applicable to the used space.

Let S be the set of objects, E the set of adequate splitting objects and $d: E \times S \mapsto \mathbb{R}_{\geq 0}$ the belonging distance function, which is efficiently computable.

Definition 2.1. *A Monotonous Bisector* Tree (MBT)* is an ordered binary rooted tree with the set of marked nodes $V \cap \mathbb{N}_0 = \emptyset$ in the following way[1]:

1. To each node $v \in V$ there is related an 8-tuple of the form

$$(\text{lSplit}(v), \text{lRadius}(v), \text{lSon}(v), \text{lObject}(v),$$

$$\text{rSplit}(v), \text{rRadius}(v), \text{rSon}(v), \text{rObject}(v))$$

   with $@\text{Split}(v) \in E$, $@\text{Radius}(v) \in \mathbb{R} \geq 0$, $@\text{Son}(v) \in V \cup \mathbb{N}_0 \cup \{nil\}$, $@\text{Object}(v) \subset S$.
2. Tree structure:
   An implicit representation of the tree structure is given through

$$\text{lSon}(v) \begin{cases} \in \mathbb{N}_0 \cup \{nil\}, \text{if } v \text{ has no left successor} \\ \in V, \text{namely the left successor of } v, \text{ otherwise} \end{cases}$$

   (analogously for $\text{rSon}(v)$).
3. Objects are stored in leafs only:
   If $@\text{Son}(v) \neq nil$ for $v \in V$, then it is always $@\text{Object}(v) = \emptyset$.
4. Heredity of split values:
   For all $v \in V$ with $@\text{Son}(v) =: \tilde{v} \in V$ there is $@\text{Split}(v) = \text{lSplit}(\tilde{v})$.
5. Numbering of the partitions:
   For all $v, \tilde{v} \in V$ with $@\text{Son}(v), \tilde{@}\text{Son}(\tilde{v}) \neq nil$ there is $@\text{Son}(v) \neq \tilde{@}\text{Son}(\tilde{v})$.

If $\text{lSon}(v)$ resp. $\text{rSon}(v) \in V$, then they point to the sons of v in MBT. If they are in $\mathbb{N}_0$, then v hasn't got a left resp. right son. In this case they contain the number of the next node in the C-tree, which is related to their partition. Finally if they are nil, then $\text{lObject}(v)$ resp. $\text{rObject}(v)$ is the set of represented objects in this partiton. At this point only the structure of the MBT is defined, but not the corresponding semantics (the way objects are assigned to partitions and the form of the partitions). This will be done subsequently.

---

[1] Always textually replace $@$ by the letter "l" and "r". In the following we will use this shorter notation.

The MBT is mainly used for describing the partitions in the C-tree, since the objects should be stored as deep as possible (in the leafs) in the C-tree[2]. The demand to have an 8-tuple for each node in the MBT is not a restriction. An empty tree can be constructed choosing an artificial split value for rSplit(root) and set rSon(root):= nil, rObject(root):= $\emptyset$.

For a simpler notation we give the following two definitions.

Definition 2.2. For $e \in E$ and $r \in \mathbb{R}_{\geq 0}$ we denote with $Sphere(e,r) := \{s \in S | d(s,e) \leq r\}$. $e \in E$ is called a split value or a cluster center.

Definition 2.3. For a *MBT* with nodes V we define:

1. $v \in V$ is called @ *Partition* if and only if @Son$(v) \in \mathbb{N}_0$.
2. $v \in V$ is called @ *Leaf* if and only if @Son$(v) = nil$.

To be able to page objects, they should fit on a page of bounded size. Therefore we specify the objects manageable by the C-tree. Originally only convex polygons could be represented. This is where the term "cell" comes from. Let $k \in \mathbb{N}$ be given:

Definition 2.4. A *cell* is an object, that can be described with k real parameters using a fixed modeling.

We can model for example all convex polygons with at most $\lfloor k/2 \rfloor$ vertices in $\mathbb{R}^2$ by the convex hull of their vertices. The type of modelling is arbitrary, but equal for all objects. In fact there may only be a unique key for each object. Only the distance function uses the modelling. For realizing the C-tree it is only important to be able to store the objects.

Let S be such a universe of cells. We will now define the C-tree.

Definition 2.5. The *C-tree* (CT) for the parameter $P \in \mathbb{N}$ is a leaf-oriented rooted tree with nodes $V = \{0, 1, \ldots, N\}$ to represent a set $C \subset S$ of cells, with the following properties:

1. Paging:
   (P) Each node fits on a page of capacity P in the secondary memory.
2. Structure:
   (S1) Each node $v \in V$ represents a MBT(v) with nodes $V(v)$.
   (S2) The edges of the C-tree are implicitly given by
   v'_{CT} is father of v_{CT} if and only if
   $\exists v_{MBT} \in V(v'_{CT})$: v_{MBT} is @ Partition in MBT (v'_{CT}) and @Son$(v_{MBT}) = v_{CT}$.
   (S3) Heredity of split values:
   If v'_{CT} is father of v_{CT} and $v_{MBT} \in V(v'_{CT})$ with @Son$(v_{MBT}) = v_{CT}$, then @Split(v_{MBT}) = the left split value of the root of MBT(v_{CT}).

---

[2] In the Günther-Celltree originally there was a sharp distinction between nodes to store objects (the leafs) and directory nodes (internal nodes). The in fact growing main memories provide the design of larger pages. In this approach this would force large pages for objects. These would not be able to be very distinctive any more. We avoid this problem through mixing directory and objects on one page. Furthermore underflowing object buckets can be handled cheaper.

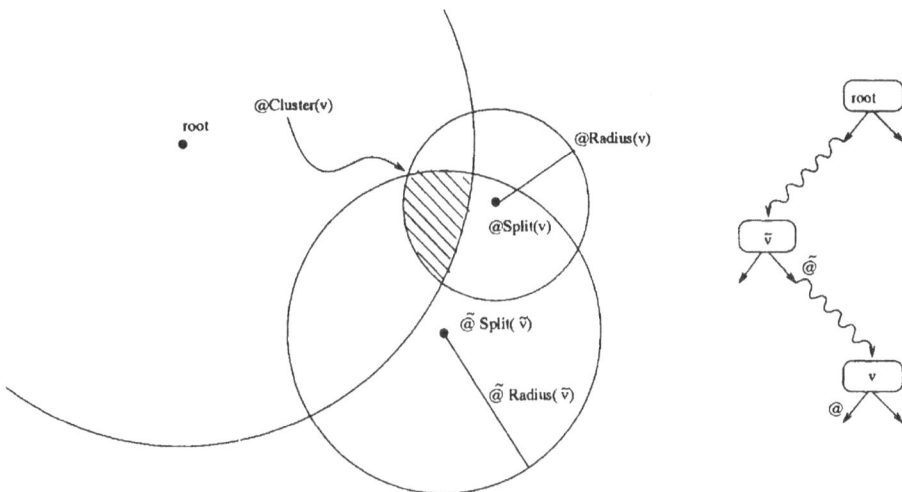

Figure 1. The cluster of a node

For $v_{CT} \in V$ and $v_{MBT} \in V(v_{CT})$ we denote with $w(v_{MBT})$ the unique path consisting of MBT-nodes starting from the root of the C-tree to v_{MBT}, and with $W(v_{CT})$ the corresponding path in the C-tree.

3. Representation of objects:

For $v_{CT} \in V$ and $v_{MBT} \in V(v_{CT})$ we term $@Cluster(v_{MBT}) := \mathrm{Sphere}(@\mathrm{Split}(v),$ $@\mathrm{Radius}(v)) \cap \bigcap \{\mathrm{Sphere}(\tilde{@}\mathrm{Split}(\tilde{v}), \tilde{@}\mathrm{Radius}(\tilde{v})) \mid \tilde{v}, \tilde{@}\mathrm{Son}(\tilde{v}) \in w(v_{MBT})\}$.

(O1) Cluster cover cells:

For all $v_{CT} \in V$ and $v_{MBT} \in V(v_{CT})$ there is $@\mathrm{Object}(v_{MBT}) \subset @\mathrm{Cluster}(v_{MBT})$.

(O2) Unique search path:

For $v_{CT} \in V$ and $v_{MBT} \in V(v_{CT})$ there is $c \in @\mathrm{Object}(v_{MBT})$ if and only if:

$c \in C$ and $\forall v'_{MBT} \in w(v_{MBT}) \forall v'_{CT} \in w(v_{CT})$ holds: $d(\mathrm{lSplit}(v'_{MBT}), c) \prec$ $d(\mathrm{rSplit}(v'_{MBT}), c) \Leftrightarrow \mathrm{lSon}(v'_{MBT}) \in w(v_{MBT}) \cup W(v_{CT})$ or $v'_{MBT} = v_{MBT}, @ = \text{"}l\text{"}$.

The C-tree is a rooted tree of MBT's, describing the paging of the virtual MBT. The defined path $w(v_{MBT})$ is related to this virtual MBT. Furthermore the cluster of a node v_{MBT} is the intersection of all spheres appearing on the path from the root to v_{MBT} in the virtual MBT. This is shown in Fig. 1. (O2) expresses that one has to go "left" (resp. "right") on a search path, if the left (right) split value is nearer to the considered object.

Figure 2 shows for a scene of convex objects the MBT of the root of the C-tree. For clearance subsequently the partitioning of the scene is represented by relevant parts of the bisector of adjacent split values (compare the voronoi diagram). The MBT is partitioning the plane into clusters of convex polygons. The large degrees of freedom in choosing the MBT for a node of he C-tree provides a good balancing of the tree. The next section shows, how this can be done.

Let sizeof(cell) be the space requirement of a cell and sizeof(mbt_node) of a MBT-node. According to (P) for a fixed page size P the maximum number of sons of a node

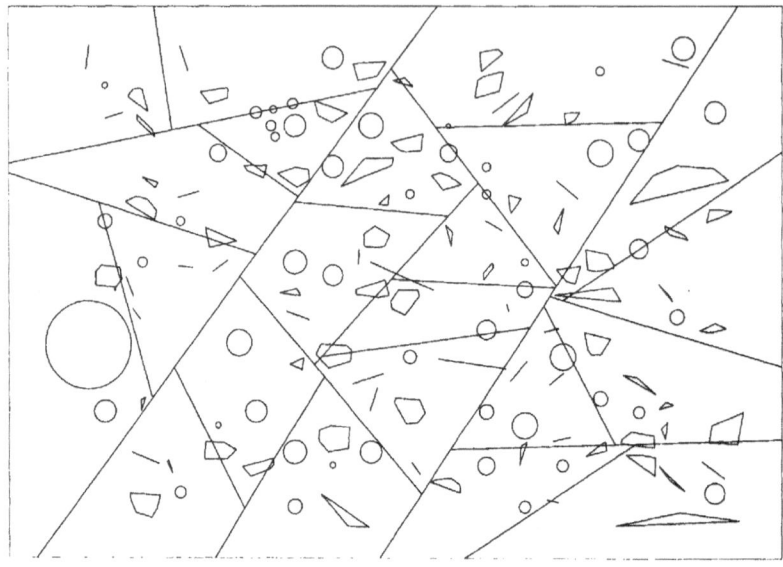

Figure 2. Partitioning of a scene through the root of the C-tree

in the C-tree is $M := \left\lceil \dfrac{P}{\text{sizeof(mbt\_node)}} \right\rceil + 1$ and $\bar{M} := \left\lceil \dfrac{P - \text{sizeof(mbt\_node)}}{\text{sizeof(cell)}} \right\rceil$ is
the maximum number of cells in a node[3]. Therefore a C-tree of height h contains at
most $\bar{M} M^h$ cells.

3. Creating the C-Tree with Estimated Height, Effort and Space Requirements

The idea is to create a Monotonous Bisector* Tree and to map it to the C-tree, i.e. to
page it. To do this, recursively a second split value belonging to the heredited split
value is chosen. The aim is to partition the set $C \subset S$ into two equal parts and
decrease the cluster radii as fast as possible. This is achieved by two algorithms
which can be applied flexible. The partitioning with bisectors automatically makes
the Monotonous Bisector* Tree flexible in respect to the distance function d.

The *balancing step* chooses for a given $C \subset S$ and $e_1 \in E$ a second split value e_2, such
that both resulting clusters C_1 and C_2 contain at least a linear portion of C. This is
very tricky, but possible in linear time $\mathcal{O}(|C|)$ for a set S of compact objects in $\mathbb{R}^N$,
$E = \mathbb{R}^N$ and d a Minkowski-metric; see [17]. This type of partition step guarantees
the balancing of the tree and therefore bounds the height of the tree logarithmically.

The *contraction step* chooses e_2 such that the cluster radii decrease as fast as possible.
Let $o_1, \ldots, o_k \in C$ be the objects with the same maximum distance of e_1. Let P be the

---

[3] Each page contains at least one MBT-node.

point on any of the o_i, $(i = 1, \ldots, k)$, taking on the distance. Now e_2 is chosen carefully to get as many objects as possible (but at least P) into his cluster. In the worst-case only P is separated from the original cluster and the cluster radius is not necessarily decreasing. Nevertheless it is possible to reduce the cluster radius to any fraction $q \in]0, 1[$ with a fixed number $l = l(q)$ of contraction steps. These steps do not have to be consecutive. Thereby the number $l \in \mathbb{N}$ of steps depends on q and the space S (in $\mathbb{R}^N$ especially on the dimension, see [17]). In our implementation we used $e_2 := \frac{2}{3}P + \frac{1}{3}e_1$. This step also costs $\mathcal{O}(|C|)$ time.

Experimental results show that both the balancing step and the contraction step are necessary to achieve a "good" tree. But the latter does not allow a statement concerning the balance. Therefore the rigid demands for the shape of the Günther-Celltree[4] can not be realized. This is the reason why the structure of the C-tree in Section 2 is designed more flexible.

The paging raises additional demands to the creation process. Paging is only sensible if the creation is feasible with bounded main memory, too. We can state that the balancing and contraction steps are executable under these restrictions, if the objects are given in a file.

If the page size P becomes too large, it is not sensible to store up to $\bar{M}$ objects in a leaf. So we introduce a bucket size B. Every bucket in the C-tree has got at most B objects: $\forall v \in V \, \forall v' \in V(v) \, \forall @ \in \{\text{"}l\text{", "}r\text{"}\} : | @ \, \text{Object}(v') | \leq B$.

The creation algorithm partitions hierarchically the set of objects using balancing and contraction step until a CT-node is filled. It then proceeds recursively with the sons. Thereby the following strategy is used: *For the sake of balanced subtrees, always the largest cluster is partitioned next.* Figure 3 shows how the successively created Monotonous Bisector* Tree is paged onto the C-tree.

The subsets of objects, that are temporarily created by the algorithm, must be held in secondary memory, too. They are termed as $1\text{Set}(v)$ resp. $r\text{Set}(v)$ according to the related MBT-node v. The set of actual leafs in $\text{MBT}(w)$ is called $V_{\text{leaf}} := \{(v, @) \in V(w) \times \{\text{"}l\text{", "}r\text{"}\} \, | \, v \text{ is } @ \, \text{Leaf}\}$. The parameter α is a load factor for internal nodes. It is left open here which value of α is best in the dynamic case. B denotes the size of all buckets in the tree.

Algorithm 3.1. Create (w, C, e, α, B)
w: root of the C-tree
$C \subset S$: the set of objects to represent
e: heredited split value
$\alpha \in [\frac{1}{2}; 1]$: filling factor for internal nodes
$B \leq \bar{M}$: bucket size

---

[4] Shape of the Günther-Celltree:
- All leafs are on the same level.
- The root has got no or more than two sons.
- Every internal node except the root has got at least m sons.

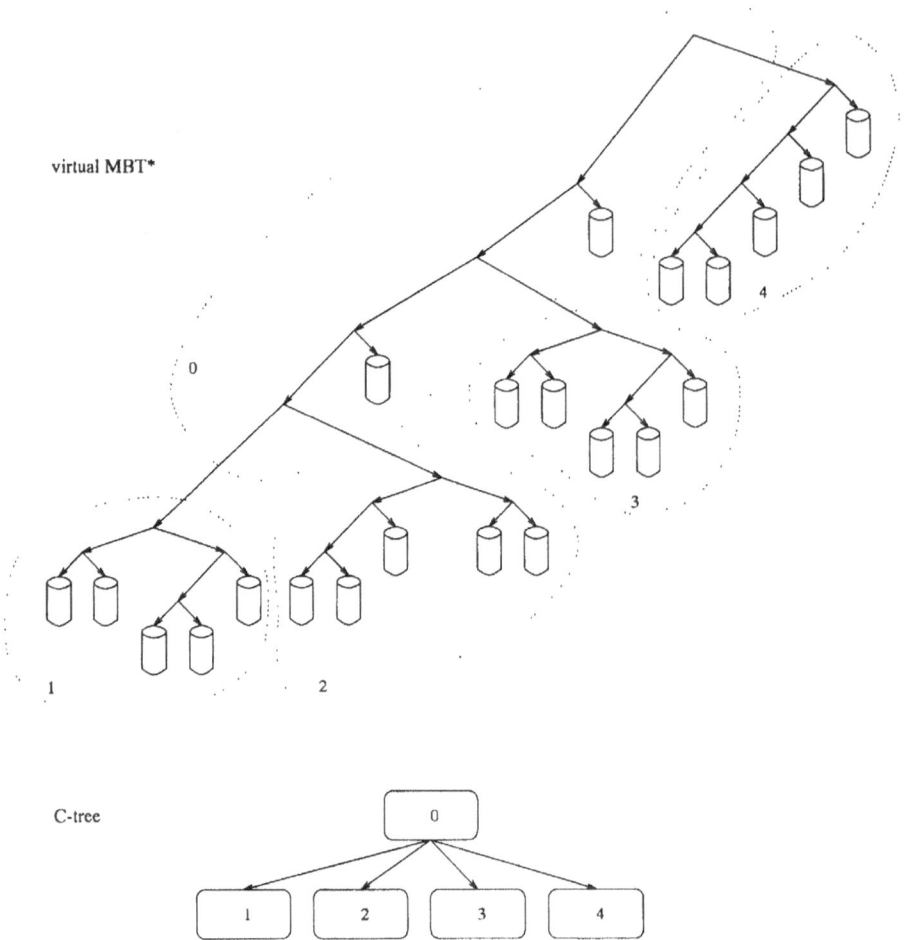

Figure 3. Paging the virtual Monotonous Bisector* Tree onto the C-tree

1. If $|C| \leq B$, then create a tree MBT(w) with only one node, that represents the objects in C.
2. Execute a partition step on e and C, which results in the second split value e'.
   Let $V(w):= \{v\}$, where v is a new MBT-node with: lSplit(v):= e, rSplit(v):= e', lSon(v), rSon(v):= nil and lObject(v), rObject(v):= $\emptyset$.
   Split C according to the split values into lSet(v) and rSet(v) and compute lRadius(v) and rRadius(v).
3. While $|V(w)| < \alpha M - 1$:
   Determine a $(\tilde{v}, \tilde{@}) \in V_{\text{leaf}}$ with $|\tilde{@}\,\text{Set}(\tilde{v})| = \max\{|@\,\text{Set}(v)|\,|(v, @) \in V_{\text{leaf}}\}$.
   (*the cluster with the largest cardinality is partitioned*)
   If $|\tilde{@}\,\text{Set}(\tilde{v})| > B$, then execute the next step, else goto 5.
4. Execute a partition step on $\tilde{@}\,\text{Split}(\tilde{v})$ and $\tilde{@}\,\text{Set}(\tilde{v})$, which results in the second split value e'. Let $V(w):= V(w) \cup \{v\}$, where v is a new MBT-node with: lSplit(v):= $\tilde{@}\,\text{Split}(\tilde{v})$, rSplit($v$):= e', lSon(v), rSon(v):= nil and lObject(v), rObject(v):= $\emptyset$.

Split $\widetilde{@}\text{Set}(\tilde{v})$ according to the split values into $\text{lSet}(v)$ and $\text{rSet}(v)$ and compute $\text{lRadius}(v)$ and $\text{rRadius}(v)$. $\widetilde{@}\text{Set}(\tilde{v}) = \varnothing$.

Last, insert the new node v through $\widetilde{@}\text{Son}(\tilde{v}):= v$. GOTO 3.

5. For all $(v, @) \in V_{\text{leaf}}$ mit $@\text{Set}(v) \neq \varnothing$ let $@\text{Son}(v):= $ new node in C-tree and execute Create $(@\text{Son}(v), @\text{Set}(v), @\text{Split}(v), \alpha, B)$.

Theorem 3.1. *Algorithm 3.1 creates a C-tree with root w, which represents the elements in C. The objects are stored in leafs and the buckets fulfill the maximum load B.*

We will now estimate the balance of the created tree. Therefore we consider the internal structure of a node. $\text{Card}(v)$ denotes the number of objects represented in the subtree with root v (v is a MBT-node or a CT-node). In other words $\text{Card}(v)$ *is the cardinality of the cluster of v.*

The theory of Monotonous Bisector* Tree [17] states that for the balancing step (BS) holds: $\text{Card}(@\text{Son}(v)) \geq \left\lceil \dfrac{\text{Card}(v)}{2} \right\rceil$, if the cells are convex polygons in $\mathbb{R}^N$ and the distance function d is any L_p-metric. Unlikely for the contraction step (KS) no such statement can be made. Using the KS, one of the two sons may have less than half of the original cells. We term these kind of sons as *underfilled*.

While creating the tree we are able to choose sensitively (according to the development of cluster radii and cardinality) the BS or KS partition step. *Subsequently we assume that at most every second step[5] is a KS.* Test of the Monotonous Bisector* Tree [12] showed that this is quite efficient. We now ask for the fraction q of filled sons of a CT-node, that is the number of filled leafs in the MBT.

An underfilled son can only be created by a KS, so: #underfilled sons $\leq$ #KS. On the other hand every BS creates at least one new filled son: #filled sons $\geq$ #BS $+ 1$.

How large can q get in the worst-case? For this the MBT must be created by as many KS as possible, but only few BS. We assume that a KS has got an underfilled son. Then this path is not used furthermore in Algorithm 3.1. So the worst-case looks like shown in Fig. 4. Hence it holds for the height $2h$ and $2h - 1$ of the worst- case:

$$\frac{q}{1-q} = \frac{\#\text{filled sons}}{\#\text{underfilled sons}} \geq \frac{\#\text{BS} + 1}{\#\text{KS}} \geq \frac{2^0 + 2^1 + \cdots + 2^{h-1} + 1}{2^0 + 2^1 + \cdots + 2^h} = \frac{2^h}{2^{h+1} - 1} \geq$$

$\dfrac{1}{2}$. So $2q \geq 1 - q$ and hence $q \geq \frac{1}{3}$.

Remark. On the average one can expect as much BS as KS and so $q \approx \frac{1}{2}$. Furthermore the KS is normally balanced, if the objects are distributed equally. In general we do have a trade-off between a larger q and a "better" tree (cluster radii descent faster). This can be controlled through the number of contraction steps used.

---

[5] On a path from the root to a leaf at most every second.

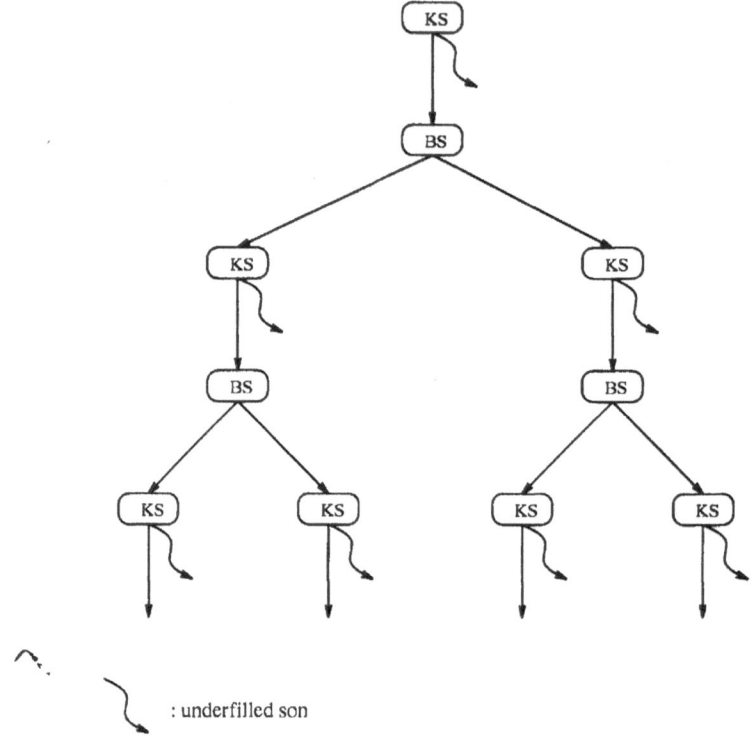

: underfilled son

Figure 4. Worst-case for q

We now know that $v \in V$ has got at least $M_{\alpha q} := \lceil (\lceil M\alpha \rceil + 1)q \rceil$ filled sons (denoted now $V' \subset V$), which represent at most $\text{Card}(v)$ objects. The creation procedure partitioned in each case the largest cluster, so it is:

$$\max_{v' \in V'} \text{Card}(v') \leq 2 \min_{v' \in V'} \text{Card}(v') \tag{1}$$

$$\max_{v' \in V'} \text{Card}(v') \left(1 + \frac{1}{2} + \cdots + \frac{1}{2} \right) \leq \text{Card}(v)$$

$$(M_{\alpha q} - 1)\text{-times}$$

and hence:

$$\max_{v' \in V'} \text{Card}(v') \leq \frac{2\text{Card}(v)}{M_{\alpha q} + 1}$$

This provides a guaranteed reduction factor for $\text{Card}(v)$. With this factor we are able to estimate the height of the tree:

Theorem 3.2. Let $M_{\alpha q} := \lfloor (\lceil M\alpha \rceil + 1)q \rfloor$, $q = \frac{1}{3}$, the cells be convex polygons of $\mathbb{R}^N$ and d any L_p-metric $(1 \leq p \leq \infty)$. Then Algorithm 3.1 creates a C-tree with the height *of at most*

$$\begin{cases} \left\lceil \log_{\frac{M_{\alpha q}+1}{2}} \frac{|C|}{B} \right\rceil, & \text{if } |C| > B \\ \qquad\qquad 0, & \text{if } |C| \le B \end{cases}$$'

if at most each second step is not balanced and $M_{\alpha q} \ge 2$.

Theorem 3.3. *Under the assumptions of Theorem 3.2, Algorithm 3.1 creates a C-tree in $\mathcal{O}(|C|\log|C|)$ time.*

Proof: The virtual Monotonous Bisector* Tree can be constructed in $\mathcal{O}(|C|\log|C|)$ time: The balancing step provides logarithmic height. The effort follows directly out of this fact and the partitioning effort of $\mathcal{O}(|C|)$ in each step. We will now show, that the additional effort for paging the Monotonous Bisector* Tree is also $\mathcal{O}(|C|\log|C|)$. Therefore we organize V_{leaf} (the set of actual leafs) as a maximum-heap in respect to the cardinalities $|@\text{Set}(v)|, (v, @) \in V \times \{$"$l$", "$r$"$\}$. The size of the heap is at most M. Hence deleting the maximum element and inserting needs $\mathcal{O}(\log M)$. Therefore step 3 causes an additional effort of $\mathcal{O}(M \log M) = \text{const.}$, if M is fixed. For large page sizes P, also B and M may be large. But the theorem holds even if $M \in \mathcal{O}(|C|)$. □

Theorem 3.4. *Under the assumptions of Theorem 3.2 the C-tree created by Algorithm 3.1 requires $\mathcal{O}(|C|)$ space (in secondary memory). More precisely the following bounds hold: For all internal nodes the load of pages is $\ge \alpha$, up to at most one exception on a path from the root to a leaf.*

Proof. From step 3 of the algorithm we conclude that the iterated partitioning in step 4 breaks off only if the load factor is fulfilled or each of the sons fits into one bucket. Therefore the load α is reached for all internal nodes except the last one on the path to a leaf. Since the Monotonous Bisector* Tree itself has got only linear size, the above load on an average shows that the space requirement is $\mathcal{O}(|C|)$. □

We now have reached the aim of an average page load. We have seen that like a rigid tree shape, the average load prevents from degenerating and is more flexible. Moreover on any search path to a leaf at most two underfilled nodes are visited. Therefore underfilled pages can not cumulate on a search path.

We can improve the bounds of the above theorems in the case $B \ll \bar{M}$, if we try to put more buckets in one page. More precisely from the guaranteed reduction factor in Theorem 3.2 we can derive a number $\bar{M}_B$, which is the maximum number of objects a C-tree with one node can be constructed for (step 1. in Algorithm 3.1). Instead of B in step 3, we have then the threshold $\bar{M}_B$. This saves the load factor for internal nodes and raises the load of leafs to $\dfrac{\bar{M}_B}{2}$. The bounds in Theorem 3.2 are improved to

$$\left\lceil \log_{\frac{M_{\alpha g}+1}{2}} \frac{|C|}{\bar{M}_B} \right\rceil.$$

From the theory of Monotonous Bisector* Tree we are able to transfer the following result:

Theorem 3.5. *Under the assumptions of Theorem 3.2 the cluster radii on a path from the root to a leaf can be estimated by a geometrically decreasing sequence (geometric k-step development).*

4. Queries

The C-tree provides the same queries as the Monotonous Bisector* Tree:

- nearest neighbor queries
- fixed-radius-near neighbor queries
- ray-shooting queries
- range queries
- points/objects in polygon retrieval
- objects hitting polygon retrieval
- objects hitting curve retrieval
- hidden-line/surface retrieval
- special problems in motion planning.

The way queries are performed is common to all rooted trees. Starting at the root all sons are recursively searched, if their subtree is relevant for the query (pruning of subtrees). This can easily be checked examining the cluster of the sons. Furthermore exact matching leads to a unique search path, because of (O2). The geometrically decreasing radii guarantee an efficient execution of the queries. On the contrary the Günther-Celltree only supports range and point queries.

When implementing a query, one has to remember, that the position in the tree is determined by a CT-node $v_{CT} \in V$ and a MBT-node $v_{MBT} \in V(v_{CT})$. To avoid page faults it is best to search an entire CT-node first, before recursively loading any successor page. Therefore the recursive calls must be stored in a stack.

Because the creation of the C-tree is sensitive in respect to the distance function d, also queries that depend on d (nearest neighbor, fixed-radius-near neighbor) are supported. Thereby a query object must be comparable to an object in S (cluster object). Let generally be Q a set of query objects and $\bar{d}: Q \times E \mapsto \mathbb{R}_{\geq 0}$, $\tilde{d}: Q \times S \mapsto \mathbb{R}_{\geq 0}$ the distance functions between query objects and split values resp. cluster objects. Then we additionally demand two triangle inequalities. For all $q \in Q$, $e \in E$, $s \in S$ holds:

$$\tilde{d}(q, s) \leq \bar{d}(q, e) + d(e, s) \tag{2}$$

$$\bar{d}(q, e) \leq \tilde{d}(q, s) + d(e, s) \tag{3}$$

In the case $Q = E = \mathbb{R}^N$, $S \subset \mathscr{P}(\mathbb{R}^N)$ the demanded properties reduce to the given distance function d. Inequality (2) allows to completely accept a subtree and (3) to prune a subtree.

First we consider the search for the fixed-radius-near neighbor of $q \in Q$ and the radius MAXDIST. That means we want to retrieve all $s \in S$ with $\tilde{d}(q, s) \leq$ MAXDIST. Let $(S'. e')$ be the cluster S' of the actually tested subtree with the cluster center e'. With

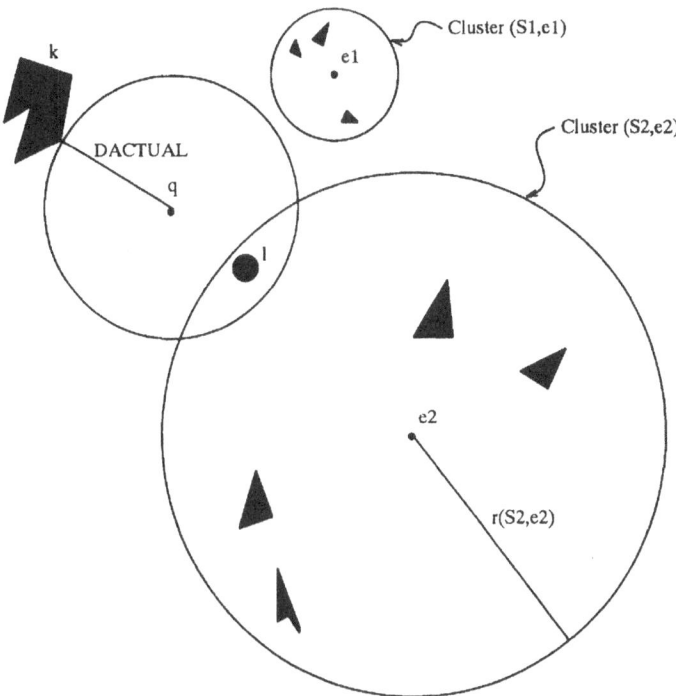

Figure 5. Pruning of subtrees when searching the nearest neighbor

$r(S', e')$ we denote the radius of the Cluster. Then it follows through (2):
$\forall s \in S': \tilde{d}(q, s) \leq \bar{d}(q, e') + d(e', s) \leq \bar{d}(q, e') + r(S', e')$. If we now have the situation that:

$$\bar{d}(q, e') + r(S', e') \leq \text{MAXDIST},$$

then $\tilde{d}(q, s) > \text{MAXDIST}$ for all $s \in S'$. So the entire subtree can be pruned.

For the search of the nearest neighbor of $q \in Q$ the following approach can be used: When already a good candidate $k \in S$ is derived (e.g. through searching the leaf with the nearest cluster center to q and taking any of the objects stored there), we can set DACTUAL:$= \tilde{d}(q, k)$. Then we can use the same arguments like before. A subtree can be pruned, if $\bar{d}(q, e') - r(S', e') \geq$ DACTUAL, where e' denotes the cluster center and S' the represented objects of the actual node. Here we have $\geq$ instead of $>$, because objects with the same distance aren't better candidates. Figure 5 shows an example in $\mathbb{R}^2$ equipped with the euclidean metric. The subtree related to the cluster (S_2, e_2) must be searched, whilst the subtree of cluster (S_1, e_1) can be pruned although e_1 is nearer to q than e_2 is. In fact for $l \in S_2$ there is $\tilde{d}(q, l) < \tilde{d}(q, k)$. If we find a better candidate than k, then we have to reset DACTUAL and continue the search.

5. Dynamic Operations: Inserting and Deleting Objects

The C-tree should support insertion and deletion of cells without degenerating, i.e. Theorem 3.2 and Theorem 3.4 are valid in the dynamic case with perhaps worse constants. To prove this we start with a definition:

Definition 5.1. An internal node where all sons are leafs, is termed *B-node*. All other internal nodes are termed *I-node*.

When we view a tree without its leafs, then B-nodes are exactly the new leafs and I-nodes the new internal nodes.

The derivation of Theorem 3.2 bases on the fact, that for each I-node exists at least $M_{\alpha q}$ filled sons $V' \subset V$ with (1) holds. If $\mathrm{Card}(v)$ denotes the cardinality of the objects in the subtree with root v, then:

$$\max_{v' \in V'} \mathrm{Card}(v') \le 2 \min_{v' \in V'} \mathrm{Card}(v')$$

We now demand the weaker statement that after each operation still holds: For each I-node $v \in V$ there is a subset $V' \subset N(v)$ of the sons of v, with:

$$|V'| \ge M_{\alpha q}$$

$$\min_{v' \in V'} \mathrm{Card}(v') \ge \max_{v \in N(v) \setminus V'} \mathrm{Card}(v) \qquad (4)$$

$$\max_{v' \in V'} \mathrm{Card}(v') \le 4 \min_{v' \in V'} \mathrm{card}(v')$$

instead of the number 4 in the last inequality, we could have chosen any other real number > 2. The idea is to give each node a linear clearance, where it is still filled.

From (4) it follows, that each I-node has got at least a load of αq. This is all we want to demand, since the subtree of underfilled nodes may be deleted without notice. It is only important that always $M_{\alpha q}$ filled sons are present.

We start with a C-tree, which is constructed by algorithm 3.1. So (4) is fulfilled. Now we must ensure that (4) is still valid after each dynamic operation. This provides the logarithmic height and a minimum load on the average. To be able to test (4), we have to store $\mathrm{Card}(v)$ for each node v within the tree.

For rebalancing we use the static creation Algorithm 3.1, which fulfills (4). Since α is still unused, we can set $\alpha := 1$.

Algorithm 5.1. Balance (v)

1. Store all objects in the subtree of v in $\mathrm{Set}(v)$ and delete the entire subtree except the root v itself.
2. Create $(v, \mathrm{Set}(v), e, \alpha, B)$ where e is the split value of the MBT-father of v.

At this point we have the opportunity to enlarge the clearance of balancing, if we choose $\alpha < 1$. If the page v which caused the error is not full, we may only rebalance the subtree in $\mathrm{MBT}(v)$ instead of the entire CT-node. Using very large pages this reduces the rebalancing effort. But it does not affect the following worst-case results.

So we are now able to explain the *algorithms for inserting and deleting objects.* Because they are very similar, we will describe the insertion procedure. The variant of deleting objects is shown by the changes in brackets.

Algorithm 5.2. Insert [Delete] (w, o)
w: root of the C-tree
o: cell to be inserted [deleted]

1. $(O2)$ provides the unique search path W to $@\text{Object}(v_{\text{MBT}})$ in v_{CT}.
   If $o \in [\notin]$ $@\text{Object}(v_{\text{MBT}})$ is already present [not present], then STOP.
2. Insert [Delete] o in [from] $@\text{Object}(v_{\text{MBT}})$.
   (*this may cause a temporary overflow [underflow]*)
3. Actualize the cluster radii on the search path W.
4. Starting at the root check all I-nodes $v \in W \setminus \{v_{\text{CT}}\}$:
   If (4) is violated by v, then Balance(v) and STOP.
5. If an overflow [underflow] of v_{CT} happened, i.e. $|@\text{Set}(v_{\text{MBT}})| > B \ [= 0]$, then: if
   Father(v_{CT}) is a B-node,
   - then Balance $(\text{Father}(v_{\text{CT}}))$
   - else Balance (v_{CT}) [if v_{CT} is empty, then delete v_{CT}].

In the undesired case that the object determing the cluster radius of $v \in V$ is deleted, the new farthest object must be found. Therefore we may use the given spatial index. But in the worst-case the effort is still $\mathcal{O}(\text{Card}(v))$ for each node on the search path. This summarizes to $\mathcal{O}(|C|)$ for actualizing the cluster radii on the search path. Another possibility is to compute the cluster radii delayed in the background. Thus after deletion they may be temporarily too large. But the effort in the amortized case would then be equal to the insertion effort.

Step 3 of the algorithm shows that $(O1)$ is fulfilled. From 4. it follows (4) and from step 5. (P). So we have:

Theorem 5.1. *Let* $M_{aq} := \lfloor (\lceil M\alpha \rceil + 1)q \rfloor, q := \frac{1}{3}, \alpha := 1,$ *the cells be convex polyhedrons in* $\mathbb{R}^N$ *and* d *any* L_p-*metric. After an arbitrary sequence of insertions and deletions the C-tree has got a height of at most*

$$\begin{cases} \left\lceil \dfrac{\log M_{aq} + 3}{4} \dfrac{|C|}{B} \right\rceil, & \textit{if } |C| > B \\ \qquad\qquad 0, & \textit{if } |C| \leq B' \end{cases}$$

if at most each second partition step is not balanced and $M_{aq} \geq 5$.

Theorem 5.2. *Under the assumptions of Theorem 5.1 after an arbitrary sequence of insertions and deletions the C-tree has got* $\mathcal{O}(|C|)$ *space requirement in secondary memory. More precisely the following bounds hold: For all internal nodes the load is* $\geq aq$, *up to one exception on a path from the root to a leaf.*

Since inserting or deleting an object may cause a complete creation of the tree, we have:

Theorem 5.3. *The effort to insert or delete in the C-tree is* $\mathcal{O}(|C| \log |C|)$ *time in the worst-case.*

We will now examine insertion and deletion more precisely, to get a statement about the *amortized costs*. In Algorithm 5.2 the search path in step 1, 3 and in test (4) in step

4 is passed once in $\mathcal{C}(\log|S|)$. When deleting we actualize the cluster radii immediately in $\mathcal{C}(|S|)$ or delayed in $\mathcal{C}(\log|S|)$. The steps 2 and 5 can be executed in $\mathcal{C}(1)$, since the page size is fixed. Altogether the effort to insert or delete with immediate (resp. delayed) actualizing of the cluster radii is logarithmic (resp. linear), if in step 4 no rebalancing is necessary. The following theorem now shows how expensive this rebalancing is.

Theorem 5.4.

1. *The amortized effort for inserting in the C-tree is $\mathcal{C}(\log^2|C|)$.*
2. *The amortized effort for deleting in the C-tree is $\mathcal{C}(\log^2|C|)$, if cluster radii are actualized delayed.*
3. *The amortized effort for deleting in the C-tree is $\mathcal{C}(|C|)$, if cluster radii are actualized immediately.*

Proof: It suffices to examine the additional costs for rebalancing in step 4. Therefore we only have to consider the I-nodes.

Let $v_{CT} \in V$ be a I-node. How many insertion and deletion operations $o(v_{CT})$ are at least necessary to force v_{CT} to be rebalanced? After executing Balance(v_{CT}) it is according to (1) for the M_{aq} largest sons V' of v_{CT}:

$$\max_{v' \in V'} \text{Card}(v') \leq 2 \min_{v' \in V'} \text{Card}(v').$$

If we now have to rebalance v_{CT} for the first time *after an arbitrary sequence of e insertion and l deletion operations* (i.e. $o(v_{CT}) = e + l$), then (4) is violated. For the worst-case (inserting in the largest and deleting in the smallest cluster) there is:

$$\overset{(4)}{\max_{v' \in V'} \text{Card}(v')} + e > 4\left(\min_{v' \in V'} \text{Card}(v') - l\right) \overset{(1)}{\geq} \max_{v' \in V'} \text{Card}(v') + 2\min_{v' \in V'} \text{Card}(v') - 4l$$

$$\Leftrightarrow 4o(v_{CT}) \geq e + 4l > 2\min_{v' \in V'} \text{Card}(v') \geq \frac{2\,\text{Card}(v_{CT})}{\alpha M}.$$

This means after at least

$$o(v_{CT}) = \frac{\text{Card}(v_{CT})}{2\alpha M} \tag{5}$$

operations a rebalancing of v_{CT} with effort $A_{v_{CT}} = \mathcal{C}(\text{Card}(v_{CT})\log\text{Card}(v_{CT}))$ may occur. The amortized effort for this is therefore $\mathcal{C}(\log\text{Card}(v_{CT}))$.

We now consider the *complete effort for a worst-case sequence of length $o(v_{CT})$ which forces v_{CT} to be rebalanced*. Let $N := \text{Card}(v_{CT})$. How large is the rebalancing effort for the sons of v_{CT}? Let v_i ($i = 1, \ldots, m$) be the son of v_{CT}, which is affected when the ith rebalancing of a son of v_{CT} occurs. Let $N_i := \text{Card}(v_i)$. Then the complete effort for all sons of v_{CT} is:

$$A_s = \sum_{i=1}^{m} \mathcal{C}(N_i \log N_i)$$

The guaranteed reduction factor in the dynamic case is $K := \dfrac{M_{aq} + 3}{4}$. Thus for all

$i = 1, \ldots, m$ it is: $N_i \leq \dfrac{N}{K}$. Since $v_1, \ldots, v_m$ were rebalanced, we have:

$$\sum_{i=1}^{m} N_i \overset{(5)}{=} 2\alpha M \sum_{i=1}^{m} o(v_i) \leq 2\alpha M o(v_{cr}) \overset{(5)}{=} N$$

and therefore

$$A_S \leq \sum_{i=1}^{m} \mathcal{O}\left(N_i \log \frac{N}{K} \right) \leq \mathcal{O}\left(N \log \frac{N}{K} \right).$$

So the complete effort for the subtree of v_{CT} with height h is determined by:

$$A = A_{v_{CT}} + A_S + \cdots$$

$$\leq \mathcal{O}\left(N \log N + N \log \frac{N}{K} + N \log \frac{N}{K^2} + \cdots + N \log \frac{N}{K^h} \right)$$

$$= \mathcal{O}\left(N \left(\log N + \log \frac{N}{K} + \log \frac{N}{K^2} + \cdots + \log \frac{N}{K^h} \right) \right)$$

$$\leq \mathcal{O}\left(N \left(\log N + h \log \frac{N}{K} \right) \right)$$

$$\leq \mathcal{O}\left(N \left(\log N + \log \frac{N}{B} \log \frac{N}{K} \right) \right)$$

We conclude that the amortized effort for balancing is $\mathcal{O}(\log^2 N)$.

Concerning point 1 and 2 of the theorem we have proven that the effort for deletion with delayed actualizing of cluster radii and for insertion is (inserting + rebalancing):

$$\mathcal{O}(\log N) + \mathcal{O}(\log^2 N) = \mathcal{O}(\log^2 N).$$

Concerning point 3 we showed that the effort for deletion with immediate actualizing of cluster radii is:

$$\mathcal{O}(N) + \mathcal{O}(\log^2 N) = \mathcal{O}(N). \qquad \square$$

6. Concluding Remarks

In [16] we report about an implementation of the C-tree with different scenes and queries. It has been tested extensively and compared to the Monotonous Bisector* Tree. These tests showed that the Monotonous Bisector* Tree and C-tree are flexible and robust indices for managing geometric spatial data. Nevertheless both data structures are not competing, but supporting each other in different aims and applications. Both data structures are applicable for different types of geometric queries, where we have observed optimal query times in tests.

The Monotonous Bisector* Tree is especially applicable in small scenes ($\mathcal{O}$(main memory)), because it is efficient and easy to implement. When managing very complex objects the Monotonous Bisector* Tree is advantageous, too.

But in very large scenes the C-tree is clearly superior. The field of applications is only restricted through the available secondary memory. Furthermore the C-tree is distinguished through its very *cooperative* character. In today's multi-tasking environments it does not compete with other processes, because it requires only a constant amount of main memory.

In our future research we will explore the application of spatial indices in motion planning. Therefore we will first investigate a combination of a C-tree (as a filter for locally relevant data) with common motion planning strategies.

References

[1] Bentley, J. L.: Multidimensional binary search trees used for associative searching. Comm. ACM *18*, 509–517 (1975).

[2] Bayer, R., McCreight, E.: Organization and maintenance of large ordered indices. In: Proceedings of the ACM-SIGFIDET workshop on data description and access, pp. 107–141, Houston 1970.

[3] Fuchs, H., Kedem, Z., Naylor, B.: On visible surface generation by a priority tree structure. Comput. Graphics *14*, 124–133 (1980).

[4] Günther, O., Bilmes, J.: The implementation of the cell tree: design alternatives and performance evaluation. Inf.-Fachber. *204*, 246–265 (1989).

[5] Günther, O., Noltemeier, H.: Spatial database indices for large extended objects. In: Proceedings IEEE—7th international conference on data engineering, Kobe (Japan), April 1991.

[6] Günther, O.: Efficient structures for geometric data management. Lecture Notes in Computer Science, Vol. 337. Berlin Heidelberg New York Tokyo: Springer 1988.

[7] Guttmann, A., C-Trees: A dynamic index structure for spatial searching. In: Proceedings of the ACM SIGMOD, pp. 47–57 (1984).

[8] Heusinger, H. Cluster-Verfahren für Mengen geometrischer Objekte. PhD thesis, Universität Würzburg 1989.

[9] Lomet, GP-tree (grow and post-tree methods). In: Advances in spatial databases. Berlin Heidelberg New York Tokyo: Springer 1991.

[10] Nievergelt, H., Hinterberger, H., Sevcik, K. C.: The grid file: an adaptable, symmetric multikey file structure. ACM Trans. Database System *9*, 38–71 (1984).

[11] Nievergelt, J.: 7 ± 2 Criteria for assessing and comparing spatial data structures. Lecture Notes in Computer Science, Vol. 409, pp. 3–27. Berlin Heidelberg New York Tokyo: Springer 1990.

[12] Noltemeier, H., Verbarg, K., Zirkelbach, C.: A data structure for representing and efficient querying large scenes of geometric objects: MB* Trees. Computing [Suppl.] *8*, 211–226 (1993).

[13] Ooi, B. C.: Efficient query processing in geographic information systems. Lecture Notes in Computer Science, Vol. 471. Berlin Heidelberg New York Tokyo: Springer 1990.

[14] Preparata, F. P., Shamos, M. I.: Computation geometry—an introduction. Berlin Heidelberg New York Tokyo: Springer 1985.

[15] Samet, H.: The design and analysis of spatial data structures. Reading: Addison–Wesley 1990.

[16] Verbarg, K.: Räumliche Indizes—Celltrees: Analyse und experimenteller Vergleich mit monotonen Bisektorbäumen. Master's thesis, Universität Würzburg 1992.

[17] Zirkelbach, C.: Geometrisches Clustern—ein metrischer Ansatz. PhD thesis, Universität Würzburg 1992.

K. Verbarg
Institut für Informatik
Universität Würzburg
Am Hubland
D-97074 Würzburg
Federal Republic of Germany

Computing Suppl. 10, 341–356 (1995)

Piecewise Linear Approximation of Trimmed Surfaces

M. Vigo and **P. Brunet**, Barcelona

Abstract. Stereolithography applications require a surface model of the modeled object consisting of a mesh of triangular facets. This model can also be used for mechanical analysis through finite element methods. In this paper, a new algorithm for the piecewise linear approximation of trimmed surfaces is presented. The algorithm generates a triangulation that approximates the initial surface within a pre-defined tolerance. The approximation is conformal, without cracks in edges: a closed polyhedron is obtained in the case of a closed initial surface. The algorithm first builds a quadtree-structured bound on the patch curvatures for every surface patch, and then works by first discretizing trimming curves and afterwards relaxing the location of a sufficient number of vertices inside the trimmed region in every patch. The resulting triangulation satisfies the max-min criterion in parametric space.

Key words: Trimmed surfaces, Delaunay triangulations, surface approximation.

1. Introduction

Many application areas in CAD/CAM are concerned with discretization and triangulation of surfaces. We can quote, among others, finite element methods (FEM), stereolithoghraphy (SLA) and surface rendering. Classical problems such as the Delaunay triangulations and the triangulation of polygons have been well studied by computational geometry and are a good theoretical basis for the approximation and discretization of surfaces. Several papers deal with implicit surfaces ([1, 2] etc.), but their main drawback is that their results are not directly applicable to the parametric case, because of their different mathematical treatment.

Because finite element methods work with discrete entities, when objects described by surfaces are to be dealt with, they must be approximated with some kind of linear elements: triangles are a straightforward choice. In this case, the basic requirement is to have element shape regularity, node constraint capability and density control [8]. This leads to a small amount of triangles that are as well shaped as possible.

Stereolithography is a process that obtains prototype parts in a short time; the input for machines that perform this task is a description of the object in terms of triangles, which should be as exact as possible.

Finally, when surfaces are to be visualized, the speed of the algorithms involved is crucial. As hardware can render polygons very quickly, one solution consists of triangulating the surface and passing the results to the computer. Algorithms for this conversion must also avoid cracks within neighboring patches or surfaces, as discontinuities would produce unpleasant visual effects.

In all these applications, the initial CAD surface model must be approximated and converted into a discrete mesh of triangles. The starting model is usually a standard file description (VDA, IGES) of the geometry. Some of the basic requirements of the conversion process,

- The triangulation must be a good approximation of the initial surface: maximum distance in R^3 between the surface and the triangle mesh must be less than a pre-defined tolerance ε;
- The triangulation must be conformal, without cracks between triangles in neighbour patches and within patches;
- Triangles must be well-shaped, the optimum being a max-min angle optimal triangulation
- The number of triangulation vertices must be minimal;

Besides inherent problems related to standardization and input formats, one of the main difficulties of this surface approximation problem is the multiplicity of goals. Up to now, published works only address these requirements in a partial way, as will be shown in the next section. In the present paper, a new algorithm for the piecewise linear approximation of trimmed surfaces is presented, with the aim of addressing in a nearly optimal way the whole set of above requirements. It focusses on stereolithography applications, although it could be used in other related areas. After presenting some of the most relevant existing schemes in the next section, the proposed algorithm is presented in Sections 3, 4 and 5. Section 6 presents and discusses several practical examples.

2. Previous Work

In this section, five representative papers of the research done in the same direction as our work are briefly presented and discussed; some of them show several similarities with our approach. The first one presents a general purpose method, the second and third were proposed for rendering purposes, the fourth one is aimed at SLA applications, and the last one comes from the FEM field.

Filip et al. [4] presented some theorems giving bounds on the maximum deviation between parametric surfaces and linear approximations. They used them for the calculation of piecewise linear approximations of non-trimmed patches, by uniformly discretizing in u and v directions in the parametric plane. The resulting algorithm produces cracks between patches, and does not take advantage of the internal distribution of curvatures in the patch. Triangles always correspond to right triangles in the parametric plane.

Herzen and Barr [6] triangulate (not trimmed) surfaces, building what they call a restricted quadtree in parametric space based on a bound of the patch curvature. Once the quadtree is constructed, the points corresponding to the vertices of each node are evaluated in order to obtain a triangulation in Euclidean space. This approach has the advantage of being adaptative, and also right triangles in the parametric plane are always obtained. However, it produces cracks between neighboring patches, it generates a large number of triangles, and the resulting triangulation is only suitable for rendering purposes.

Rockwood et al. [7] deal with trimmed surfaces, and they also work in parametric space. First, a bound based on the surface curvature is found for each patch. This bound determines the size of the rectangles that will fit the surface. Next, the interior of the trimmed patch is tiled with rectangles of this size. The rectangles are bisected onto right triangles in the parametric plane. The trimming boundary curves are also discretized according to the curvature. A polygonalization that is connected with the interior quadrilateral mesh is obtained by a coving method. The main drawbacks of this scheme are that it obtains very odd-shaped triangles close to the trimming curves and that it does not handle cracks between surfaces. In addition, the algorithm for obtaining the polygonalization of the boundaries is quite complex: it works by finding extrema of trimming curves in order to subdivide them into uv-monotone segments.

Sheng and Hirsh [9] triangulate trimmed surfaces for stereolithography applications. They take special care of the edges of the solid that is being discretized, because the resulting mesh must represent a valid object. Their method can be summarized as follows:

- find bounds for curvature of the curves and surfaces that describe the object;
- construct a polygonal approximation of the trimming curves, based on these bounds;
- merge the neighboring polygonal approximations obtained in the previous step;
- start with a Delaunay triangulation of each face based on the vertices of the polygonal approximations of the trimming curves in parametric space;
- refine the triangulation, subdividing the triangles depending on the curvature bounds for the surface;
- generate the final triangulation model by obtaining the R^3 vertices from vertices of the parametric triangulation.

This method has the disadvantage that it is not adaptative (global bounds are found for every patch) and that it does not care about the shape of the resulting triangles. The consequence of not being adaptative is that the number of required triangulation vertices is too large, as will be shown in Section 6. On the other hand, the algorithm guarantees the absence of cracks.

Shimada and Gossard [8] generate adaptive 2D/3D meshes given a boundary representation of the geometry of a space region and a density distribution over that region. They also allow added constraints to control the position of the final nodes. The resulting mesh has more points where the density function is higher, and the shape of the elements is as regular as possible. The method is based on a physically-based approach that simulates the attraction/repulsion forces between bubbles covering the region:

- discretize the bounds of the region;
- calculate the amount of bubbles necessary to cover the region;
- place bubbles inside the region;
- simulate bubble motion and solve the equations of motion, given the initial configuration of the bubbles;

• connect the center points of the bubbles by Delaunay triangulation of a valid mesh.

Both equations and density distribution function are in parametric space, so forces have to be projected onto the surfaces or curves. Although it is a good approach and guarantees a conformal triangulation with well-shaped triangles, the model for interbubble forces is somewhat empirical and quite complex, and solving the equation system can become inefficient. On the other hand, the discretization of the region boundaries is not related to neighbour patches, only the single-patch problem being addressed. As a consequence, cracks between patches are not avoided.

Mesh simplification is another field directly related with triangulations. Although it does not deal with the exact representation of the surfaces, methods developed in this area are a good reference for linear discretization algorithms [10].

3. Bounds for Linear Approximations

Given a general C^2 parametric curve $f(t)$ and two parameter values a, b, the maximum distance between $f(u)$ and the linearly parametrized line segment $l(u)$ (Fig. 1) is bounded by an expression that depends on the second derivatives [4]:

$$\sup_{a \leq u \leq b} \| f(u) - l(u) \| \leq \tfrac{1}{8}(b-a)^2 \sup_{a \leq u \leq}{}^b \| f''(u) \|$$

In a similar way [9], given a parametric patch $S(u, v)$ and an arbitrary triangle T with vertices (A_p, B_p, C_p) in the parametric space (Fig. 1), the maximum distance between $S(u, v)$ and the linearly parametrized triangle $l(u, v)$ satisfies

$$\sup_T \| S(u, v) - l(u, v) \| \leq \tfrac{2}{9} \Omega^2 (M_1 + 2M_2 + M_3)$$

Ω being the maximal edge length of the triangle T in the parametric space, and M_1, M_2, M_3 upper bounds of the norms of the second derivatives of $S(u, v)$ (with respect to u, u and v, and v respectively) in the domain $(u, v) \in T$. It must be observed that in the general case bounds on the distance to linear approximations can be wrong for non-functional (parametric) models [3]. In this case, however, this problem is avoided by weakening the bounds and considering distances between the images of the same parametric point (u, v), $S(u, v)$ and $l(u, v)$, [4].

A direct consequence of the last equation is that an upper bound for the maximal edge length Ω exists when the distance between the surface $S(u, v)$ and its linear

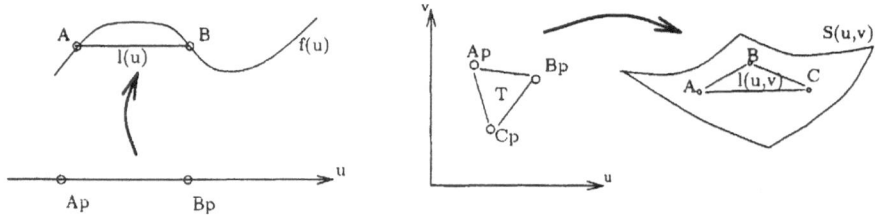

Figure 1. Linear approximation of curves and trimmed surfaces

approximation $l(u, v)$ must be under a certain pre-defined tolerance ε:

$$\Omega \leq 3 \sqrt{\frac{\varepsilon}{2M_1 + 4M_2 + 2M_3}}$$

Following the proof in [4], it can be seen that the expression $2M_1 + 4M_2 + 2M_3$ can be replaced wherever it appears by

$$\max\{M(p)\} = \max\left\{2\frac{\partial^2 S}{\partial u^2}(p) + 4\frac{\partial^2 S}{\partial u \partial v}(p) + 2\frac{\partial^2 S}{\partial v^2}(p)\right\}$$

where p is the point with parametric coordinates (u, v). We can define the $\Omega(p)$ function for each point in parametric space:

$$\Omega(p) = 3\sqrt{\frac{\varepsilon}{M(p)}}$$

A segment AB in parametric space will be said to be admissible (Fig. 1) if

$$\Omega(p) \geq \|AB\| \quad \forall p \in AB$$

and a triangle T is admissible if

$$\Omega(p) \geq L_m \quad \forall p \in T$$

where L_m is the length of the largest edge of the triangle. These definitions ensure that any admissible triangle will approximate the surface with the given tolerance, ε. We define a second function, $R(p)$, for each point in parametric space:

$$R(p) = \min\{\Omega(q)\}\forall q; \quad \operatorname{dist}(p, q) \leq \Omega(p)$$

This function allows us to redefine the admissibility of segments and triangles taking into account only their bounds. A segment AB will be admissible if $\|AB\| \leq R(A)$ or $\|AB\| \leq R(B)$, and a triangle is admissible when all its edges are admissible.

A single lower bound on the Ω values of the patch (or on the R values) could be considered in order to test the admissibility of the triangulation edges. However, in order to adapt the triangle sizes to the curvature changes within the patch and to minimize the number of necessary triangulation vertices, a quadtree structure of Ω bounds and R bounds is built in the parametric space during a preprocess step. We will call these quadtrees the omega-quadtree and the R-quadtree in the rest of the paper. Both quadtrees subdivide the parametric space (u, v) until regions with similar values of the corresponding bounds are obtained.

In order to build the omega-quadtree, the patch is first subdivided up to a certain pre-defined level, and the Ω bound is computed for every subpatch using the above equations. The quadtree is then obtained through compactation of this uniform mesh of subpatch bounds: every set of four contiguous nodes with similar Ω bounds is compacted on a single mode of the immediate upper level.

Finally, in order to build the R-quadtree, the influence area for every terminal node of the omega-quadtree is first computed. The influence area is the locus of the points being at a distance from the node of less than the Ω value of the node. Then, the $R(P)$ value of every parametric point P is the minimum value of the influence areas that contain P.

4. Overall View of the Algorithm

The proposed surface discretization algorithm works in seven sequential steps:

STEP 1: VDA information is processed, and a boundary representation is generated

STEP 2: The geometry of the vertices connecting patches is determined

STEP 3: The Ω-quadtree and the R-quadtree are computed for every patch

STEP 4: Trimming and boundary curves are discretized for every patch

STEP 5: A sufficient number of interior vertices is placed randomly in every patch

STEP 6: A relaxation algorithm is started in order to find an admissible location of interior vertices for every patch. The final triangulation in parametric space is the Delaunay triangulation of interior vertices and vertices of the discretized boundary and trimming curves – restricted to be consistent with these trimming boundaries

STEP 7: Triangulations in parametric space are mapped on R^3

The goal of Step 1 is to validate the input information and create topological links between geometric entities. Then, Steps 2, 4 and 6 create consistent geometry shared by adjacent patches. After Step 2, vertices joining three or more patches are precisely located. They have a unique geometric representation which can be accessed from all joining patches, and their distance to the patches is not greater than the predefined tolerance ε. The computations in Step 3 (see previous section) guarantee that all triangles will be admissible and that their size will automatically adapt to curvature changes along the patches. Furthermore, at the end of Step 4, trimming and boundary curves that close faces in patches are discretized, being approximated by a polyline which is unique and shared by the two adjacent patches. The distance from this polyline to both curves on surfaces is not greater than the predefined tolerance ε, and the polyline ends are the vertices obtained in Step 2. Every polyline subsegment is furthermore guaranteed to be admissible with respect to the R-quadtrees of the patches sharing the discretized curve. Now, after Step 5, the number of interior vertices at every trimmed face of every patch guarantees that it is possible to redistribute them and obtain an admissible triangulation with respect to the R-quadtree bound. The goal of Step 6 is to reach this redistribution of the internal vertices.

This set of postcondition requirements for the different steps of the algorithm ensures that: 1) the final discretization will be a good approximation of the initial surface with respect to the pre-defined tolerance ε (this is guaranteed by Steps 2, 4, 5 and 6); 2) the triangulation will be conformal, without cracks (this is the goal of unicity the Steps 2 and 4); 3) triangles will be well-shaped (this is due to the relaxation in Step 6 and to the later Delaunay triangulation); 4) the number of triangulation vertices will be small, due to the estimation computed in Step 5 and to the adaptation to curvature changes through the R-quadtree of curvature bounds.

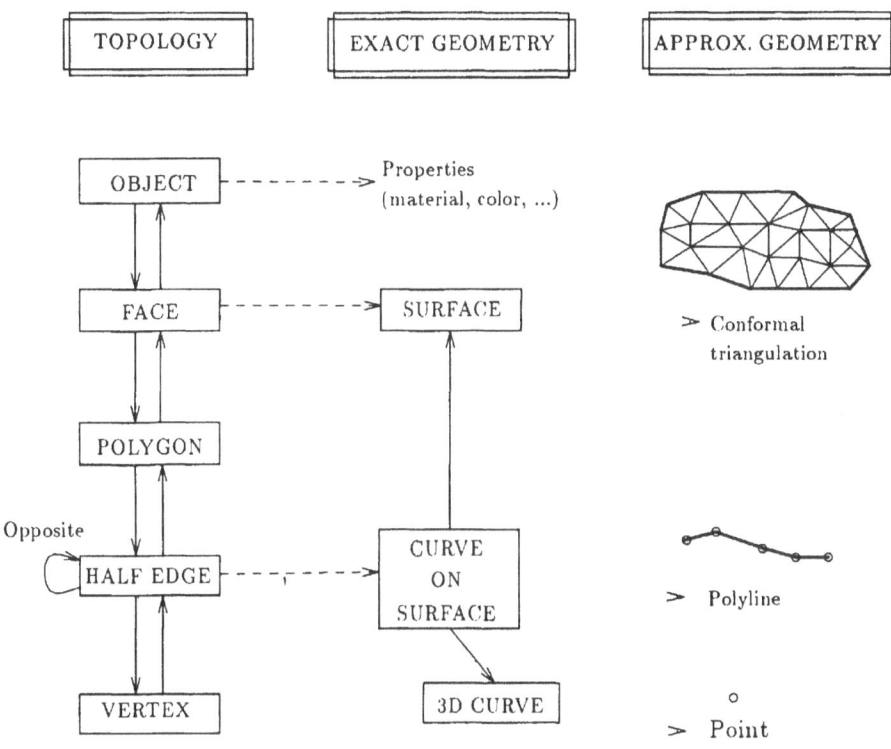

Figure 2. Intermediate boundary representation for trimmed surfaces

The rest of this section is devoted to the description of Steps 1, 4 and 5. Step 3 has already been discussed in Section 3, and Step 6 will be described in the next section.

STEP 1: VDA validation and conversion onto a boundary representation.

In this initial step, VDA information is validated while it is integrated in a complete boundary representation (Fig. 2). VDA geometry (surface or patch equations, parametric representations of trimming curves on surfaces, 3D representations of these trimming curves) are complemented by (usually) new information on topology and relationships between adjacent elements. The geometric entities in our boundary representation are (Fig. 2):

- Objects: set of faces; other specific information.
- Face: a trimmed part of a patch; contains pointers to the surface patch equation, to its final approximate geometry – the triangulation – and to the surrounding polygons – boundary, holes.
- Polygon: list of half edges.
- Half edge: the exact geometry is a parametric curve on a surface, and it also has a pointer to the polyline that will approximate it after Step 4; its opposite half edge has the pointer to the curve on the adjacent patch.
- Vertex: it points to its approximate geometry, computed in Step 2.

During the validation, every half edge is required to have its curve on surface, the 3D curve and the curve on surface of its opposite half edge closer than the tolerance ε. In fact, this is the criterion that is used for creating new "opposite" links. At the end, if the VDA information represented a closed object, every half edge would have an opposite different from nil.

STEP 4: trimming and boundary curve discretization.

Discretization of a 3D curve between two faces in different patches must be unique, and coherent with the curvature bound quadtrees of both patches. This is accomplished by using two proximity maps between points of the 3D curve and points of the corresponding curves on surfaces (Fig. 3):

$$F_1: C_{3D} \rightarrow C_{2D1} \text{ such that } A \rightarrow F_1(A)$$

$$F_2: C_{3D} \rightarrow C_{2D2} \text{ such that } A \rightarrow F_2(A)$$

where $F_i(A)$ is defined by

$$\text{dist}(S(F_i(A), A) = \text{dist}(S(C_{2Di}, A)$$

A segment (A, B) of the C_{3D} curve is admissible only if the segment $(F_1(A), F_1(B))$ is admissible with respect to the quadtree bound on patch P_1 and the segment $(F_2(A), F_2(B))$ is admissible with respect to the quadtree bound on patch P_2 (Fig. 3). This test is recursively repeated in order to ensure that every linear segment in the discretization of C_{3D} is admissible: the whole curve is first tested, and recursively bisected when its chord is not admissible.

STEP 5: Placing a sufficient number of interior vertices in every patch.

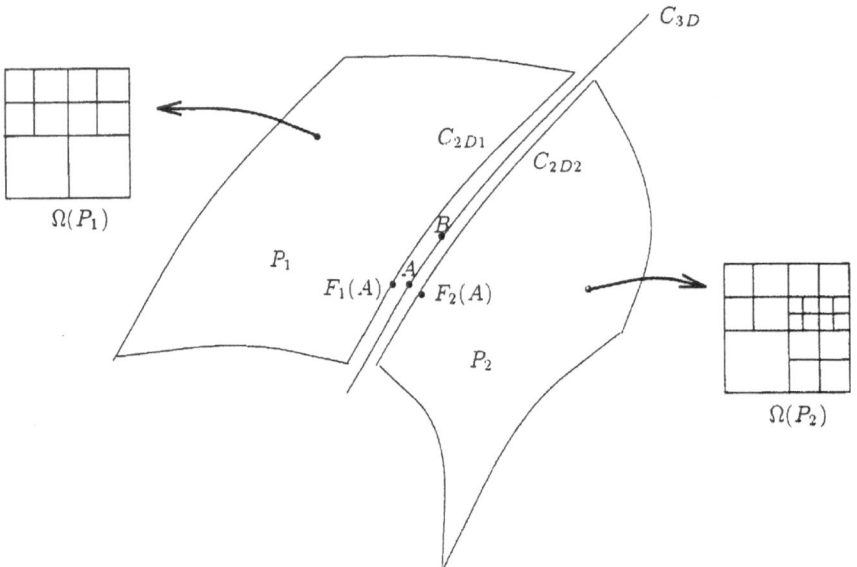

Figure 3. Discretization of trimming curves

In order to compute a bound on the number of interior vertices in the trimmed region of a patch, let us assume that we already have an admissible Delaunay triangulation, with vertices $V_1, \ldots, V_{nv}$ such that $\| V_i V_j \| \leq \max(R(V_i), R(V_j)) \forall i, j$. On the other hand, let us assume that vertices V_i are not too close, and there exists a value $\alpha, 0 \leq \alpha \leq 1$ such that $\| V_i V_j \| \leq \alpha R(V_j) \forall i, j$. Furthermore, let us now consider a circle C_i centered at every vertex V_i with a radius $\alpha R(V_i)/2$. This circle is completely contained in the Voronoi tile T_i of the vertex V_i, as $\alpha R(V_i)$ is a lower bound on the distance between adjacent admissible vertices of the triangulation. As a consequence, the surface area of the circle C_i is always less than the surface area of the Voronoi tile T_i of its vertex V_i:

$$\text{Sur } f(C_i) < \text{Sur } f(T_i) \forall i$$

This equation can be used to compute an upper bound on the number of triangulation vertices nv. If the total surface of the trimmed patch (face) is S_t, the surface area of the circle centered at V_i is $\frac{\pi}{4} R(V_i)^2$, and we can write:

$$S_t > \sum_{i=1}^{nv} \text{Sur } f(T_i) > \sum_{i=1}^{nv} \text{Sur } f(C_i) = \pi \frac{\alpha^2}{4} \sum_{i=1}^{nv} R(V_i)^2$$

The first inequality comes from the fact that tiles belonging to vertices in the trimming and boundary curves are not considered in the summation. In the particular case of a constant-valued R over the trimmed patch, the summation equals $nv \cdot R^2$, and

$$S_t > \pi \frac{\alpha}{4} nv \cdot R^2$$

yielding to the bound

$$nv < \frac{4S_t}{\pi \alpha^2 R^2}$$

In the general case, the final expression of the upper bound on nv is obtained [5] by using the preceeding bound on every R-quadtree node and summing up the partial bounds,

$$nv < \frac{4}{\pi \alpha^2} \sum_k \frac{\text{Sur } f_k}{R_k^2}$$

Where k traverses the set of R-quadtree terminal nodes, Sur f_k is the surface area of the visited node after being clipped by the trimming curve, and R_k is the R value associated with the node.

As a consequence, randomly placing a number of $\frac{4}{\pi} \sum_k \frac{\text{Sur } f_k}{R_k^2}$ vertices inside every trimmed face in every patch guarantees that it is possible to redistribute them and obtain an admissible triangulation with respect to the R-quadtree bound. In our implementation we first triangulate the vertices from the discretization of trimming curves, and then use a random weighted choice from among all the triangles. The approach is similar to that of [10], but in our case the R-quadtree bound for each triangle is used together with its area as the weighting factor.

5. Relaxation of Internal Vertices

Methods based on relaxation are usually based on attraction-repulsion forces between vertices [10], [8]. They are quite attractive, because of their similarity to natural fenomena. However, they have several drawbacks. Force models are rather empirical, and several parameters must be tuned according to particular cases. Also, either they take into account only neighbour points, or they can become very computationally expensive ($O(n^2)$ complexity for the force computation). Furthermore, in the present problem there is no way of ensuring that the final triangulation will be admissible without testing it: long, non-admissible edges can appear in the triangulation of an apparently admissible distribution of points; the triangulation can generate edges connecting points classified as non-neighbours by the force model.

The relaxation approach that we propose is based on the explicit computation of the Dealunay triangulation of the vertices at each relaxation step. This ensures the goal of obtaining an admissible triangulation. The computational cost of the triangulation is $O(n \log(n))$, but the analysis of the new placement of each vertex is now only based on the study of its triangulation edges.

Having already placed a sufficient number of vertices in the patch trimmed face (Step 5), at every relaxation step we first triangulate the interior and boundary vertices using a constrained Delaunay triangulation. Then, for every vertex V_i its triangulation edges are analysed. Edges $V_i V_j$ longer than $\max(R(V_i), R(V_j))$ would produce non-admissible configurations, and therefore they must be shortened. In this case, a move of the vertex shortening this particular edge is decided. Also, edges shorter than $\alpha R(V_i)$ can be lengthened, α being a user-adjustable parameter in the range of 0.8–1.0. In this case, a move of the vertex increasing the length of the edge is decided. The move step is fixed in all cases. The final movement of the vertex is computed through the vector sum of all moves induced from its edges. Several experiments have shown that α can be very close to the unity. Lower values of α obviously tend to produce bad shapes in the resulting triangles.

The relaxation ends when there is no edge longer than $R(V_i)$ and vertices are stable. The resulting triangulation is admissible by definition. During the relaxation process, several vertices can be thrown outside the trimmed region of the patch. This is always the case when the nv bound computed in Step 5 is too large. In our implementation, expelled vertices are kept in a stack, and they can be re-inserted only if long and stable edges are detected. Long edges are thus bisected. This scheme is especially useful for trimmed faces with almost disconnected regions.

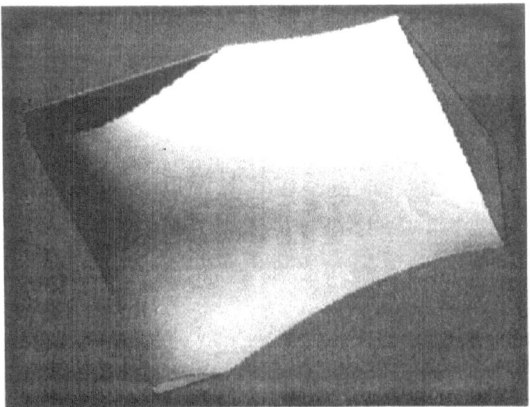

Figure 4. Test patch

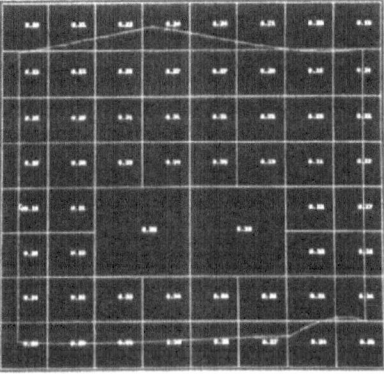

Figure 5. Quadtree of curvature bounds in parametric space. Discretization of trimming curves, and associated triangulation

6. Results and Discussion

The algorithm presented has been implemented on a Sun SPARC station. Some results are explained in this section.

Figure 4 shows a solid whose top face (the non-planar one) has been chosen as a test patch. As explained, first the trimming curves of the face are discretized, and the quadtree of curvature bounds is computed. These results are shown in Fig. 5 in parametric space, together with the triangulation associated with the curves discretization. The values written in each terminal node are the maximum allowed length of any triangle edge that will approximate the region. The algorithm proposed in [9] was implemented in order to compare it with the scheme presented in this paper; Fig. 6 shows the resulting triangulation for this test case. Notice the irregular shape of the triangles, imposed by the initial Delaunay triangulation of the face. The next

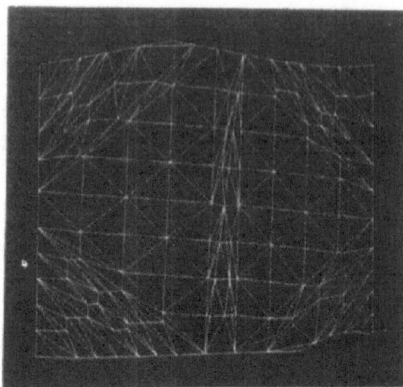

Figure 6. Sheng and Hirsch triangulation in parametric space

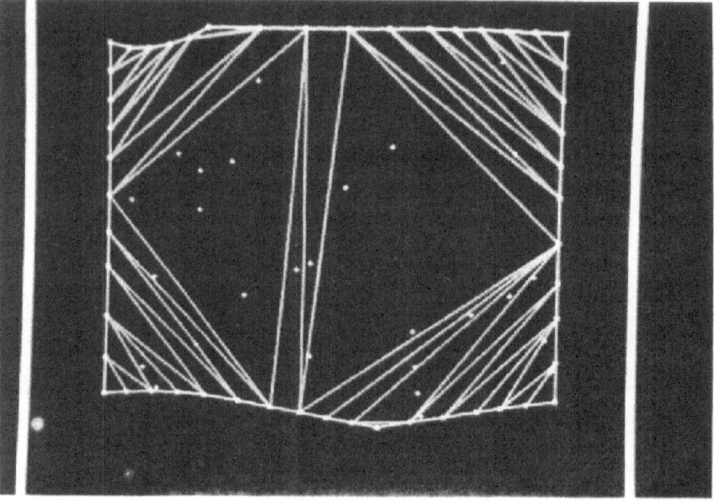

Figure 7. Interior vertices before relaxation

step consists in placing interior vertices randomly inside the face (Fig. 7), and then applying the relaxation process. Figures 8a and 8b show the interior vertices while the relaxation is taking place and once it has converged respectively. As can be seen, most of the triangles are close to equilateral, and the number of interior nodes is lower than in the [9] scheme.

A number of tests have been applied to evaluate the performance of our solution. The first test consists in placing a wrong number of points inside the face, and applying the relaxation method. The results show that when more points than needed are used, the spare ones are pushed outside the face, so that the resulting

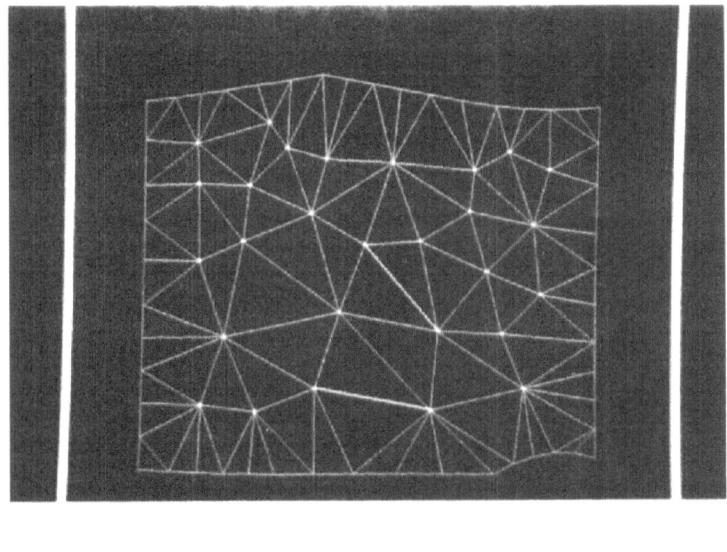

a

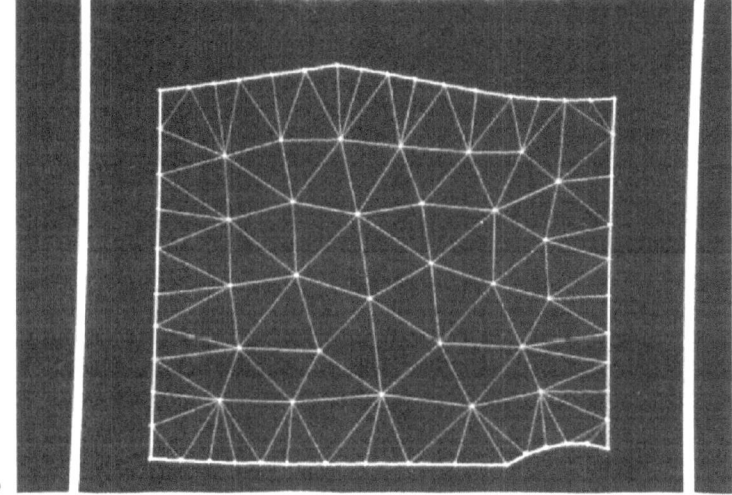

b

Figure 8. Interior vertices and triangulation. **a** Intermediate iteration during the relaxation. **b** Result after the relaxation

triangulation contains nearly the correct number of interior points. Obviously, if too few points are placed in a face, the resulting triangulation contains edges larger than those allowed, but the triangles are still well-shaped. The second test consists in starting the relaxation with different initial point configurations. As was expected, the method converges after more iterations when the distribution of initial points is not homogeneous.

Finally, relaxation when the surface shows strong curvature variations, or the trimmed patch has pronounced features on its bounds (large concavities or holes)

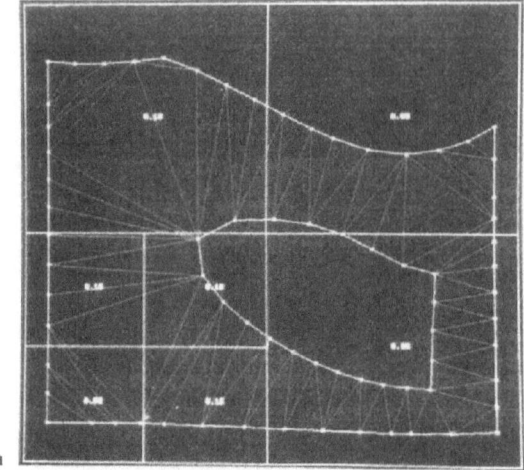

a

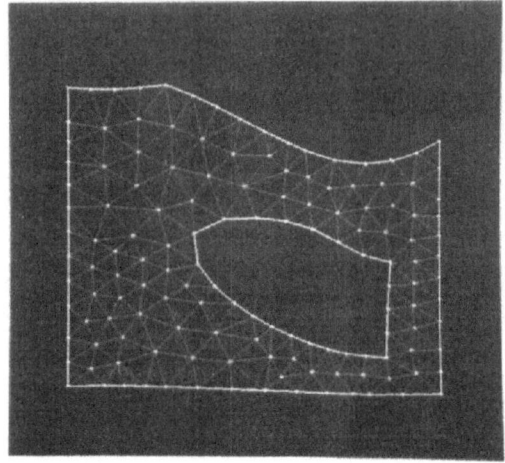

b

Figure 9. Linear approximation of a different patch. **a** Discretization of its trimming curves and the quadtree of curvature bounds. **b** Interior vertices and associated triangulation after the relaxation

was tested. The results show that the method still works well for these cases (see Figs. 9a, 9b, 10a, and 10b). While the algorithm converges in 20 iterations for a normal patch, these special cases require up to 50 iterations. Table 1 shows execution times of the relaxation process for each case:

7. Conclusions

A new scheme for the generation of planar approximations of trimmed surfaces has been proposed. The method is based on a tolerance bound and produces

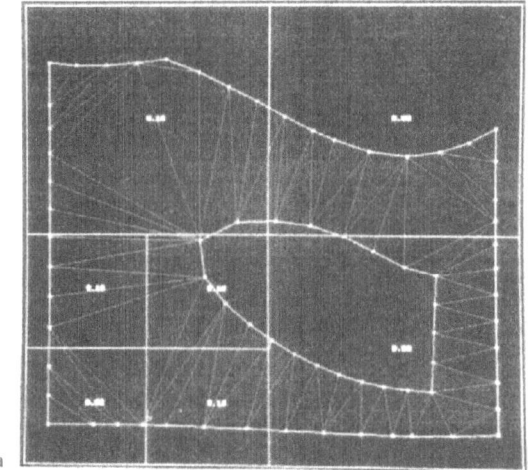

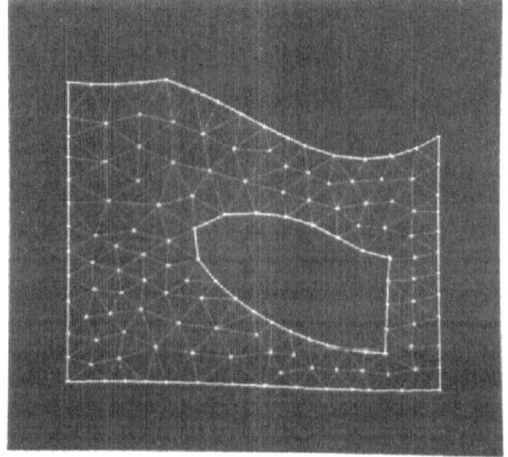

Figure 10. Linear approximation of a third patch. **a** Discretization of its trimming curves and the quadtree of curvature bounds. **b** Interior vertices and associated triangulation after the relaxation

Table 1

Surface	Interior points	Boundary points	Iterations	CPU time (seconds)
Patch 1 (Fig. 8)	28	51	50	1.90
Patch 2 (Fig. 9)	65	40	50	2.86
Patch 3 (Fig. 10)	59	73	50	3.43

a conformal triangulation containing a limited number of triangles. The max-min angle criterion is used to optimize the shape of the resulting triangles.

Further work includes testing the optimality of the proposed solution, and finding lower bounds for the number of points in the triangulation. More precise bounds on the surface curvature should also produce better planar approximations.

Acknowledgements

The authors would like to thank Nuria Pla for her contribution to the development of the bounds, and W. L. F. Degen for his useful suggestions on the curvature bounds for parametric patches. This work was partially supported by the CEC, under Brite project BR-5478. Marc Vigo was also supported by an FI grant from the Generalitat de Catalunya.

References

[1] Allgower, E. L., Gnutzmann, S.: Simplicial pivoting for mesh generation of implicitly defined surfaces. Comput. Aided Geom. Des. *8*, 305–325 (1991).
[2] Brown, J. L.: Vertex based data dependent triangulations. Comput. Aided Geom. Des. *8*, 239–251 (1991).
[3] Degen, W. L. F.: Über den maximalen Sehnenabstand bei Kurvenbögen mit beschränkter Krümmung. Personal communication, 1993.
[4] Filip, D., Magedson, R., Markot, R.: Surface algorithms using bounds on derivatives. Comput. Aided Geom. Des. *3*, 295–311 (1986).
[5] Pla, N., Vigo, M., Brunet, P.: Adaptative conformal triangulation of trimmed surfaces. Research Report, Dept. of Software, UPC 1993.
[6] Von Herzen, B., Barr, A.: Accurate triangulations of deformed intersecting surfaces. Comput. Graphics (SIGGRAPH'87) *21*, 103–110 (1987).
[7] Rockwood, A., Heaton, K., Davis, T.: Real-time rendering of trimmed surfaces. Comput. Graphics (SiGGRAPH'89) *23*, 107–116 (1989).
[8] Shimada, K., Gossard, D. C.: Computational methods for physically-based FE mesh generation. In: Human aspects in computer generated manufacturing (Olling, G. J., Kimura, F., eds.), pp. 411–420. Amsterdam Elsevier 1992.
[9] Sheng, X., Hirsch, B. E.: Triangulation of trimmed surfaces in parametric space. Comput. Aided Des. *24*, 437–444 (1992).
[10] Turk, G.: Re-tiling polygonal surfaces. Comput. Graphics (SIGGRAPH'92) *26*, 55–64 (1992).

Dr. M. Vigo
Prof. Dr. P. Brunet
Universitat Politècnica de Catalunya
Dept. L.S.I., Diagonal 647 (Edifizi ETSEIB)
E-08028 Barcelona
Spain

Computing Suppl. 10, 357–361 (1995)

Computing
© Springer-Verlag 1995

An Efficient Algorithm for Evaluating Polynomials in the Pòlya Basis

J. Warren, Houston

Abstract. A new $O(n)$ algorithm is given for evaluating univariate polynomials of degree n in the Pòlya basis. Since the Lagrange, Bernstein, and monomial bases are all special instances of the Pòlya basis, this technique leads to efficient evaluation algorithms for these special bases. For the monomial basis, this algorithm is shown to be equivalent to Horner's rule.

Key words: Polynomial bases, evaluation.

1. Pòlya Basis Functions

Let $n_k(t)$ be the restriction of the standard B-spline basis functions [3] of degree n over the nondecreasing knot sequence $t_1, t_2, \ldots, t_{2n}$ to the interval $t_n < t < t_{n+1}$. The Pòlya basis functions are the unique polynomial functions $d_k(t)$ that satisfy Marsden's identity [5].

$$(x - t)^n = \sum_{k=0}^{n} d_k(t) n_k(x).$$

The Pòlya basis functions of degree n can be written explicitly as

$$d_k(t) = \prod_{i=1}^{n} (t_{k+i} - t). \tag{1}$$

(See Barry and Goldman ([2], pp. 28) for more details.)

Using the explicit definition of Eq. (1), the Pòlya functions can be defined directly for arbitrary sequences $t_1, \ldots, t_{2n}$. Under this definition, the Pòlya functions includes several important bases as special cases. These bases are:

- The Lagrange basis,
- The Bernstein basis,
- The monomial basis.

Given a set of $n + 1$ distinct values $u_0, \ldots, u_n$, The Lagrange basis functions $l_k(t)$ [1] of degree n are the $n + 1$ polynomial basis functions that satisfy

$$l_k(u_i) = \delta_{ki}.$$

The kth Lagrange basis function can be explicitly written as

$$l_k(t) = \frac{1}{\alpha_k} \prod_{i \neq k} (u_i - t)$$

where $\alpha_k = \prod_{i \neq k}(u_i - u_k)$. The Lagrange basis may be viewed as an instance of the Pòlya basis. Consider the Pòlya basis with knot sequence

$$t_1 = u_1, t_2 = u_2, \ldots, t_n = u_n,$$
$$t_{n+1} = u_0, t_{n+2} = u_1, \ldots, t_{2n} = u_{n-1}.$$

The Pòlya basis function $d_k(t)$ is exactly the Lagrange basis function $l_k(t)$ multiplied by the constant α_k.

The Bernstein basis functions of degree n are

$$b_k(t) = \frac{n!}{k!(n-k)!} t^k (1-t)^{n-k}.$$

These basis functions are the building blocks for one of the most popular curve representations in geometric design, Bézier curves [4]. Specializing the knot sequence to

$$t_1 = 0, t_2 = 0, \ldots, t_n = 0,$$
$$t_{n+1} = 1, t_{n+2} = 1, \ldots, t_{2n} = 1$$

yields a Pòlya basis of the form

$$d_k(t) = (-t)^{n-k}(1-t)^k.$$

Thus, the Pòlya basis function $d_k(t)$ is exactly the Bernstein basis function $b_{n-k}(t)$ multiplied by the constant $(-1)^{n-k}\dfrac{n!}{k!(n-k)!}$.

Finally, the monomial basis, $\{1, t, t^2, \ldots, t^n\}$ is probably the most commonly used basis in mathematics. To express the monomial basis as a Pòlya basis, one must first develop a notion of a knot at infinity. The affine knot t_i can be homogenized to yield the homogeneous representation $(t_i, 1)$. The linear factors in the Pòlya basis can be expressed in the form $(t_i * 1 - 1 * t)$. If $t_i = \infty$, then the corresponding homogeneous knot is $(1, 0)$. The related linear factor is $(1 * 1 - 0 * t)$, the constant 1. (For information on this interpretation, see [2].) Thus, specializing the knot sequence to have values

$$t_1 = 0, t_2 = 0, \ldots, t_n = 0,$$
$$t_{n+1} = \infty, t_{n+2} = \infty, \ldots, t_{2n} = \infty$$

yields Pòlya basis functions of the form

$$d_k(t) = t^{n-k}.$$

These basis function are exactly the monomial basis function after renumbering.

2. Evaluation in the Pòlya Basis

Let $d(t)$ be a polynomial of degree n in the Pòlya basis.

$$d(t) = \sum_{k=0}^{n} P_k d_k(t). \tag{2}$$

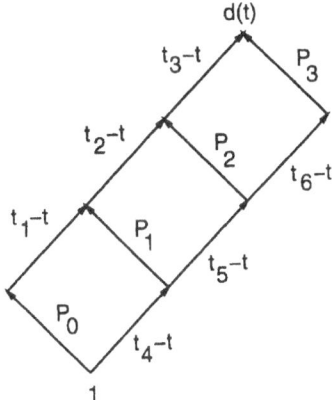

Figure 1. The ladder recurrence for $n = 3$

We next describe an algorithm for evaluating $d(t)$ that requires $O(n)$ arithmetic operations per evaluation. The following recurrence describes such an algorithm.

$$L_0 = 1,$$
$$U_0 = P_0,$$
$$L_{i+1} = L_i * (t_{i+n+1} - t),$$
$$U_{i+1} = U_i * (t_{i+1} - t) + P_{i+1} * L_{i+1}. \tag{3}$$

This recurrence can be expressed as a ladder shaped diagram in Fig. 1. Values are associated with the intersection of arrows and flow up the ladder. A value at an intersection is passed along an outgoing edge and multiplied by the label of that edge. The resulting value is then added to the value at the end of edge. The $L_i's$ lie along the lower right portion of the ladder. The $U_i's$ lie along the upper left portion of the ladder.

Theorem 1. *If $d(t)$ is defined as in Eq. (2) and U_n is defined in Eq. (3), then*

$$U_n = d(t).$$

Proof: Obviously, L_j satisfies

$$L_j = \prod_{i=1}^{j} (t_{i+n} - t). \tag{4}$$

We claim that U_j satisfies

$$U_j = \sum_{i=0}^{j} P_i \left(\prod_{\alpha=i+1}^{j} (t_\alpha - t) \right) \left(\prod_{\beta=1}^{i} (t_{\beta+n} - t) \right).$$

The proof is by induction on j. For the base case $j = 0$, this expression evaluates to P_0 exactly as specified by the recurrence. If this expression is true for U_j, then we next show that also is true for U_{j+1}. By Eq. (3),

$$U_{j+1} = U_j * (t_{j+1} - t) + P_{j+1} * L_{j+1}.$$

Using the inductive hypothesis and replacing L_{j+1} by the right hand side of Eq. (4) yields

$$U_{j+1} = \left(\sum_{i=0}^{j} P_i \left(\prod_{\alpha=i+1}^{j} (t_\alpha - t) \right) \left(\prod_{\beta=1}^{i} (t_{\beta+n} - t) \right) \right) * (t_{j+1} - t)$$

$$+ P_{j-1} * \left(\prod_{\beta=1}^{j+1} (t_{\beta-n} - t) \right).$$

After simplification, this relation establishes the inductive hypothesis for U_{j+1}.

To conclude the proof, we note that if $j = n$, then U_n agrees exactly with the definition of $d(t)$. □

3. Comparison with Other Methods

Given a set of $n+1$ coefficients for a polynomial in the Pòlya basis, evaluating this recurrence requires $O(n)$ additions and multiplications. Note that this method avoids any divisions by functions of t. To evaluate a polynomial in the Lagrange basis using this recurrence, one must first divide the kth coefficient in the Lagrange basis by α_k and then apply the recurrence. All of the α_k's can be precomputed once in an $O(n)$ preprocessing step and subsequently looked up. The speed of this method compares favorably with that of Neville's algorithm [1], the standard evaluation recurrence for the Lagrange basis that requires $O(n^2)$ operations per evaluation.

The standard evaluation method for polynomials in the Bernstein basis is the deCasteljau algorithm [4]. This algorithm requires $O(n^2)$ operations per evaluation. To evaluate a polynomial in the Bernstein basis using the ladder recurrence, the coefficient of $b_k(t)$ is first multiplied by $(-1)^{n-k}(n!/k!(n-k)!)$ and then used as input to the ladder recurrence. Since these binomial coefficients can be precomputed in $O(n)$ time, evaluation using the ladder recurrence requires only $O(n)$ time per evaluation.

In the case of the monomial basis, the knots $t_{n+1}, \ldots, t_{2n}$ were placed at infinity. The linear factors corresponding to these knots were the constant functions 1. Therefore, all the terms L_i in the ladder recurrence were also one and the lower left side of the recurrence in Fig. 1 could be deleted. The resulting recurrence on the U_j's satisfies

$$U_0 = P_0,$$
$$U_{i+1} = U_i * t + P_{i+1}.$$

This recurrence is exactly Horner's rule for evaluating the polynomial

$$\sum_{j=0}^{n} P_{n-k} t^k$$

in the monomial basis [1].

4. Conclusion

The existence of $O(n)$ evaluation methods for polynomials in the Lagrange and Bernstein basis is not surprising in itself. For the Lagrange basis, building such a method is simple if one is allowed to resort to division by $(u_k - t)$. In the case of the Bernstein basis, a table of various powers of t and $(1 - t)$ can be used to evaluate the basis functions efficiently. The beauty of this result lies in the fact that the ladder recurrence provides a natural generalization of Horner's method for the monomial basis to polynomials in the Lagrange and Bernstein basis. The recurrence provides an efficient method for evaluating a polynomial in the Pòlya basis without resorting to division by polynomial functions in t (and the numerical difficulties associated with such an approach). For the Lagrange basis, the recurrence is asymptotically faster than Neville's algorithm. For the Bernstein basis, the recurrence is asymptotically faster than the DeCasteljau algorithm. One remaining question is the stability of the ladder recurrence, especially in relation to the methods described above. The author hopes to address this problem in future work.

References

[1] Burden, R., Faires, J. D.: Numerical analysis. New York: Prindle, Weber & Schmidt 1985.
[2] Barry, P., Goldman, R.: Algorithms for progressive curves. In: Knot insertion and deletion algorithms for B-spline curves and surfaces (Goldman, R., Lyche, T., eds.), pp. 11–63. New York: SIAM 1993.
[3] de Boor, C.: A practical guide to splines. Berlin Heidelberg New York: Springer 1972.
[4] Farin, G.: Curves and surfaces for computer aided geometric design: a practical guide. New York: Academic Press 1988.
[5] Marsden, M. J.: An identity function for spline functions with applications to variation-diminishing spline approximation. J. Approx. Theory 3, 7–49 (1970).

J. Warren
Department of Computer Science
Rice University
Houston, TX 97251-1892
U.S.A.

Springer Titles in Computer Science

R. Albrecht, G. Alefeld, H. J. Stetter (eds.)
Validation Numerics
Theory and Applications
1993. 23 figures. IX, 291 pages. (Computing / Supplement 9)
Soft cover DM 160,–, öS 1120,– *. ISBN 3-211-82451-0

G. Farin, H. Hagen, H. Noltemeier (eds.)
in cooperation with W. Knödel
Geometric Modelling
1993. 175 figs. and 6 plates. VIII, 316 pages. (Computing / Supplement 8)
Soft cover DM 160,–, öS 1120,– *. ISBN 3-211-82399-9

G. Tinhofer, E. Mayr, H. Noltemeier, M. M. Syslo (eds.)
in cooperation with R. Albrecht
Computational Graph Theory
1990. 68 figures. VII, 282 pages. (Computing / Supplement 7)
Soft cover DM 148,–, öS 1036,– *. ISBN 3-211-82177-5

U. Kulisch, H. J. Stetter (eds.)
Scientific Computation
with Automatic Result Verification
1988. 22 figures. VIII, 244 pages. (Computing / Supplement 6)
Soft cover DM 128,–, öS 900,– *. ISBN 3-211-82063-9

K. Böhmer, H. J. Stetter (eds.)
Defect Correction Methods
Theory and Applications
1984. 32 figures. IX, 243 pages. (Computing / Supplement 5)
Soft cover DM 72,–, öS 504,– *. ISBN 3-211-81832-4

R. Albrecht, U. Kulisch (Hrsg.)
Grundlagen der Computer-Arithmetik
1977. 13 Abbildungen. IX, 150 Seiten. (Computing / Supplement 1)
Broschiert DM 69,–, öS 480,– *. ISBN 3-211-81410-8

* 10% price reduction for subscribers to "Computing"

Prices are subject to change without notice

Springer-Verlag Wien New York

Sachsenplatz 4–6, P.O.Box 89, A-1201 Wien · 175 Fifth Avenue, New York, NY 10010, USA
Heidelberger Platz 3, D-14197 Berlin · 3-13, Hongo 3-chome, Bunkyo-ku, Tokyo 113, Japan

Springer Journals

Surveys on Mathematics for Industry

Editorial Board:

H. Engl (Managing Editor), T. Beth, C. Cercignani, M. Deistler, R. E. Ewing,
A. Fasano (representative of ECMI), D. Ferguson, A. Friedman, A. Gilg,
R. Glowinski, M. Grötschel, H. Hagen, R. Janßen, K.-H. Keil, T.-T. Li,
B. Lindorfer, A. Louis, P. Markowich, R. Mennicken (representative of GAMM),
H. Neunzert, J. Periaux, P. Rentrop, A. A. Samarskii, A. Tayler, A. Tesei (representative
of SIMAI), W. Törnig (representative of DMV), I. Troch (representative of ÖMG),
M. Yamaguti (representative of JSIAM)

Editorial Assistant: A. Neubauer

The main goal of this journal is to bridge the gap between university and industry
by the presentation of mathematical methods relevant for industry and the expo-
sition of industrial problems which are of interest to mathematicians.
To achieve this goal, the journal publishes (exclusively in English): surveys on new
mathematical techniques, surveys on established mathematical techniques with a
new range of applications, surveys on industrial problems for which appropriate
mathematical models or methods are not yet available, articles comparing mathematical models or methods for particular industrial problems, articles describing
mathematical modelling techniques, broad historical surveys, articles of general
interest about the use of mathematics in industry, occasional book reviews and
reports about conferences in the field of Industrial Mathematics.
Papers may be submitted to Managing Editor or will be solicited by member of the
Editorial Board. The Managing Editor also welcomes suggestions for possible topics by prospective authors with a short outline of the intended paper for a first evaluation of the suitability of the topic for the journal. All papers will be refereed.

Subscription Information:

1995. Vol. 5 (4 issues):
for institutional subscribers: DM 268,–, öS 1876,–, plus carriage charges
for individual subscribers: DM 144,–, öS 1008,–, plus carriage charges
Special rates for individual members of DMV, ECMI, GAMM, JSIAM, ÖMG, and
SIMAI: DM 80,–, öS 560,–, plus carriage charges. Orders must be placed directly
with the respective society or Springer-Verlag Wien.
ISSN 0938-1953, Title No. 724

Springer-Verlag Wien New York

Sachsenplatz 4–6, P.O.Box 89, A-1201 Wien · 175 Fifth Avenue, New York, NY 10010, USA
Heidelberger Platz 3, D-14197 Berlin · 3-13, Hongo 3-chome, Bunkyo-ku, Tokyo 113, Japan